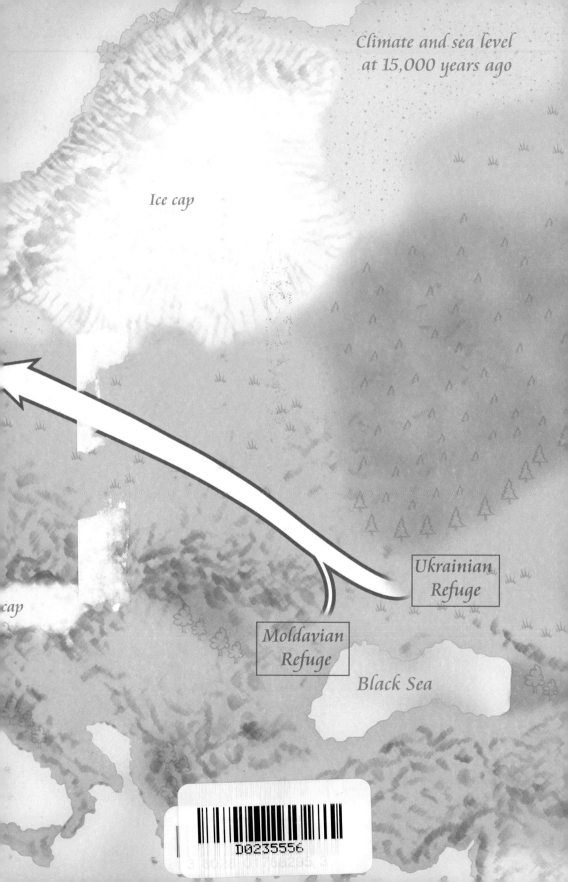

Climate and sea level
at 15,000 years ago

Ice cap

Ice cap

Ukrainian
Refuge

Moldavian
Refuge

Black Sea

The Origins of the

BRITISH

A GENETIC DETECTIVE STORY

By the same author

Eden in the East:
The Drowned Continent of Southeast Asia

Out of Eden:
The Peopling of the World

The Origins of the

BRITISH

A GENETIC DETECTIVE STORY

STEPHEN OPPENHEIMER

CONSTABLE • LONDON

Constable & Robinson Ltd
3 The Lanchesters
162 Fulham Palace Road
London W6 9ER
www.constablerobinson.com

First published in the UK by Constable,
an imprint of Constable & Robinson Ltd, 2006

A copy of the British Library Cataloguing in Publication Data
is available from the British Library

ISBN-13: 978–1–84529–158–7
ISBN-10: 1–84529–158–1

Printed and bound in the EU

In memory
of
David Oppenheimer
classicist, composer, clinician
and
father

CONTENTS

ACKNOWLEDGEMENTS

I started thinking about the peopling of the British Isles in terms of sources, numbers and timing of founders over five years ago, when I heard Barry Cunliffe talk about the ancient unity of cultures along the Atlantic coast. I was attending an excellent series of talks put on by Linacre College, Oxford. These talks were later published under the title *The Peopling of Britain: The Shaping of a Human Landscape* (edited by Paul Slack and Ryk Ward), but I picked up ideas from other fascinating lectures in the series by archaeologists Heinrich Härke, Paul Mellars and Clive Gamble and the demographer Richard Smith.

Much of *Origins* results from new analysis that I have carried out on existing genetic data; parts of the story are retold from other people's published work and synthesized from the relevant disciplines, with a focus on genetic heritage. The bibliography measures my attempt to cover different points of view and the perspectives of the different disciplines – the study of human history, archaeology, genetics, language, climate and sea level. For instance, as with my last book, I have found Jonathan Adams' palaeo-environmental website invaluable for maps. There are several geneticists whose laboratory work, with colleagues, over the past few years has added enormously to the genetic database for the British Isles and neighbouring countries, and should be singled out. They include Cristian Capelli, Jim Wilson, Michael Weale, Mark Thomas, Martin Richards, Zoë Rosser, Siiri Rootsi, Kristiina Tambets, Luísa Pereira, María Brion, Antonio Torroni, Peter Forster and Chris Tyler-Smith, among many others. As a rule (and there are few exceptions), geneticists generously make their raw data available in the public domain on publication of their analyses. This has allowed me (and others) to pool similar genetic data, re-analyse and make inter-regional comparisons.

But I have not done this in isolation. I have had much assistance and advice from many of the people mentioned above. Orcadian geneticist Jim Wilson has been enormously helpful in pointing me towards appropriate Y-chromosome datasets, making his own data available, giving critical advice, warning of pitfalls and suggesting alternative methods of analysis, even how to get the best out of Excel, not to mention e-mailing me relevant publications. The same applies to Martin Richards, Peter Forster and Vincent Macaulay, who have given me much of their time on the phone and insights into their own work in progress. Martin's publications, with colleagues, on the impact of the Neolithic on Europe from the perspective of mitochondrial DNA form a benchmark in the use of the phylogeographic method in European prehistory. Peter Forster and I share several perspectives on English genetic ancestry, and we even came up with the same linguistic questions by different routes. I have also benefited from discussion and comparing methods of analysis with Bill Amos. Other geneticists who have given me their time in discussion include Cristian Capelli, Siiri Rootsi, Kristiina Tambets, Mark Thomas, Michael Weale and Antonio Torroni.

I have also received constructive criticism and advice from archaeologists, including Barry Cunliffe, who kindly read and commented on the entire book, Nick Barton, who concentrated on the Late Glacial and Early Holocene sections, and Helena Hamerow, who read the Anglo-Saxon parts. Others who generously gave me their time and help include Jill Cook, who provided the photo-micrograph of the Creswell horse engraving, Simon James, Beatrice Clayre and Graeme Barker.

I am very grateful for time and constructive criticism given me by linguists, in particular Patrick Sims-Williams, who was a great source of advice and references, including many he had written himself. He also read through and commented on the language sections. Other linguists – April McMahon, John Penney, Katherine Forsyth and Wendy Davies – gave their time and advice, as did members of the English Place-name List on the Internet. I also received very useful material and insight from non-linguists working on language trees and networks, including Peter Forster, Quentin Atkinson, Geoff Nicholls and Russell Gray.

Acknowledgement of the above does not necessarily indicate their agreement with the views expressed in this book.

As ever, my agents, Carole Blake and Julian Friedmann, always go the extra mile and more. I count myself very lucky to be on their list.

I have had an enormous amount of support from the team at my publisher, Constable. With such a large number of figures and refer-ences, this has involved much more work than the average popular science book. I would like to thank particularly Pete Duncan (my supportive editor), Bill Smuts (illustrator), John Woodruff (science editor), Katharine Timberlake (genetics editor) and many others, including Jan Chamier, Andy Hayward, Sarah Moore, Bruce Connal, Charlotte Deane and Beth Macdougall, not forgetting Nick Robinson, who runs the whole show with such enthusiasm.

I thank friends and family, who have put up with my grumpy behaviour over the months leading up to publication, and pointed out that life need not centre around a laptop. In particular, I thank my friends from the Bradshaw Foundation, Damon De Laszlo, John and Peter Robinson, and Jean Clottes, who arranged diversions such as visiting a pristine, protected section of the Magdalenian cave at Niaux in the lower Pyrenees. There, in the crypt of our ancestors, we saw immaculate, confident charcoal cartoons of a horse and a stoat and fifty metres of clear, fresh footprints made by three immortal children, two of them running hand in hand, now all over 13,000 years old. Surely, one of the artists in this Ice Age refuge had British cousins, who carried on the equine tradition in Robin Hood Cave, at Creswell Crags.

Finally, thanks to my long-suffering wife Freda and my children Maylin and David.

PREFACE

As a child, I sometimes wondered why people told jokes about Englishmen, Irishmen, Welshmen and Scotsmen. Why should our origins and differences matter? Part of growing up was realizing that they do matter and trying to understand why. We do not benefit from ignoring our own cultural identity and physical origins or by casually derogating those of others.

The English have traditionally had uneasy relations with their 'Celtic' neighbours. Part of the unease has to do with history, part with nationhood and part with territory. Perhaps I could illustrate this by a mid-twentieth-century anecdote:

Welsh idyll

I was taken as a babe-in-arms for my first holidays to a cottage on the side of a mountain in North Wales. Sixty years ago, my aunt had 'inherited' some sticks of furniture and the informal lease of the cottage, at a peppercorn rent, from an elderly English headmistress. More of a shepherd's hut built of massive granite blocks, the cottage had one small living room and two tiny bedrooms. Yet, for several decades it housed many members of our extended family over Easters and summers.

I made annual pilgrimages to this magic hut over the next twenty years. There was lots of nothing for kids to do in the cottage. On a small shelf on a convertible sofa were mildewed books and torn society magazines collected during the 1930s. My elder brother and sister played with the farmer's daughter while my younger sister and I tried to trail after. In my late teenage and early university days in the sixties, I regularly dragged my Oxford friends up to the cottage in Wales. I remember once the farmer's daughter, who was by now married, making a sharp comment about one of my guests' open-air washing habits. I enquired about this from my friend, a rather posh debutante, who explained innocently that she had given up washing in the chipped enamel bowl in the shed and started going straight to the spring. Here she had regularly bared her lovely body to the cows and sheep in the hillside field. Unfortunately, the spring overlooked the bus stop and the milk churns on the coastal road, and so more critical eyes from the village enjoyed the treat.

Towards the end of my clinical studies, things changed abruptly. My elderly aunt received her written marching orders from the farmer's daughter, who on the death of her own mother had inherited the farm. As my cousins were leaving after their last holiday there, they noticed a contractor's van on the hill. Workmen had come to install an electricity transformer beside the cottage. My aunt was very upset. Above all, she felt a sense of betrayal that the capital of friendship she had built up over such a long time had evaporated with the death of the farmer's wife. She felt that the old lady would never have allowed it. Of course, whatever her dubious squatting rights, my aunt had no moral right to such expectations or feelings of betrayal. Our large English extended family had enjoyed exclusive use of the cottage for a negligible rent. I was very fond of my aunt, but her view of her relationship with the farmer's family was rose-tinted. Like my own mother, she treated the farmer's daughter in the same bossy way as she did her nephews and nieces. This, combined with a conviction that she was right, usually left the opposition speechless.

Tight-lipped but courteous was indeed how I usually found the Welsh-speaking farmer's family on the occasions I ventured into their parlour. The farmer's wife may just have been too polite, or at a loss to know how to deal with this prolonged unremunerative tenancy. Her daughter, with more education and a different perspective, must have been waiting some time. There was obviously a pressing reason for terminating the tenancy: to improve the cottage and let it out at a more commercial holiday rate. But the body language I saw at the time also indicated a deep sense of resentment and outrage. We were foreigners, we had strange ways and foreign English names – even Germanic, like my father's – and we had unjustified control of their property.

Stirring it

There should be no doubt as to the very real power and impact of perceived national and regional identity. There is certainly no mistake about this among the modern populations of the British Isles* outside England. Disputes, attacks, mutilations, murders and occasional bombings are committed annually in the name of such divisions. A wide range of overt and hidden considerations of ethnicity affect the regional allocation of UK Government spending, and Britons are constantly reminded, both in their history books and in the modern history of their political institutions, of the

---

*I should say from the start that throughout this book I use the term 'British Isles' (and, only very occasionally, 'British') in the traditional sense: to refer to *all* the islands in the immediate vicinity of Britain, including Ireland, Shetland and Orkney (and, exceptionally, more distant ones such as the Channel Islands which were included in the genetic dataset I have used). I appreciate that many Irish do not regard themselves as British, with good reason. My inclusion of Ireland as part of the British Isles is only to avoid repeating the geographic reality, and not to make any contemporary political statement. I also use 'Britain' in the Roman sense to refer to the 'big island' in place of the cumbersome political term *Great Britain* with its overtones of the Act of Union.

importance of regional division. Football teams are named after ethnic labels, and fans fight under their flags. No British person hears more than a sentence from another before mentally placing them in their regional context.

I argue in this book that these divisions are real and consistent. They have a deep and very ancient history, which can be traced both culturally and genetically to two widely separated European regional origins for our ancestors. Even the 'myths' of Celtic and Anglo-Saxon ethnicity represent the same two real divisions. As with all folklore, however, these labels ultimately mean something quite different, and may refer to older events. The distortions of history, propaganda and misunderstanding need to be viewed in the context of other evidence if we are to separate actual events from fiction.

Why do I feel the need to write a book which emphasizes national and ethnic divisions, when these have caused such grief and ought to be buried? Well, first I believe in celebration of our diversity rather than cultural levelling. Pretending that differences should not exist is a political fib. I just do not accept that a sense of pride in culture and diversity is the primary cause of nationalistic crimes. The causes of such crimes can be more clearly identified in the deliberate agenda set by some politicians – as can be seen from the recent history of Germany and the Balkans. Interest in our cultural and biological diversity is sinister only when the agenda are competitive, derogatory or exclusive. Of course, such sinister agenda are not exclusive to individuals in power, but people of different cultures usually manage to co-exist without genocide until stirred up by ambitious politicians.

The Origins of the British describes the story of the peopling of the British Isles since their recolonization after the end of the last Ice Age, from about 15,000 years ago. I combine genetics, climatology, geology, archaeology, linguistics, culture and history to reconstruct and explain our roots and differences. Most speculation on our roots centres on resonant ethnic labels, such as Celts, Anglo-Saxons,

Vikings and Normans. This book is no exception to that emphasis, but those labels do not mean quite what we have been led to believe.

There is a rising trend of scepticism about the validity of terms such as 'Celtic' and 'Celts', which is apparent in several recent books written by archaeologists. Despite this, it is indeed true that there are systematic genetic differences between the so-called Celtic regions of the western British Isles and England, and that there are some parts of the British Isles which show close genetic links to Scandinavia. Perceptions of 'Celtic' and 'Viking' ethnicity held by people living in, say, Ireland and York are not simply meaningless, and should not be flushed away by the bathwater of academic Viking- and Celto-scepticism. But while traces of invasions, from the historical period and just before, do exist in today's regional genetic patterns, the overall picture of deep genetic divisions between England and the British Atlantic coasts and islands is much more ancient than is implied by the story which tells of how Anglo-Saxons ethnically cleansed other Britons from their land. Some geneticists still actively promote the Anglo-Saxon wipeout view. Equally, documentaries that invite us to join a quest to trace the 'Blood of the Vikings' seem to miss the point that our relationships with Scandinavians are much older than the Viking raids.

Likewise, the perception of genetic and cultural differences between 'Celtic' regions and England has a basis in reality, but has little to do with the nineteenth-century orthodoxy, still current today, of Celtic origins in Iron Age central Europe. Rather, the regions of Wales, Cornwall, Ireland and western Scotland have for many thousands of years shared genetic and cultural links with Iberia and the French Atlantic coast. I hope to establish more clearly than ever before the true genesis of the peoples we currently call insular Celts: the Scots, Welsh, Cornish and Irish.

From the Neolithic period, we shall travel from megalithic complexes like Stonehenge and massive passage-tombs like New-grange in Ireland to the extraordinary preserved Neolithic village of

Skara Brae in Orkney (Plate 5). During the Bronze and Iron Ages, we can marvel at the swirly designs of British gold and bronze artefacts, previously thought to support Celtic origins in Central Europe, to the precious ornaments found in exotic Saxon burials such as the one at Sutton Hoo. In each case, the British have continually adopted and developed new ideas, language and cultural practices from the recurrent invasions from across the North Sea and the English Channel, while managing to retain most of their prehistoric genetic heritage. That process of cultural borrowing from new visitors continues today.

The book is based on similar research to that employed for my previous book, *Out of Eden*: I have used the so-called phylogeographic approach to follow and date gene flow from different parts of the Continent into the British Isles. Details of the methods I have used can be found in the appendices. However, I should mention here that, as in the previous book, I have used personal names as nicknames or aides-mémoires for the major gene groups. In each case, I have chosen a personal name, from the appropriate region, which starts with the first letter of the technical name (i.e. from the consensus scientific nomenclature). So, for example, I have called the male gene group R1a1, which is strongly associated with the Ukraine, 'Rostov'. As before, this is intended not to personalize such small elements in our genome, but to help the reader keep track of the migrations of different gene groups. And also as before, there will be the odd academic reviewer who still regards this practice as a familiarity and trivialization unworthy of their genes and their discipline. Too bad. I agree with my publishers and other reviewers that it helps a general readership.

Genesis: think not what we can do for the media, but what they might do to us

In 2004, I gave a talk on *Out of Eden* at the Edinburgh International Science Festival. The topic was a genetic perspective on the peopling of the world by humans. Arrangements for the lecture at the excellent Royal Museum seemed largely to be run by bright medical students. The talk went well, but the morning before I had made the mistake of responding to a request from the publicity lass to provide more of a local angle for the Scottish newshounds. Although there was no mention of Scotland in any of my books, I unwisely agreed to be available from my hotel room to offer some opinions on where the different regional populations of the British Isles had come from and on the ancient division of English and Celtic speakers. I say 'unwisely' because of what subsequently happened in the media.

I gave several interviews to reporters from my Edinburgh hotel-room phone on the Sunday morning, and then gave my book talk that evening in the grand lecture theatre of the Royal Museum. I saw no journalists. I signed books, answered keen, well-informed questions, and then ate duck in a French restaurant with my wife. Later we crossed the Grassmarket, back to our hotel, to sleep. On the bank holiday morning of our drive back to England, *The Scotsman* appeared under my hotel-room door.

I couldn't miss the front-page article, titled 'Scots and English aulder enemies than thought',[1] with the accompanying editorial leader 'Relatively well connected'. I went to a newsagent. *The Herald* also ran the story, giving it more space on an inside page: 'English-Scots split goes back 10,000 years: genetic proof of Celts' ancient ancestry'.[2] The *Independent* ran a front-page article under the heading 'Celts and English are a breed apart? Absolutely, says Professor'.[3] Other Scottish and English national papers ran the story in various forms and under different bylines. I could tell that they were all syndicated from just one of the interviews I had given, since multiple errors were repeated, such as a fabricated description and a title of an imaginary post in Oxford that might have been taken from a Roald Dahl book. Other persistent errors included the erroneous claim that all of this was in my book and had been discussed in my lecture of the previous day.

The syndicated story then appeared on the newspapers' own Internet pages, from where it spread in ripples to a wide spectrum of other websites. For example, a BBC news page announced that 'Scientist mulls Anglo-Scottish split'.[4] A page run by a coven of witches found interest in the idea of the English–Scottish split; various genealogy web-groups conducted fierce discussions, to which I was invited to contribute, and did. A Fascist site calling itself White Stormfront took up the story, and one of their brave anonymous contributors depressingly volunteered that 'the English have always been mongrels'. In fact, the latter was a true enough statement of ancient admixture, but made racist simply by choice of a derogatory noun. Over a hundred secondary reports and blogs arose in the following week, and my book sales rocketed on Amazon.

Mixed feelings? Certainly. Although pleased with the sales, I was annoyed and embarrassed by the errors and by the news editors' titles, which scratched away at old cultural wounds. The use of the term 'Celts' in this way actually perpetuated the old myths. Of course, I immediately wanted to correct the errors – although not in the newspapers. Instead, I contacted my editor Pete Duncan at Constable & Robinson. After I had unloaded my feelings to him, he challenged me to tackle the errors and myths in book form. I did, and here it is.

What amazed me most was the extraordinary public interest in the question of who are the Celts and the English. There are millions of people, both in the British Isles and in the English-speaking diaspora in North America, Australia, New Zealand and South Africa, who visit genealogy websites and are fascinated and eager for more information on their own origins, their 'British roots'. It is easy to dismiss the exact geographic and temporal origins of the British Celts and the English as questions for the academics, but that is not how most of us see them.

Ethnic identity, Celtic vs English, highland vs lowland, is real for millions. Consequently, there is a huge media market ministering to such perceptions. But the problem of ancient rivalry is not easily confined to such simple labels. If we take just one of the most intractable ethnic feuds in modern British community life, that of Catholics vs Protestants, and then look at its representation in one of our best-known forms of ritual warfare, football, we find the Celtic term creeping in with fresh spin. At one time the football clubs Everton and Liverpool were identified as Catholic and Protestant rivals in Liverpool, a former Viking colony and one of England's largest Irish colonies. Farther north, in Scotland, Glasgow still sports exactly the same rivalry and violently polarized allegiances, but between clubs named Celtic and Rangers.

PROLOGUE

Facing the Atlantic

Many people regard the different regional populations of the British Isles as 'races apart'. Words like 'race' have in general little validity or utility, but it is certain that Ireland, Scotland, Wales, Cornwall and England all have different cultural histories. The idea of genetically putting Brits in their places was not completely off the top of my head and, although never attempted before in popular science writing, has been an interest of mine for some time. Five years ago, I attended a wonderful lecture on the prehistory of the peoples of the European Atlantic coast by archaeologist Professor Barry Cunliffe of Oxford University. I was so inspired by his passionate talk and novel angle on West European prehistory that I immediately set out to study the genetic and archaeological evidence. I continued that interest over the years, while subsequently writing a book on another topic. Cunliffe illuminated a story of cultural continuity of the Atlantic coast peoples, stretching from the west of Scotland, down through Wales, Ireland and Cornwall, then across the Channel and south through Normandy and Brittany, right down to Spain and Portugal. The cultural unity and trade links of this coastal strip had somehow persisted as a binding force, from the Late Mesolithic, over 7,000 years ago, when the sea level had risen enough for colonists to penetrate between Ireland and Wales from the south, through to modern times.

This cultural continuity of the Atlantic coastal strip overlays the extraordinary millennial cycles of change and influence coming in from elsewhere in Europe. The coastal network was the main thrust of Cunliffe's lecture, but I am a daydreamer, and several other gems attracted me while listening to his talk. These were two observations, probably related. One, more a realization on my part, was that when Britain and Ireland were first recolonized in the Late Mesolithic, they were still connected (until 8,500 years ago) by dry land across the North Sea to the Continent, thus geographically filling in Cunliffe's image of the Atlantic coast's cultural continuity. The second observation was that En   d had repeatedly missed out on the cultural fashions that peri ically swept up the Celtic Atlantic coast from the south over the last 10,000 years. Rather, England tended to link culturally with north-west Europe, on the other side of the North Sea.

Even today, with the landbridge gone, the tip of Brittany is closer to Cornwall than to anywhere else in France or up the Channel. So, it seemed that, for cultural reasons, the natural route of explorers and traders from Spain, the French coast and the Mediterranean, through the ages, was towards the west coast of Britain, not the east. The cultural-geographical link directly across the southern entrance to the Channel between Brittany and Cornwall, and on to Wales, Ireland and Scotland, remained intact whether it was the introduction of styles of megalithic monuments to Ireland, or Tristan's to-and-fro seafaring peregrination between these places. He voyaged by sea to all these 'Celtic' countries of the Atlantic façade, in different versions of the original Dark Ages romance, while he wandered blindly towards betrayal of honour and trust, in his love for Isolde.[1]

Of course, my interest was pricked by the possibility that there might be real genetic parallels for the recurrent cultural movements into the British Isles. These, according to my reading of Cunliffe's story, should have been from *two* sources, one up the Atlantic coast from the south and the other from north-west Europe, indicating that people may have migrated (or invaded) from these two directions. This was a hope both vain and sanguine, since dramatic movements inferred from the archaeological record always tend to have a much fainter and more conservative genetic parallel; and even conquests of the historical period may have represented no more than the imposition of an elite minority rather than a mass influx of a different nation (think of the Norman conquest – in the long run our conquerors became anglicized; we did not become French). In other words, fashion and culture move faster and more comprehensively than the people who carry them, and genetic traces of individual cultural sweeps may be disappointingly faint.

On the other hand, my hope was that the conservatism of inter-regional gene flow also allowed those same genetic traces to persist for thousands of years in the same communities. The one great advantage of the British Isles for genetic study over continental Europe (or any other continent, for that matter), apart from their isolation, is that it was a landscape empty of people after the Ice Age. At its worst, half of Britain was covered in an ice sheet and the rest was polar desert. This left a clean genetic sheet, a blank slate, until about 15,000 years ago, with no confusing genetic traces remaining from any hunter-gatherers who may have lived there before the ice.

The reality of the genetic picture was potentially much more illuminating than I had hoped. At the time I attended Barry Cunliffe's Linacre Lecture there were already enough genetic data in the literature to underwrite his cultural view of a unique and ancient Atlantic coastal community. Even by using just a crude marker system such as ABO blood grouping, a line could be drawn north–south along the Welsh border or its physical embodiment, Offa's Dyke. The physical line separates the Welsh from the English, as Saxon King Offa had intended. But it also effectively points to the genetic links that non-English regions of the Isles have with one another – and ultimately, as I shall explain, more with Spain and the Basque Country than with the immediately adjoining European lands of north-west Europe.

Other more specific markers, such as mitochondrial DNA (mtDNA) and the Y chromosome, gave clear confirmation of this two-source picture, but with subtle differences corresponding to characteristic male and female migratory patterns. MtDNA (only passed down through our mothers) is most useful for dating and recording initial colonizations and true migrations, for instance where whole communities move from one region to another. The Y chromosome (held only by males) in general gives a much sharper geographical pattern but tends to overemphasize male-dominated migrations such as conquering elite-invasions. There is also less consensus, at present, on the calibration of the male-chromosome clock. These male/female differences are valuable in illuminating the reasons for new cultural waves in the archaeological record.

Origins of the Celts: Central or Southern Europe?

One observation shines bright from the genetics. The bulk of informative male gene markers among the so-called Atlantic Celts are derived from down in south-west Europe, best represented by people of the Basque Country. What is more, they share this Atlantic coastal link with certain *dated* expansions of mtDNA gene groups, representing each of the main, archaeologically dated, putative colonization events of the western British Isles. One might expect the original Mesolithic hunter-gatherer colonists of the Atlantic coast, over 10,000 years ago, to have derived from the Ice Age refuges of the western Mediterranean: Spain, south-west France and the Basque Country. And that was indeed the case: shared genetic elements, both in the British Isles and Iberia, did include such Mesolithic mtDNA founding gene lines originating in the Basque region.

Perhaps more surprising and pleasing was the identification, among 'Atlantic Celts', of gene lines which arrived later, in the British Neolithic period, deriving ultimately from the very first farming communities in Turkey. The British Neolithic began over 6,000 years ago, but the archaeological and genetic evidence points to two separate arms, or pincer routes, of Neolithic migration into the British Isles from different parts of Europe, each with its own cultural precursors and human genetic trail markers. Most Neolithic migration is apparent rather than real, but in this instance migration is supported by genetic evidence.

One of these migrations may have come up the Atlantic coast and into Cornwall, Ireland and Wales, preceded in France by the arrival of a particular pottery type known as Cardial Impressed Ware. Cardial Ware had in turn spread mainly by sea, west along the northern Mediterranean coast via Italy and the Riviera, and then across southern France to arrive near Brittany by around 7,000 years ago. In parallel with this cultural flow, specific gene lines appear to have travelled along the northern Mediterranean coast, round Spain and directly through southern France to the British Isles. In the case of this real Neolithic migration, however, the Basque Country seems to have been bypassed. The other Neolithic migration went up the Danube from the Black Sea to Germany and the Netherlands (but more of that later).

What is truly remarkable about the Mediterranean coastal Neolithic spread, as sketched by genetics and archaeology, is that there is another parallel trail, one which may explain the origins of the Celtic languages. New evidence places the split that produced the Celtic branch of the Indo-European language family rather earlier than previously thought. Dating of this branch split could put Celtic linguistic origins at the start of the European Neolithic, consistent with the separate southern Neolithic expansion round the coast of the Mediterranean. The final break-up of the Atlantic coast Celtic languages may have been as early as 5,000 to 3,000 years ago, during the Neolithic period in the British Isles.

While the genetic evidence for an ancient southern origin for the ancestors of modern British Celts provides a ringing echo to Cunliffe's archaeological vision of the Atlantic cultural network, it is very different from the familiar scene painted in history books, and from nineteenth-century romantic re-creations, of a once vast Celtic empire in Central Europe. There is another rider, since most of those southern 'ancestors' arrived even earlier than the Neolithic.

The last three hundred years have seen the construction of the orthodox picture of the Celts as a vast, culturally sophisticated but noisy and warlike people from Central Europe who invaded the British Isles during the Iron Age, around 300 BC. Central Europe during the last millennium BC was the time and place of the exotic and fierce *Hallstatt* culture and, later, the *La Tène* culture, sporting their intricate, prestige, Iron Age metal jewellery wrought with beautiful, intricately woven swirls. Hoards of such weapons and jewellery, some fashioned in gold, have indeed been dug up in Ireland, seeming to confirm Central Europe as the source of migration. The swirling style of decoration is immortalized in a glorious illuminated Irish manuscript, the *Book of Kells*, evoking the western British Isles as a small surviving remnant of past Celtic glory. This view of grand Iron Age Celtic origins on the Continent and progressive westward shrinkage since Roman times is still held by many archaeologists. It is also the epistemological basis of strong perceptions of ethnic identity held by millions of the so-called *Celtic diaspora* now residing in the former British Empire and America.

Not all archaeologists see it that way. Dissidents include Colin Renfrew, who in his landmark *Archaeology and Language: The Puzzle of Indo-European Origins*, published in 1987, questioned the evidence for this whole perspective of invading 'Celtic ethnicity'. In fact there is now a growing consensus view of the *lack* of evidence, both in the archaeology and in early historical documents, for any large-scale pre-Roman Iron Age invasions of the British Isles, apart from shared Belgic tribal names across the Channel (of which more below).

Although mainstream archaeologists, on principle, do not refer to it much, there is a large corpus of Irish legendary-historical records, written down and collated by various cleric-academics over the past 1,500 years, which echoes the views of the dissident archaeologists. These unique texts fail to support the concept of any military invasions of Ireland after those of the Late Bronze Age. The latter invasions, stretching back from the Bronze Age to the Late Neolithic, are all explicitly recorded in the Irish Kingship Lists as coming from the Mediterranean region, in particular from or via Spain and even from Greece, suggesting an alternative legendary reconstruction of Gaelic history.

More recently, Simon James has been more outspoken than his fellow-archaeologists. In his book *The Atlantic Celts* he describes the story of the Iron Age Celtic invaders of the British Isles from Central Europe as just that – a story. In particular, he unravels a modern myth created in the early eighteenth century by a Welsh antiquarian, Edward Lhuyd. The term 'Celtic' had never been applied to inhabitants of the British Isles until the time of Lhuyd, who correctly identified the relatedness of languages spoken today in Brittany and throughout the western British Isles. Lhuyd was not the first to arrive at that conclusion, and he believed in successive Celtic waves, but he erred in conflating this linguistic unity with the Roman non-linguistic term 'Celtae'. The latter was used during classical times, often rather loosely, to describe tribes somewhere in Western Europe, in much the same ill-defined way that some people nowadays speak of 'Asians'. Unfortunately, the pseudo-ethnic terms 'Celtic' and 'Celts', with their Central European Iron Age baggage, have stuck since the nineteenth century.

James speculates that Lhuyd, living in the eighteenth century, preferred the term 'Celtic' as a language label to the more geographically appropriate 'Gallic' for obvious nationalistic reasons. The connection, or further conflation, of Atlantic Celts with the Iron Age Hallstatt and La Tène cultures has no basis in direct

linguistic evidence,[2] and came with nineteenth-century archaeology. By this time, Lhuyd's linguistic idea 'Celtic' had matured into
a rich story of ethnic identity with strong nationalistic overtones
telling of ancient Celtic invaders. The Romantic-heroic image was
seized upon by people living in the non-English-speaking parts of
the British Isles. They had plenty of real history to remember, of
English oppression over the previous 1,500 years.

The English

This is where one of the most deeply embedded of British roots
myths comes in: namely, that the English story starts late in the day
with Angles, Saxons and Jutes, as inferred from the illuminated
writings of the Dark Age clerical historians Gildas (sixth century
AD) and the Venerable Bede (seventh century). Here the label 'myth'
is mine, rather than that of any dissident archaeologist.

I agree that much of the unique genetic, cultural and linguistic
identity of the English did come from the nearby continent of
north-west Europe, but I contend that this process started, not as
some blitzkrieg during the Dark Ages, as we learn from our history
books, but long before the arrival of the Romans. What is more,
Beowulf – our first written poem *and* the only surviving complete
saga in Old English – used a Germanic language, one of whose
ancestors could have arrived in England even before the Romans
made their mark in Britain.

Apart from the etymology of our country's name, how did the
conventional view of the English as descended from recent Saxon
invaders come about? The Saxon story goes right back to the Dark
Ages. Bede and St Gildas tell respectively of fierce invading Angles
from Angeln (in Schleswig-Holstein, north-west Germany), or
Saxons from Saxony and Jutes from Jutland over the fifth and sixth
centuries AD. And then there is the well-documented history of
Anglian and Saxon kingdoms covering England for half a
millennium before the Norman invasion. The Saxon suffix '-sex',

for example (Sussex means 'south Saxons'), is plastered all over our English shire-names.

So, who were those Ancient Britons and their descendants remaining in England to be slaughtered when the legions finally left? For recent scholars, the presence in Roman England of some Celtic personal and place-names suggests that occupants of England were Celtic-speaking at that time. This argument could gain further support from the story of Iron Age Celtic invasions driving through England, if that were true. However, there is a reasonable linguistic presumption that there were 'Celts' living in England before and during the Roman occupation. But then, in the absence of any other linguistic evidence, this firms up to the modern linguistic view that before the Roman invasion *all* rather than *some* Ancient Britons were 'Celts' and Celtic-speaking.

It is natural to conclude that something cataclysmic happened in England during the Dark Ages. Many think, for instance, that the Celts were totally eradicated – culturally, linguistically and genetically – by invading Angles and Saxons. This sort of logic derives partly from the idea of a previously uniformly 'Celtic' English landscape, together with the clear evidence of uniformly Germanic or Norman modern English place-names today, and the preponderance of Germanic words in modern English.

Now, Gildas and Bede painted a grim picture, but neither actually specified complete ethnic cleansing. Some geneticists, and rather fewer historians and archaeologists, however, still believe that these invasions were massive and involved the influx of whole communities from Germany. In the extreme view, invaders were thought to have swept across a defenceless and largely depopulated England and to have replaced all the remaining 'Celts' in the country. Such complete replacement, not only of a people but of their presumed ancient English Celtic linguistic and cultural heritage, would have to be explained in the context of the lack of any Celtic linguistic substratum in English of any period.

How sure can we be that England was universally Celtic before? Roman writers, for instance Strabo, explicitly exclude Celtic affinities of the English on various grounds, such as greater size and not so yellow-haired. Unfortunately we have little to go on as to what Romans actually meant by 'being like Celts'. From his own stated view of 'Celtic', Strabo would have meant south-west rather than Central European. Tacitus, on the other hand, felt that those Britons living near Gaul were more like the Gauls physically and linguistically as a result of migration, although it is probable that he meant the Belgic and not the Celtic Gauls (see below). He was more explicit about some other Britons that we now choose to call Atlantic Celts. Referring to the Welsh, whom he calls 'a naturally fierce people', he states: 'The dark complexion of the Silures, their usually curly hair, and the fact that Spain is the opposite shore to them, are an evidence that Iberians of a former date crossed over and occupied these parts.'[3] This observation can hardly be support for the notion that the other parts of the British Isles were necessarily the same or had a common ancestry even at that time, and merely reinforces the new genetic evidence I shall present in this book.

So, if not 'Celtic' by Strabo's description, but rather Gaulish according to Tacitus, who were the Britons occupying England at the time of the Roman invasion? The Belgae of northern Gaul (Belgium and France north of the Seine) had tribal namesakes in England during Caesar's time (e.g. there were tribes called Belgae and Atrebates around Hampshire as well as in Gaul). Tacitus, like Caesar, reported that between Britain and Gaul 'the language differs but little'.[4] As we know from Caesar's famous opening paragraph of the *Gallic Wars*, which begins 'All Gaul is divided into three parts',[5] 'Gaul' included the Belgae in northern Gaul, a region that stretched from the Rhine as far south as the Seine and Paris. However, unlike the Celtae of the middle part of Gaul, who he said identified themselves as Celtic in their own language, Caesar did not specify the language of the Belgae – stating repeatedly, however, that they had more in common with the Germani.

The history of early coins in Britain reveals a pre-Roman influence that is predominantly derived from north Gaul. The earliest coins to circulate in south-east England, c.150 BC, were made in Gaul and were produced by the Belgae. The richest Iron Age treasure ever discovered in Britain was unearthed at Snettisham in Norfolk. A burial date of c.70 BC is suggested by coins found in the majority of such hoards as grave goods, along with bronze, silver and gold torcs (Plate 16). Coins were subsequently produced locally throughout southern England, but not in contemporary Cornwall, Wales, Scotland or Ireland.

Even farther north, the curious Iron Age culture of East Yorkshire known as the Arras Culture, characterized by chariots and square burial barrows, lasted for four hundred years until the Roman invasion and showed cultural links with northern Gaul. The fact that the Romans would call the inhabitants of East Yorkshire 'Parisii', a name also given to the tribe who went on to found Paris, has led some to speculate that these people were immigrants from northern France.

So, one might surmise that the 'common language' referred to by Tacitus as being spoken on both sides of the Channel was not Celtic, but was similar to that spoken by the Belgae. From present linguistic geography, and from numerous hints dropped in Caesar's *Gallic Wars* about languages spoken in northern (Belgic) Gaul, the language shared across the Channel is more likely to have been of the Germanic group (see Part 3). If so, it might have been a member of the West Germanic branch of Indo-European (i.e. something like Dutch, Flemish or, more likely, Frisian) rather than Atlantic Celtic (Gaulish). In other words, a Germanic-type language or languages could already have been indigenous to England at the time of the Roman invasion. In support of this inference there is some recent linguistic analytic evidence, which I shall discuss, that the date of the split between Old English and Continental Germanic languages goes much further back than the Dark Ages, and that English may owe more to Scandinavian languages. But such speculation merely adds to the confusion that standard comparative linguistic analysis *already* places Old English (the language of *Beowulf*) on its own separate branch, and closer to Frisian than to Saxon. The last observation is clearly inconsistent with the orthodoxy of Angles and Saxons replacing Celts, quite apart from the near-complete absence of a Celtic substratum in either Old or Modern English.

Modern popular images of the sort of English people the Romans met on their arrival, and left on their departure, vary. They range from the dark, feral, woad-painted savages, gibbering a version of Cornish, depicted in the recent Hollywood movie version of King Arthur's story, to the more honest admission of ignorance shown by French cartoonists René Uderzo and Albert Goscinny in their famous *Asterix* comic-strip adventures of Celtic-speaking tribes in Brittany. In *Asterix Goes to Britain*, the cross-Channel connection is caricatured in the form of a moustachioed, spindly-legged toff in plus-fours, a fraffly polite British cousin of the Gaulish Asterix. Named Anticlimax, he comes over from England to Asterix's village in Armorica to ask for help fighting the Romans. In sketching the latter portrait they create their hallmark mixture of slapstick and modern contemporary lampoon, underwritten by a canny reading of the classics.

Uderzo and Goscinny incidentally come much nearer the mark than Hollywood. In one frame, Asterix echoes Tacitus' comparison in telling his friend, the huge, dull Obelix (who has just violently misunderstood the purpose of a handshake with Anticlimax), that '*they* don't talk quite the same as us'. If the cousinship and the people and language links had been Belgic rather than Gaulish, the sketch would, in my view, have been very close to reality. The first coins struck in Britain, around 40 BC, bear the name Commios. It is believed that this might well be Commius, a king of the Belgic

Atrebates, who fled to Britain in 51 BC after rebelling against Julius Caesar.

My unorthodox view of English roots does not deny the historical significance of the imposition on the indigenous population of an Anglo-Saxon ruling elite. There are ample historical records for the establishment of Saxon kingdoms in England – Wessex, Essex, Kent, Sussex, East Anglia, Northumbria and Mercia – and of violent internecine warfare, but that may have been carried on against a pre-existing English cultural and genetic background.

The English maternal genetic record (mtDNA) denies the Anglo-Saxon wipeout story. English females almost completely lack the characteristic Saxon mtDNA marker type still found in the homeland of the Angles and Saxons. The Y chromosome evidence is potentially more informative, but the same data have been used by researchers variously to 'prove' either a wipeout or slightly less than 50% replacement. In this book I shall show that although there is some evidence for invasion in the first millennium AD, the 'replacement' was a mere 5%. So what does the Y chromosome say about English links with the Continent? A picture emerges that is surprisingly similar to that provided by mtDNA. There are general English similarities with Frisia, but these result mainly from common colonization history and intrusions in Neolithic times. Interestingly, the sixth-century writer Procopius of Caesarea mentions the Frisians as Dark Age invaders of the Isles, but his second-hand report is highly suspect. Specific genetic links do exist between the English and the European source regions suggested by Bede, but they do not support the wipeout theory.

A picture thus comes into focus of the 'Anglo-Saxon invasions' as less of a replacement and much more akin to the later Norman conquest: that of battles for dominance between various chieftains, all of ultimately Norse origin. Frisian, Jute, Angle, Saxon, Norman: each invader shared much culturally and, except in the last case, linguistically with their newly conquered indigenous subjects.

So, how far back in time can we trace the linguistic-genetic links across the North Sea? Or to put this question another way, for how long have the English been different from their Atlantic Celtic neighbours to the west in Wales, Ireland and Cornwall, and to the north in Scotland? The cultural record indicates that the continuous trade relationship between England and the nearby Continent carried on in parallel with substantial incoming gene flow back well over 5,000 years before the 'Saxon Advent'. The line separating the English from the rest of the Isles is repeated in different expressions of the cycles of glorious artistic and funerary traditions over the same period. Examples are trade networks of various kinds, distribution of megaliths and passage graves and portal tombs, pottery beakers and their exclusive trade networks, and gold torcs.

This story of two sets of Britons has echoes back in the introduction of farming – which, incidentally, arrived first in Ireland. The spread of farming to the western British Isles by the Mediterranean route, as mentioned above, contrasts with a parallel but separate spread into north-west mainland Europe up the Danube from the Black Sea, arriving in the Netherlands by 7,300 years ago. The north-westerly Neolithic cultural expansion was hallmarked by the spread of another type of pottery, known as *Linearbandkeramik*. From archaeological evidence, northern Neolithic traditions, although not *Linearbandkeramik*, may have spread across the North Sea to Norfolk and Cambridge by 6,200 years ago (see Chapter 5), thus possibly taking the roots of English separate identity back over 6,000 years.

A note of caution

Academe is naturally conservative. Added to this, the excesses of Nazi 'racial anthropology' shocked scholars elsewhere into moving towards extremes of political correctness. Migrations became equated with racial 'Aryan invasions'. As a consequence, anything that sails too close to overt 'migrationism' has been frowned upon in

English archaeology for the past fifty years. Likewise, Colin Renfrew, in his revolutionary popular book on Indo-European origins, *Archaeology and Language*, peppers his essentially migrationist message with qualifications and caveats against such practice.

As explained above, I strongly endorse such caution about seeing migrations where there is evidence only for cultural spread. I also argue in favour of the conservative nature of prehistoric gene flow in both of my books on genetic trails (*Eden in the East* and *Out of Eden*). Again, in this book, I argue that those who arrived first on the physical landscape tend to dominate the modern genetic one. But this does not mean ignoring genetic evidence for real migrations when it is there. Three aspects are special about the British story. First, the British Isles were cleared of people after the last Ice Age, thus avoiding the complication of a Palaeolithic gene pool. Second, and uniquely for such a small region in Europe, there is a deep genetic line dividing the English from the rest of the British. Third, the separate genetic source regions for the English and the Atlantic Celts are clear and distinct, and correspond to specific interpretations of the cultural evidence.

Part 1

THE CELTIC MYTH: WRONG MYTH, REAL PEOPLE

1

'CELT': WHAT IT MEANS TODAY, AND WHO WERE THE CLASSICAL HISTORIANS REFERRING TO?

Insular Celts: a modern myth?

Since how we view Celtic cultures today is probably most important for how we view them in the future, we should start with current perceptions. Nearly all adults in the British Isles will at least have heard of the terms 'Celts' and 'Celtic', and will have some opinion. Many of those living in non-English parts of the British Isles probably even recognize one another's descriptions of the terms, but opinions on their meaning will vary.

I recall vague descriptions from my own reading: 'A once great people, with no written history, speaking distinctive archaic tongues, now beaten back to their last strongholds in the western parts of the British Isles; brave, clannish, warlike, disunited, makers of fine jewellery and beautifully decorated weapons; poetic but illiterate creators of some of the most haunting oral European legends of magic, bravery and tragedy' (Plates 1–3). Other random images that might spring to mind are of the French cartoon characters

Asterix and Obelix (whose only fear was the sky falling on their head), of football clubs and T-shirts, or perhaps references to Ireland and faeries, as in Yeats's *Celtic Twilight*.[1]

How has this picture been built up over the years? In its 1913 edition, Webster's dictionary confirmed these stereotypes, defining 'Celt' thus:

> **Celt** ... *n*. [L. *Celtae*, Gr. *Keltoi* ...] One of an ancient race of people, who formerly inhabited a great part of Central and Western Europe, and whose descendants at the present day occupy Ireland, Wales, the Highlands of Scotland, and the northern shores of France.

This was a definite improvement in tone on the entry to the 1828 edition of Webster's, which dismisses Celts as 'One of the primitive inhabitants of the South of Europe.' But this earlier record contains an interesting change of European territory, reflecting a nineteenth-century move in archaeological perception towards a Celtic homeland in Central Europe. Not everyone sees Celts that way today, and not everyone defined them so specifically in the past. The Greeks and Romans, respectively, used the terms *Keltoi* and *Celtae*, and were, after all, contemporaries of the people *they* called Celts; but they never mentioned any connection with the British Isles.

Language is regarded as extremely important in modern perceptions of Celtic identity and ethnicity. One linguist, Myles Dillon, even insists that language is *the* test[2]; and that the only agreed definition of Celts should be people who spoke Celtic dialects. By 'Celtic' he presumably means the branch of Indo-European languages called Celtic by modern linguists. This view is at odds with classical descriptions of Celts, which were not primarily based on language, so it does not seem helpful. After all, the ancients, not linguists, introduced the term 'Celt'.

Classical commentators, following the excellent example of the first known historian, Herodotus, generally gave quite detailed and broad-based accounts of regional populations, many of which are still of great interest. Unfortunately for the preoccupations of modern archaeologists, they paid rather little attention to language. Also, as we shall see, Herodotus and those who followed him were less than specific or consistent about whom they meant by 'the Celts', let alone what language they spoke. This means that there is potential for doubt as to whether the modern Celtic languages have any connection at all with classical Celts, let alone the sort of identity that is claimed today. So it is rather important to be sure that there is no confusion about what is meant by the term 'Celtic'. Needless to say, confusion is just what has happened.

The only parts of Europe that now speak what modern linguists call 'Celtic languages' are the British Isles and Brittany. This creates difficulties in linking those languages with the putative origins of Celtic culture in Austria and southern Germany, an area of Central Europe that is German-speaking today. As we shall see, there is clear evidence for the presence of Celtic tongues in ancient times in parts of France, northern Italy and Spain (i.e. in south and south-west Europe), but during Roman times they were largely replaced by local hybrid Romance languages – French, Italian and Spanish. There is no such evidence for Celtic languages ever having been spoken in a 'Celtic homeland' in Central Europe, and therefore no reason to argue that Romance languages replaced them there. So if prehistorians and linguists of the last 150 years wanted to find a convincing homeland for Celtic languages, why on earth were they looking in Central rather than south-west Europe? The short answer is that Herodotus, in his identification of the geographical location of the *Keltoi*, mistakenly thought that the Danube rose somewhere near the Pyrenees rather than in Germany (but more of that below).

Celto-sceptics

Some archaeologists have, over the last couple of decades, become quite red-faced about the whole issue of Celts. They warn against the dangers of *racial migrationism* and point to the lack of archaeological evidence for mass migrations into the British Isles during the Iron Age. They further question the relevance and meaning of Celtic ethnicity. Their reasoning is that whatever the term 'Celt' may have meant to the ancients, it was not based on a clearly defined language group and thus does not amount to an adequate ethnic description.[3] Furthermore, they argue that classical Celts bear little relation to the modern imagined picture of the origins of Atlantic coastal Celts. Following this argument through, they give the modern construct of the romantic Celtic story the mantle of a myth with the apparent intention, in one case, of invalidating any use of the word.[4]

There are two problems with this attack on the commonly used words 'Celt' and 'Celtic'. First, such academic arguments will not make the words go away or stop being used. Classical authors used the terms for a thousand years. Second, the term 'ethnicity' has no better claim than 'Celtic' to a clearly defined usage. While dictionaries still conservatively define 'ethnicity' with reference to 'race' and language, anthropologists have driven current usage much more towards softer concepts of perception, affiliation and self-identification. A common mother-tongue is not a prerequisite for this.

Debunking the myth of the Central European Celtic linguistic and cultural homeland is a long overdue task, but we should not lose the baby with the bathwater, and it is important to separate the fallacy of the 'Celtic homeland' from the possibility that the 'Celtic' language story may still have something to tell us. To make that differentiation, we have to look at the evidence for the origins of Celtic culture and of modern Celtic languages in rather more depth.*

---

*To avoid having endlessly to repeat the distinction between, on the one hand, the terms 'Celt', 'Celtic', 'Celtae' and 'Keltoi' as used by whomever and however loosely, and, on the other hand, the modern definition of 'celtic languages', from here on I shall use lower-case 'c' for the celtic languages and a capital for everything else Celtic (except in passages quoted from other sources). The reason for this particular distinction is the sceptics' doubt that Celt and celtic languages have any solid or meaningful connection.

Words from the past: Celtic philology

In their debunking of the modern Celtic story, archaeologists such as Simon James (in *The Atlantic Celts: Ancient People or Modern Invention*) seem to blame early linguists for the persistence of logical errors in the Celtic myth.[5] I think this is unfair: they should really be blaming recent generations of their own profession for constructing the myth of a Celtic Iron Age homeland in Central Europe.

The person whose name features most prominently in these Celto-sceptic polemics died a long time ago and was neither a trained linguist nor an archaeologist. He belonged to a breed of general scholar that has all but died out in the last hundred years. He was an antiquarian named Edward Lhuyd who lived in Oxford in the early eighteenth century, and was keeper of one of the oldest museums of all, the Ashmolean. Lhuyd could also be called other names, although such labels were yet to be coined. He has been called a Welsh nationalist,[6] and also a philologist, on the basis that he was one of those who founded the discipline. But above all he was a persistent, innovative, hard-working and self-motivated scholar. (Historical linguists, or philologists as they used to be called, were an early product of the Enlightenment, preceding archaeologists. Antiquarians became self-aware as professional 'archaeologists' only towards the end of the nineteenth century.)

The accessibility of written and spoken European texts, both ancient and modern, combined with the rational clarity of thought and enquiry encouraged by founding Enlightenment philosophers, provided Lhuyd with a powerful cultural microscope, a time machine requiring no equipment save pen and paper, access to a library and hard work. But one does not have to be a scholar to see that some languages, such as English and German, show systematic and measurable links as soon as one looks at them. For Lhuyd and other philologists, the excitement was electrifying. It must have seemed as if the dusty craft of words was providing a new window into the past that required only a little cleaning to remove all opacities.

Lhuyd put years of his life into field and library research on the syntax and lexicons of the languages of Wales, Ireland, Scotland, Cornwall and Brittany, both historical and modern. Actually he owed some of his impetus and inspiration to the work of a Breton, Paul-Yves Pezron, who published a book entitled *Antiquité de la nation et de la langue des Celtes, autrement appelés Gaulois* in 1703.[7] Lhuyd apparently helped as an entrepreneur in arranging swift translations of Pezron's work into English and Welsh. The English translation appeared in 1706 under the title *The Antiquities of Nations, More particularly of the Celtae or Gauls, Taken to be Originally the same People as our Ancient Britains*, only a year before Lhuyd's own magnum opus, *Archaeologia Britannica*.[8]

The creative extension of Pezron's French title in its translation into English is telling, and summarizes the two basic hypotheses shared by the two men. One of these extended the application of the 'Celt' terms, beyond the meanings given them by the ancients, to include the people and languages of British Isles. In other words, these eighteenth-century bookworms created the concept of *insular Celts*.[9]

The other, less justified hypothesis was the implicit assumption that all 'Ancient Britons', including those previously living in England, were celtic-speaking. Crucially, no mention was made of a Central European homeland. Although their work was a milestone in historical linguistics, it should be remembered that neither of the hypotheses was new; the Scottish scholar George Buchanan had suggested both ideas previously in outline as early as 1582.[10] Unlike Pezron and Lhuyd, however, Buchanan had made a more objective and systematic analysis of all the classical sources, and had even tried to marshal evidence for an ancient celtic-speaking England, from place-names and standard Roman sources such as Tacitus, Strabo, Ptolemy and Caesar.

Buchanan had also linked north-west Spain with Ireland and northern England on the basis of the respective tribal names –

Brigantia and Brigantes – assigned by Ptolemy during Roman times. Buchanan argued for three related 'Gallic' dialects spoken during classical times in the British Isles: Belgice, spoken in northern Gaul and south-east England; Celtice, spoken in Spain, Ireland and Scotland; and Britannice, a language ancestral to Welsh.[11]

Simon James speculates at length on the reasons for Lhuyd's choice of the term 'Celtic' rather than 'Gallic', implying that this was a pivotal decision that gave rise to later myths. He suggests Lhuyd's own nationalism, the recurrent poor relations between the French and the English, and the coincidence of the date of publication of Lhuyd's own book with the Treaty of Union between England and Scotland as possible factors in this decision and in the subsequent explosion of popular interest in things Celtic.[12]

But this emphasis on the origin and effect of opting for the 'Celtic' rather than the 'Gallic' label, and Lhuyd's contribution to it, all seems to me perhaps laboured, or at least a polemic decoy. Disembodied slogans do not work alone. If the concept of an insular-Celtic or Breton Gallic heritage was a mere fad of the Enlightenment and Romantic eras, it would have died long ago and would have no continuing popular resonance, let alone the power to survive in the names of British football teams and the modern French cartoon characters Asterix and Obelix. Whatever his supposed motives or inaccuracies, Julius Caesar had already made the connection between language and the terms 'Celtae' and 'Galli', nearly two thousand years before, in the first paragraph of his famous campaign epic *Gallic Wars* (see p. 4). Both Lhuyd and Pezron seem to have been even-handed, whether one agrees with them or not, in using the terms 'Celtic' and 'Gallic' interchangeably. Given the classical texts available then, as today, the hypothesis of the geographical relationship between Ancient Celts and the celtic languages of Brittany and the British Isles was a reasonable provisional interpretation of new linguistic data.

What has always been lacking is a systematic testing of the linguistic model against alternative scenarios for the geographical origins of the Ancient Celts and the cultural relevance of celtic languages. The dominant view over the past 150 years has been that Celts had their origins in the Iron Age cultures of Central Europe. Although apparently the more glaring howler, this does provide such an alternative scenario for systematic comparison. James[13] is less ready to name and shame his archaeological colleagues and forebears of the nineteenth and twentieth centuries for perpetuating that leap of imagination than he is to lampoon Lhuyd. The Central European homeland theory cannot be put at Lhuyd's door, although he did believe in Iron Age migrations to Britain. Luckily, this alternative model is independent of the linguistic question, thus reducing the overlap of evidence and the opportunity for circular reasoning.

After many pages on Lhuyd, Simon James has this to say on the Central European Iron Age connection:

> During Victorian times, as scientific excavation began to develop, major discoveries in mainland Europe were ascribed, with considerable confidence, to the continental Celts or Gauls of the classical texts. Of particular importance, were the finds in Hallstatt in Austria and La Tène in Switzerland. At Hallstatt, many richly furnished Early Iron Age graves were excavated. The finds proved to be related to material from a wide region north of the Alps, and seemed to correspond in time and place to the earliest Greek references to *Keltoi* (around the sixth century BC).[14]

He goes on to describe similarities between some of these cultural finds and burials in eastern France and the Rhine basin, and adds:

> Again on grounds of date and geographical location, these remains were identified with the Celts or Gauls which classical sources reported had poured into Italy from just these areas around 400 BC. The unique traits of the artefacts ... became identified with Celtic Gaulish peoples, and the areas ... thought of as 'Celtic homelands'.[15]

Now James is clearly not impressed by the quality of evidence for these connections, but he devotes much less space to exploring the story of this archaeological inference than he does to Lhuyd. Nor does he cite the relevant classical references used for constructing this part of the Celtic myth. From my reading of his argument and those of other Celto-sceptic debunkers, however, the Central European homeland story was the weakest link in the chain of the insular-Celtic identity construction, when compared with the linguistics and requires most careful testing.

Locations for the Celts given by the classical writers

Since evidence for the supposed homeland of the Celts is central to the story of British origins, I rang Simon James to ask him how on earth anyone could have come to the conclusion that the Celts originated in Central Europe. More specifically, I mentioned Herodotus and asked him which classical sources, imputed in his book, had been used by archaeologists to place Celts north of the Alps in Central Europe.

Herodotus

As I already suspected, the main sixth-century BC 'source' was the grand historian himself, Herodotus, who, in a discussion on the difficulty of measuring the length of the Nile, demonstrated in a passing comment how little he knew about the source or course of the Danube:

> [The Nile] starts at a distance from its mouth equal to that of the Ister [Danube]: for the river Ister begins from the Keltoi and the city of Pyrene and so runs that it divides Europe in the midst (now the Keltoi are outside the Pillars of Heracles and border upon the Kynesians, who dwell furthest towards the sunset of all those who have their dwelling in Europe).[16]

In this quote Herodotus is clearly talking about Iberia in south-west Europe, but mistakenly thinks that it held the source of the

Danube. In published reconstructions of Herodotus' distorted view of the route of the Danube through Europe, historical geographers have uniformly taken 'Pyrene' to refer to a Pyrenean rather than Central European location. Avenius (Proconsul of Africa, AD 366) in his description of the Atlantic and Mediterranean coasts of south-west Europe (in which he also discusses the Celts),[17] mentioned the port of Pyrene as being near Marseilles. Livy also mentioned 'Portus Pyrenaei'.[18] Both writers were probably referring to the Roman port later popularly known as Emporiai (meaning 'markets') situated at the eastern end of the Pyrenees – now the Spanish archaeological site of Ampurias.[19] This is the simplest explanation, consistent with three of four geographical locators, for Herodotus' western or Iberian source of the Danube, implicit in his statement. These three locators are the Latin name for those mountains, Montes Pyrenae, and Herodotus' two statements that 'Keltoi are outside the Pillars of Heracles' – i.e. beyond Gibraltar – presumably on the Atlantic coast of Iberia – 'and border upon the Kynesians, who dwell furthest towards the sunset of all those who have their dwelling in Europe' – i.e. the Kynesians (referred to in other writings as the Kynetes, Cynesians or Conios) lived at the westernmost point of Europe, in south-west Iberia. This third locator fits with other authors' statements that the Kynetes were neighbours of Tartessus (i.e. north-west of Gibraltar in the Gulf of Cadiz).[20]

The fourth and least credible locator is, of course, the geographical source of the Danube, which, from the first three locators, Herodotus incorrectly thought to rise in the Pyrenees. Nineteenth-century historians were well aware that the source of the Danube is in Germany and should have recognized the inconsistency. Unsurprisingly, Herodotus did locate the mouth of the Danube correctly – discharging into the Black Sea, much nearer to his own home – and did acknowledge that his information for the rest of the Danube's long course in Europe was second-hand from other sources (Figure 1.1).

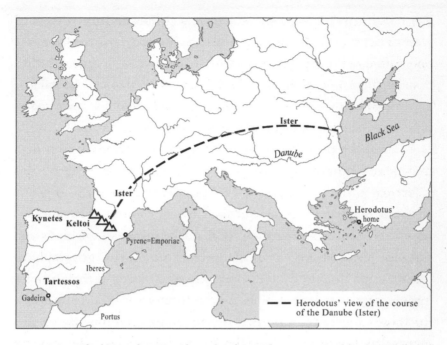

Figure 1.1 Which Danube? Herodotus lived in Halicarnassus, and his knowledge of Western Europe was sketchy. Although he realized that the Celts were far to the west, near the Pyrenees, he mixed this location up with the actual source of the Danube in Central Europe. This mistake unwittingly spawned the nineteenth-century myth of Celtic origins in Iron Age Central Europe.

It is clear that the inference from this passage in Herodotus that Celts came from Central Europe is at best wishful thinking and at worst deliberate distortion. The more rational view of Herodotus' description is obviously that Celtic lands were in south-west Europe and he had got both the course and source of the Danube wrong. This puts those nineteenth- and twentieth-century archaeologists out on a very shaky limb – and not same one Buchanan, Pezron, or Lhuyd were sitting on. The view that there was a connection between celtic languages and the classical Celts is consistent with Herodotus' statement of proximity to the Pyrenees and with the good evidence that celtic languages were spoken in France and Spain during early classical times.

This alternative view led me to search for any other classical references which might give more specific or consistent early geographical locations for the Celts. Since Herodotus may have been deliberately misread, I needed to take the same legalistic approach to any original source. In this way, it might be possible to tease out a more consistent and specific picture.

During classical times, the Celts had a particular aversion to putting anything about themselves in writing. This frustrating quirk was due not to a lack of literate Celts in the Roman Empire, for they certainly wrote about other things, but apparently to a real disinclination to write about themselves. So over roughly a thousand years from the sixth century BC, we find numerous references to 'Keltoi' (Greek) and 'Celtae' (Latin) by mainly non-Celtic classical authors. This long period of use implies that the term 'Celt' was continuously regarded as a useful descriptor, whatever evolutionary changes there were in its meaning.

Several classical authors clearly used 'Celts' in the non-specific sense of 'Western Barbarians'. This lack of clarity about peoples, ethnicity and geography is still common today and should not be used as an excuse to bin all classical descriptions of Celts. Many British today refer to 'Asians' when what they mean is 'people from the Indian subcontinent', and an apocryphal story of a television quiz has an English girl identifying the language spoken by people in China as 'Asian'. This kind of loose usage obviously reflects more on the knowledge of the user than on the specificity of such terms. It cannot be used to invalidate statements by other classical authors, who were more careful with the terms they used. I do not intend to tabulate every word that was written about the ancient Celts. Others have done this at length, both arguing for, and against the Central European Celtic homeland.[21] In reviewing these commentaries, I will focus on those statements by classical authors, which demonstrate specific locators and valid context.

The orthodox geographical interpretation

When I rang Simon James with my question about original classical sources for the orthodox view of Celtic origins, he also
pointed me at two books written over the past two decades with
titles resonant of the world of Obelix and Asterix. These were
Barry Cunliffe's *Greeks, Romans and Barbarians*[22] and David Rankin's
Celts and the Classical World.[23] As an afterthought, he mentioned
John Collis's recent book, on the sceptical side of the fence.[24] As it
happened, I got access to the older 'orthodox' books several weeks
before I obtained the sceptical one. So I shall start first with my
own impressions, gained by perusing their cited evidence and that
of Barry Cunliffe's more recent views as expressed in his book
Facing the Ocean.[25]

Avenius and Himilco

In his search for Celtic origins, Rankin quotes from Avenius'
fourth-century AD poem *Ora maritima* extensively and, in my view,
inconclusively. His stated reason for using this rather late
commentary is Avenius' own claim to have had access to many
texts, some very ancient. The most important of these is the lost
text of Himilco's *Periplus of the Northern Sea*. Himilco was a Punic
(Carthaginian) admiral who explored the north-west coast of
Europe at the end of the sixth century BC. Avenius' poem gives
verifiable information on the Atlantic coasts of Spain and Portugal,
mentioning the Pillars of Hercules (Straits of Gibraltar), Cadiz,
Tartessus and the Gulf of Oestrimnicus, and the impressive
Cordillera Cantabri mountains (ridge of Oestrimnis) towering
over its southern shore.[26]

At this point, however, Avenius' account becomes more difficult
to follow. According to him, within the Gulf of Oestrimnicus arise
the Oestrimnides Islands. Since Avenius describes these islands as
rich in the mining of tin and lead and mentions the use of curraghs

(leather-covered coracles), Rankin identifies them as either the Cassiterides (the 'tin islands' used by the Phoenician traders) or just as Cornwall.[27] According to modern interpretation of other sources,[28] 'Cassiterides' is thought to refer, misleadingly, both to Cornwall and to the north-west coast of Spain, neither of which are islands in their own right. Avenius says that Tartessus traded with the Oestrimnides, which would certainly fit the archaeology of the long-term trade between the Cadiz (then known as Gadir) area and north-west Spain.[29]

One might even add the peninsula of Brittany, also a source of tin, to the 'peninsular' tin islands of the Cassiterides. The tip of the Breton Peninsula is nearer to that of Cornwall (Land's End) than it is to any other part of France outside Brittany. The lack of any reference by Avenius to the English Channel highlights this key geographical feature, which through the ages determined and directed the Atlantic coastal trade route from Spain and southern France and then along the western fringes of the British Isles (Figure 1.2).[30]

Cornwall would make sense as one of the Oestrimnides, because Avenius goes on to say that it is two days' sailing from there to the Sacred Isle (Ireland), inhabited by the race of Hiberni, with the island of the Albions (Britain) nearby. Finally getting to Celts – the ultimate point of the quote – Avenius' report of Himilco's sea voyage takes us further north. Since the preceding leg of his voyage brought Himilco to the Irish Sea, this next section of his port-by-port *periplus* (literally his captain's log) presumably sees him moving north through the Irish Sea to the west coast of Scotland:

Figure 1.2 Early voyagers up the Atlantic Coast: Pytheas and Himilco. Pytheas the Greek bypassed Spain to get to Britain in the late fourth century BC. He may have reached as far north as Shetland. Himilco, a Carthaginian admiral, is reputed to have sailed to the British Isles two hundred years before – via Cadiz.

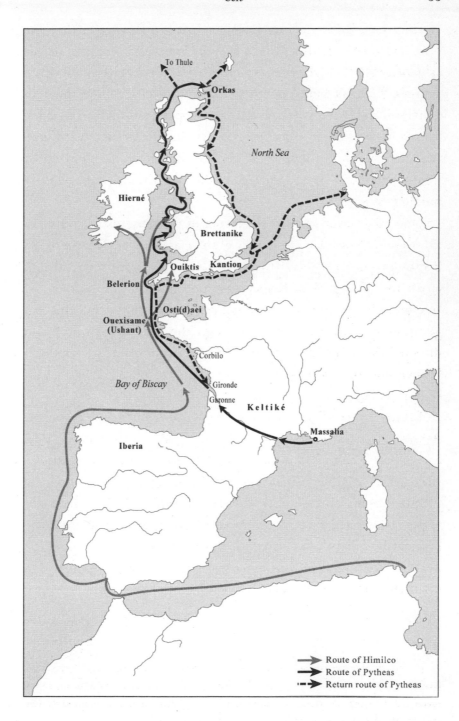

To Thule

Orkas

North Sea

Hierné

Brettanike

Ouiktis

Kantion

Belerion

Osti(d)aei

Ouexisame
(Ushant)

Corbilo

Bay of Biscay

Gironde

Garonne

Keltiké

Massalia

Iberia

Route of Himilco
Route of Pytheas
Return route of Pytheas

> ... away from the Oestrimnides
> under the pole of Lycaon (in the Northern sky)
> where the air is freezing, he comes to the Ligurian land, deserted
> by its people: for it has been emptied by the power of the Celts
> a long time since in many battles. The Ligurians, displaced, as fate
>   often
> does to people, have come to these regions. Here they hold on in
>   rough country
> with frequent thickets and harsh cliffs, where mountains threaten
>   the sky.[31]

The wild, freezing, mountainous country would certainly fit Scotland. If so, it would be the only classical reference that directly links Celts with the British Isles – a tantalizing anomaly.

Even more anomalous, however, is the confusing reference to some former inhabitants in this northern wasteland who were chased out by the Celts. Himilco called these unfortunates the Ligurians. Avenius goes on to say that the Ligurians had at first hidden themselves in the interior of this northern land before gaining suffi- cient courage to migrate south to Ophiussa (Portugal) and then to the Mediterranean and Sardinia. The reason that these lines of the *Ora maritima* are all so odd is that numerous classical references place Ligurians 'at home' in the Italian and French Riviera, and there is no other mention of a homeland near Scotland. Indeed, the Italian Riviera is today called Liguria, and has a distinct Franco-Italic dialect known as Ligurian. Several authors, including those cited by Avenius, relate that Celts and Ligurians jostled with the Greek Marseillaise for space in this region in the first millennium BC.

The simplest explanation is that Avenius, ignorant of the geographical location of Liguria, had performed a mistaken scissors-and-paste job on his many ancient source texts and confused the Riviera with Scotland. This does not wash. Given his post as Roman Proconsul of North Africa towards the end of the

Roman Empire, and the kudos he hoped to gain by his literary effort, it is highly unlikely that he was ignorant of Liguria on the Riviera over the other side of the Mediterranean. He may, however, have known something we do not about the origins of the Ligurians and their connections with Celts, and the British Isles, not to mention Himilco's throwaway remarks about Ligurians travelling south via Portugal and Sardinia – both locations on the Phoenician trade network.

Other modern readers of these lines of Avenius have placed Himilco's northern Ligurians variously in northern Spain or the Baltic coast. I think that the best approach is to accept the most parsimonious text analysis, which is that Himilco thought there were Celts and some other people in Scotland, but to be sceptical about the reference. So we are left with this mysterious suggestion of migrational links up and down the Atlantic coast.

So much for Avenius, who, whatever else he wrote, gave no comfort to the concept of a Central European Celtic homeland. Rather, he raises the possibility of an Atlantic coastal connection. A major problem in trying to untangle such later Roman accounts is their recurrent references to the same earlier Greek sources, many of which are now lost.

Strabo and Pytheas

There was another early explorer whose trip to the British Isles may partially validate or even replace some of Avenius' geographical claims. This was the explorer Pytheas, of the Greek colony of Marseilles, whose account of his own 330 BC odyssey is unfortunately lost. Instead, his text comes to us in fragments, reproduced with venomous incredulity as asides in Strabo's *Geography*, which was written in Greek around AD 20. Strabo repeats Pytheas' names of Atlantic coastal promontories, in particular Cabaeum (Brittany), with a people called the Ostimioi, and their offshore islands,[32] the outermost of which was called

Uxisame (today's Île de Ouessant, or Ushant, off the tip of Brittany). Strabo states that these were all part of Celtica, rather than Iberia. The similarity of geography and names has been used to link Himilco's Oestrimnis and Oestrimnides with the Ostimioi and thus, at least partly, with Brittany, but does not clear up the confusion of identity between the Spanish and Breton sources of tin on either side of the huge Bay of Biscay.[33]

North of Brittany, Pytheas the Greek seems to have taken much the same route as Himilco the Punic admiral, moving via Cornwall to the Irish Sea and even beyond Scotland, to Shetland and Orkney. Other accounts confirm and fill out the role of Cornwall.[34]

As an expert on the archaeology of the Atlantic coast, Barry Cunliffe has drawn some of these and other fragments and strands together with the more solid cultural and archaeological evidence. He makes a geographic reconstruction of the mid-first millennium BC tin trade between the producers on the Atlantic coast (the Atabri/Cantabri of north-coast Spain, the Cornish and the Bretons) and the middlemen in southern France and Spain and the rest of the north-west Mediterranean coast. His reconstruction identifies two separate rival networks (Figure 1.2).[35]

The Punic/Phoenician state consortium to which Himilco belonged ran the southern and older of these two networks. The locations of Punic pottery suggest that it stretched mainly from Cadiz, north along the coast just as far as north-west Spain. Since this western Mediterranean consortium also controlled the Straits of Gibraltar, the rest of the Mediterranean traders – in particular the Greeks – were prevented from using this gateway to the tin-producers.

So instead, the Marseilles Greeks seem to have used their connections with the people of 'Keltiké' (as Pytheas called southern France, according to Strabo) around Marseilles and Narbonne, who most likely used an alternative cross-country route just north of the Pyrenees to gain access to the Atlantic coast from the

Mediterranean, thus bypassing Gibraltar (Figure 1.2). This may seem quite a trek, but there are good waterways most of the way across, starting from Narbonne, moving up the river Aude and through the Carcassone gap to Toulouse, and then down the river Garonne to Bordeaux and the Gironde on the Atlantic coast. Cunliffe neatly resolves some of the reasons for Strabo's disbelief at Pytheas' travel times by using his own text to suggest that Pytheas, as a Greek pioneer, actually took this trade route across Keltiké. By his discoveries further north, Pytheas could have opened up an opportunity for the rest of the Mediterranean to bypass the Punic/Phoenician Atlantic coastal monopoly in their search for tin (Figure 1.2).[36]

However, if Cunliffe is correct, then Pytheas, with such satisfactory travel arrangements, is unlikely to have been travelling alone, without a courier, or as a novice tourist. In other words, he would have been taking advantage of a pre-existing river–land–river–sea trade route worked out by the established local inhabitants of Keltiké. According to the Central European homeland theory, these locals could not have been Celts when Pytheas made his trip in the fourth century BC since, as Cunliffe goes on to show,[37] the Celts would have been either still in their Central European homeland or moving east into northern Italy. Yet Himilco, who made his own trip up the Atlantic coast nearly two centuries before, seemed to claim that the power of the Celts was very much in evidence in northern Britain.

Again we are led down the Atlantic coast rather than into Central Europe for the Continental connections of the Celts. Strabo himself is explicit on the antiquity of the Celts in the region of Narbonne, where Pytheas might have started his journey across Keltiké:

> This, then, is what I have to say about the people who inhabit the
> dominion of Narbonitis, whom the men of former times named

'Celtae'; and it was from the Celtae, I think, that the Galatae as a
whole were by the Greeks called 'Celti' – on account of the fame
of the Celtae, or it may also be that the Massiliotes, as well as
other Greek neighbours, contributed to this result, on account of
their proximity.[38]

In this passage Strabo defines the geographical and tribal origin of the
term 'Celt' and how it then spread by some process of inclusive
labelling. If there should be any doubt about this southern centre of
gravity, there are other classical commentators who concur.
Diodorus Siculus, writing in Greek rather earlier than Strabo, states:

> And now it will be useful to draw a distinction which is unknown
> to many: the people who dwell in the interior above Massalia
> [Marseilles], those on the slopes of the Alps and those on this
> [northern] side of the Pyrenees mountains are called Celts (Keltoi),
> whereas the peoples who are established above this land of Celtica
> in the parts which stretch to the north, both along the ocean and
> along the Hercynian Mountain [today's Massif Central], and all the
> peoples who come after these, as far as Scythia, are known as Gauls
> (Galatai); the Romans, however, include all these nations together
> under a single name, calling them one and all Gauls (Galatai).[39]

This apparently independent confirmation of Strabo's geographical
identification of a Celtic heartland in the extreme south of France is
very revealing. First, it seems to limit their range 'of former times'
not to southern Germany, but to Narbonne: a small area around
Marseilles, north of the Pyrenees, west of the Alps and south of the
Massif Central, and probably east of Aquitane. Second, but equally
important for untangling the Celtic mystery, both Greek authors
feel the need to explain how the local term 'Celt' came to be
conflated by Roman writers such as Julius Caesar with the much
larger regional labels of 'Gaul' and 'Gauls'.

And others

Apart from anything else, this southern homeland would go a long way to explaining anachronistic mentions of Celtici in the south-west of Spain and Celtiberi to the east of Madrid as early as the sixth century BC.[40] This information comes from authors such as Herodotus, Eratosthenes (third century BC)[41] and Ephorus (405–330 BC), who is cited by Strabo: 'Ephorus, in his account, makes Celtica so excessive in its size that he assigns to the regions of Celtic most of the regions, as far as Gades [Cadiz], of what we now call Iberia' (see also below).[42] Diodorus Siculus, probably citing Poseidonius, states that the 'Celtiberes are a fusion of two peoples and the combination of Celts and Iberes only took place after long and bloody wars'.[43]

The Romantic mythologist Parthenius of Apamea (first century BC) gave a telling and charming version of the popular legend of the origins of the Celts in his *Erotica pathemata*,[44] which preserves the Spanish connection and even hints at Ireland. Heracles was wandering through Celtic territory on his return from a labour – obtaining cattle from Geryon of Erytheia (probably Cadiz). He came before a king named Bretannos. The king had a daughter, Keltine, who hid Heracles' cattle. She insisted on sex in return for the cattle. Heracles, struck by her beauty, had a double motivation to comply. The issue of this union was a boy and a girl. The boy, Keltos, was ancestor of the Celts; the girl was Iberos. Rankin speculates further that the homophony between 'Iberos/Iberia' and the Irish mythical ancestor, Eber, may be more than coincidence.

We can provisionally accept this literary evidence of a Celtic homeland in the south of France, but several critical questions remain. How much of the spread of 'the Celts' was due to this conflation of terms (combined with 'the fame of the Celts' and consequent Roman labelling, as Strabo speculates), and how much was due to real population migration, invasion and/or cultural expansion? This issue might be amenable to genetic study, as I show later, but it forces a reappraisal of the terrifying and documented Celtic invasions of Southern and Eastern Europe (including Italy and Anatolia) in the fourth and third centuries BC, in which they sacked Rome itself (390 BC).[45] Instead of streaming across the high passes of the Alps from Germany and Austria to the north,[46] could these Celts have been anticipating Hannibal's example, a couple of centuries later, by crossing the Alps farther south into Italy, from a homeland in the south of France?

Rankin, in his book *Celts and the Classical World*, has a whole chapter on the early and long association of the Celts with the south of France and the Greek colony of Marseilles. On the persisting assumption that the Celts came from Central Europe, he has this to say:

> It is reasonable to suppose that the Celts had arrived in the region of Southern France some considerable time before their irruption into Northern Italy. By the fifth century BC they had established themselves firmly in what was to become Cisalpine Gaul, and at the end of that century were strong enough to threaten the safety of Rome.[47]

In spite of placing Celts so early in southern France, Rankin is still clearly convinced of the primacy of southern Germany as the Celtic homeland. He describes as 'an abiding preoccupation of the Roman mind … the vulnerability … to invasion … especially from the north.'[48] Cunliffe, in his 1987 book, is more explicit, drawing maps (Figure 1.3) showing big Celtic arrows driving south from the Marne–Moselle–Bohemian region straight into Italy.[49] Yet the Celtic tribes that Livy and other Roman historians describe as taking part in these fearsome fourth-century BC invasions, the Senones and Boii, were associated, in Caesar's time, with parts of France south of the Seine.[50]

Figure 1.3 Which Celtic homeland? Iron Age in Central Europe, as argued from the Hallstatt and La Tène expansions, or a Neolithic trail from south-west Europe, as suggested by language and Atlantic cultural distributions?

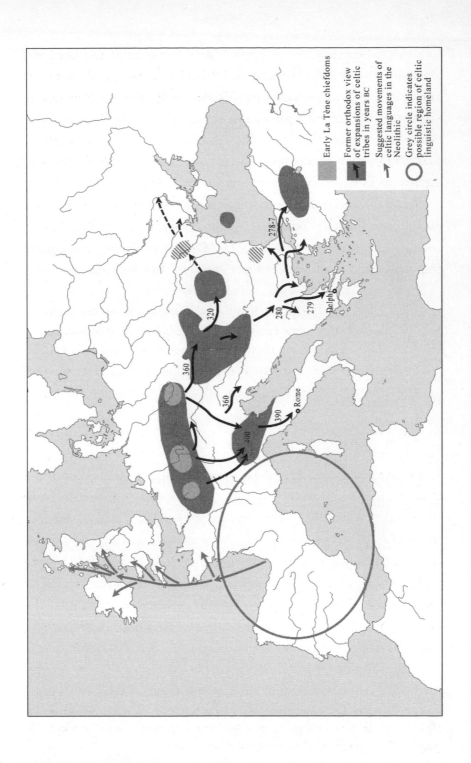

Early La Tène chiefdoms

Former orthodox view
of expansions of celtic
tribes in years BC

Suggested movements of
celtic languages in the
Neolithic

Grey circle indicates
possible region of celtic
linguistic homeland

278-7

320

360

360

280

279

Delphi

390

400

Rome

It may be claimed that several of these Celtic tribes, for instance the Boii, were subsequently associated with Bohemia. But Strabo makes it clear that Bohemia was where the Romans, who had subsequently got the upper hand over the Celts occupying northern Italy, pushed them later at pain of extinction.[51] While archaeologists base the identity of the northern invaders on types of artefact, there is very little support in the classical literature for a Celtic homeland in the north, let alone for the Celtic invasion of Italy coming directly from the north through the high passes of the Austrian and Swiss Alps.[52]

As for the third-century BC Celtic invasions of the Italian Adriatic coast and Greece and Asia Minor, Strabo is again explicit in placing their ultimate starting points in France south of the Seine. The Veneti came from Armorica,[53] and the Volcae Tectosages, who invaded Delphi in Greece and Galatia in Turkey, came ultimately from the Pyrenees.[54]

Julius Caesar

Strabo cites Julius Caesar extensively in his description of the distribution of the Celts in Gaul. So, bearing in mind Strabo's clarification on the ultimate southern origin of the term 'Celts' in Narbonne and its extension to Gauls, we might as well hear now what the iconic warrior-historian had to say himself in his memorable but lean opening lines of the *Gallic Wars*:

> All Gaul is divided into three parts, one of which the Belgae inhabit, the Aquitani another, those who in their own language are called Celts, in our [Latin] Gauls, the third. All these differ from each other in language, customs and laws. The river Garonne separates the Gauls from the Aquitani; the Marne and the Seine separate them from the Belgae.[55]

Although debunkers of the Celtic myth have argued that, in writing this document, Caesar had hidden agenda aimed at self-justification to his Roman audience, the lack of contradiction by other authors gives authenticity to this simple description (Figure 2.1a).

First, Caesar places a northern limit on the Celts within Greater
Gaul and in Western Europe as a whole. The Seine and the Marne
define this northern limit, which is much farther south than today's
French border and excludes all but a small sliver of the putative
Central European Celtic homeland. Significantly, Caesar includes
southern French locations of the tribes Livy identified as previously
having invaded Rome.[56] Caesar's use of these two rivers, as both
northern Celtic and celtic-language boundaries, is consistent with
evidence from place-name analysis. (I shall come back to this
northern Celtic boundary from another perspective in Chapter 7,
and at the same time discuss the controversial place-name evidence
in terms of where celtic languages gave way to Germanic-branch
languages in Caesar's Gaul.)

Second, consistent with Strabo's narrower south-eastern French
Celtic homeland, Caesar's Celtic zone of Gaul included, by default,
Narbonne in the south. Third, and perhaps most important in the
context of identity, Caesar, in his economic style, tells us that the
term 'Celt' is applied to one region only, and also identifies people
of this region of Greater Gaul as 'Celtic in their own language,
Gallic in ours' ('*qui ipsorum lingua Celtae, nostra Galli appellantur*').
Caesar may have conflated Celts and Gauls elsewhere, but his
meaning on this point is unambiguous, even to my rusty Latin: the
Gauls of this central region of Greater Gaul called themselves Celts
in their own language. I shall come back to the relevance of celtic
languages, but first a digression on the Celtic homeland.

Authority and the archaeological myth of the Celtic homeland

By this time in my reading of Rankin, I had seen sufficient of his
key quotes from classical sources to get a strong impression of
Celtic origins in south-west Europe, rather than the Central
European homeland he was promoting in the 1980s. I almost
began to wonder why Simon James had directed me to read such

books if the evidence used, and the classical texts referred to, were so contradictory to the orthodoxy of the Celtic Central European Iron Age homeland.

At this point I belatedly received the new book written by the Celto-sceptic John Collis: *The Celts: Origins, Myths, Inventions*.[57] Collis, Professor of Archaeology at the University of Sheffield and a leading British authority on the European Iron Age, confesses, on the first page of his acknowledgements, to having been a sceptic since his Cambridge student days in the 1960s, and perhaps even from his schooldays, when he first read Caesar and Herodotus. It is a testament to the power of academic conservatism that he has waited so long before committing himself so publicly in book form. Could this be regarded as an act of academic bravery, even in someone so senior? In Hans Christian Andersen's tale 'The Emperor's New Clothes', only a small boy had the temerity to point out that the Emperor had not a stitch on.

Having at last taken this decision, Collis makes a magnificent job of debunking 150 years of archaeological conviction-scholarship. As I mentioned above, James gives us three timorous lines of politely concealed incredulity, obliquely referring to unspecified details of 'time and place' as orthodox evidence for Celtic origins north of the Alps.[58] By contrast, the more senior author Collis devotes three and a half chapters out of eleven to deconstructing such evidence, naming and shaming the authorities responsible for this academic castle of cards.

I shall sketch this part of Collis's book here, while recommending the reader to his most thorough treatment. After covering the classical writers and early linguists, Collis enigmatically entitles his fourth chapter 'Race and time'. 'Time' here refers to the realization in the nineteenth century of the vast time depth in prehistory, in contrast to Bishop Ussher's rather shorter Biblical chronology, commencing in 4004 BC. Collis uses the term 'race' to introduce early concepts, growing during the eighteenth and nineteenth centuries, of racial differentiation and a new sense of nations, all mixed up with linguistics and comparative craniometry (head-measuring).

Collis quotes one anatomist, the otherwise brilliant Paul Broca (who gave his name to the speech area in our brains), confidently stating in 1864 that the Celts, who spoke so-called celtic languages, had arrived in central Gaul introducing the 'more civilized' Bronze Age to Europe from Asia in the East. Broca associated this event with a hypothetical change in head shape in the population. His clear picture of racial migration takes on a surreal tint, however, when he admits that opinions varied as to whether 'broad-heads' replaced 'long-heads' or vice versa.[59]

Against this background of early and distinctly flaky anthropology, elements of which were later to be developed and used by the Nazis to justify the Holocaust, the new 'Celts' were conceived and underwent a strange gestation. During the eighteenth and nineteenth centuries the term 'Celtic', unearthed initially by linguists, became re-established, after a gap of over a thousand years, as a reincarnated ethnic label – although clearly not meaning quite the same as could be inferred through the smoky window of the classical texts.

John Collis contrasts the views of mainly French authors of the nineteenth century on the origins of the Celts. The most conservative of these, Amédée Thierry, rather wisely stuck to the classical sources such as Livy, Caesar and Strabo, coming up unsurprisingly with Caesar's Middle Gaul as the original location of Gauls. In his *Histoire des Gaulois* (1827), Thierry gives the location of the Celtic homeland as south of the Seine and the Marne and well west of the Rhine, missing out most of the south-west third of France and the pre-existing Roman province of Narbonensis, presumably because of Caesar's vagueness farther south.[60] But Thierry argued for a Gallic presence in France from at least 4,000 years ago, with early invasions of Spain some time between 3,700 and 3,500 years ago, and of northern Italy by 3,400 years ago, based on some interesting interpretation of certain texts.

So far, so good. Thierry gives a text-based reconstruction that could be challenged rationally; but in the year Thierry published his book, another Frenchman, Henri d'Arbois de Jubainville, was born who subsequently had a profound effect on the nature of reconstructions of 'Celtic prehistory', replacing reason with conviction.[61]

D'Arbois de Jubainville published books and papers in the last quarter of the nineteenth century and into the twentieth, but somehow managed to infuse the rest of the twentieth century with his views on the issue of the Celts. His chronology was based mainly, but loosely, on the classics rather than the new archaeological interpretations; but his reconstructions were altogether more creative than either, accepting mythological statements in parallel with those of classical historians. He also made extensive but selective use of linguistics. D'Arbois de Jubainville saw four phases of colonization, or 'empires'. The first (undated) consisted of hunters and herders living in caves. The second empire was Iberian, dating from 6000 BC, with hunters and herders from Atlantis, speaking a non-Indo-European language, forced by the disappearance of their land to colonize new territory (i.e. Spain). He dated the third of these exotic empires at 1500 BC, formed by Ligurians (à la Himilco) as the first of the Indo-European speakers, bringing agriculture and colonizing from the East, thereby displacing the Iberians. The fourth empire is embodied by the Celts arriving from southern Germany in 500 BC, representing the second wave of Indo-European-speakers from the East.

The first three of d'Arbois de Jubainville's mythical empires seem to have been conveniently forgotten or dismissed, because, as Collis says, they were 'fanciful conjecture'. But this makes it all the more surprising that the fourth has survived in name, date and location into the twenty-first century, let alone as the orthodox paradigm of Celtic origins, since the way he abused the evidence from the classics in its creation was equally fanciful.

Collis appears to 'give face' to d'Arbois de Jubainville for the fourth
empire by saying that the reasons for his belief in the late arrival of
Celts 'need more careful consideration'. In this gentlemanly faint-
defence Collis cites the Frenchman's arguments: 'Firstly ... the
evidence of the introduction of Indo-European languages, the
assumption of their easterly origin, and their relatively late appearance
in western Europe ... secondly ... claims [of] evidence for population
change in the historical record'.[62] But when we find that the sources
for these arguments are taken from Herodotus and the *Periplus* of
Himilco as quoted by Avenius, Collis's cover as apologist is blown.

As we have seen, those tantalizing classical fragments are insufficient to form any but the flimsiest of hypotheses about the origins
and dates of classical Celts or any other West European tribes, let
alone their languages. And then we hear that this evidence was what
d'Arbois de Jubainville used to construct the story 'that the south of
France was taken over by the Gauls around 300 BC' from Central
Europe[63] – in the face of all the classical sources that give a much
earlier date (see above). We may wonder if Collis is being tongue in
cheek or just kind to his colleagues, who were all taken in by it for
the next hundred years.

After this, Collis's real sceptical stance becomes apparent. Still
citing d'Arbois de Jubainville, he continues: 'For the eastern origin
of the Celts, he uses Herodotus' assertion about the Danube rising
in the territory of the Celts [e.g. in Central Europe] as well as
Polybius' statement that the Gauls invaded northern Italy from the
other side of the Alps [e.g. from the north].'[64] As I had guessed when
reading these texts myself, Herodotus' misunderstanding of the
source the Danube is what underpins the whole house of cards of
the German homeland theory. It is further used to prop up an indef-
ensible interpretation of Polybius, since 'the other side of the Alps'
clearly refers to France and not to Germany. There is more
discussion of d'Arbois de Jubainville's distortion of his other
sources, and some description of French archaeologists who carried

this torch of Celtic origins ultimately through to this century, but Collis's message is clear.

In his fifth chapter, 'Art and archaeology', Collis explains how this basic misunderstanding of classical sources was then combined with the exciting new nineteenth-century finds of exotic jewellery and weaponry from Hallstatt and La Tène in Austria and Switzerland to create the cultural-archaeological picture of the Celts we have today. The unjustified assumption of origins of Celts (whoever they were) in Central Europe, and their late expansions into the rest of Europe from there, led to further unjustified assumptions of cultural associations, followed by inferences about migration and ethnic replacement. British archaeologists then come under the torch for conflating the glorious art and precious metalwork of Iron Age Britain (Plate 2) and Ireland with this view of invading Celts in 300 BC.

At the same time, Collis recapitulates the whole history of western archaeology in a revealing and cathartic way. At the same time as these Romantic 'evolutionary' notions of human ethnicity were developing, the principles of modern archaeology with its own folklore were being established. The concept of three stages of 'civilization' – the 'three Ages' of Stone, Bronze and Iron – was first conceived in 1823.[65] In 1863, an earlier Stone Age, or Palaeolithic, represented in the caves and river gravels of France and the more southerly part of Europe, was distinguished from the more recent or Neolithic Stone Age remains found throughout Europe.[66] These crude chronological divisions were reinforced by stone tool typology, an approach that lent itself to relative chronology. Chronology in those pre-carbon-14 days also depended on context, known historic sites and battles, dated coins and pot typology, and on careful stratigraphy and type comparison between sites.

Like most people, I find it quite amazing that so much useful information was inferred in the past – and still can be inferred – by using these practical, reason-based, pre-hi-tech methods. That is what makes field archaeology still such compulsive television viewing today, much more exciting than the whiz-bang special effects used by film producers to imply technical wizardry in their documentaries. We all have a bit of the tracker in us.

Worthy though Collis's charting of the growth of knowledge and ingenuity among archaeologists may be, it remains an inadequate veil to obscure his colleagues' persistence in sticking to d'Arbois de Jubainville's particular myth-structure of Celtic origins. Collis seems to paint a Swiftian picture of the dons of his profession as erudite, supremely articulate historian-scholars. Archaeologists, in spite of their more practical leanings, which dominate most of their day-to-day work, would seem to create their grand reconstructions in cycles of sudden leaps of conviction and group-persuasion, followed by detailed consolidation and prolonged stasis. On top of these cycles of theory are imposed further cycles of fashion and process. Fads such as migrationism (enthusiasm for movements of people) and diffusionism (enthusiasm for movements of culture) move in and out of fashion, irrespective of the weight of evidence for or against particular events. There is now a school of what is called processual archaeology, which concentrates less on cultural history and more on processes such as the way humans do things (behaviour) or the way things decay. Then there is post-processualism ... However, in spite of Collis's criticisms, their heavy load of other 'isms' and a false start on Celts, archaeologists would seem to the observer to have achieved the most extraordinary advances in European prehistory over the last hundred years.

Geneticists also come in for a deserved pasting, with well-aimed punches from Collis in his frenzy of debunking. He derides both the research in the context of the 'Celtic' question and papers published by geneticists seeking to 'prove' the later Anglo-Saxon replacement.[67] I hope to resuscitate some of the relevance of genetics from such onslaught later on.

2

CELTIC AS A LANGUAGE LABEL

Written in stone

Having demolished the evidence for a Central European homeland and any specific association of Celts with the Hallstatt and LaTène cultures, John Collis moves back briefly, in the penultimate chapter of his Celto-sceptic book, to language. Here he mainly bemoans the unhealthily close relationship between linguists and archaeologists in the nineteenth century, which has persisted in some quarters until today. Along with the obvious problem that ancient artefacts without writing do not identify the language of the maker or wearer, there is the tendency to force both language and culture into similar monolithic racial stereotypes, where the reality is of diversity, difference and mixture.

The twelve conclusions that Collis reaches in his last chapter are mainly deconstructions of the struts of the modern Celtic myth, but the only hint of reconstruction, albeit half-hearted, comes from his seventh conclusion, which refers to language:

> One interpretation of the historical and linguistic evidence also seeks the origin of the Celts in south-west Germany, but other interpretations of the classical sources are also possible, indeed perhaps more likely, and would include central and western France.[1]

This brings us conveniently back to the point in the last chapter at which I digressed into archaeological homeland myths. I had just referred to Caesar's comment that the people he called 'Celts' in Middle Gaul called themselves Celts 'in their own language' (Figure 2.1a). Since Caesar's assertion contradicts those who claim that Celts did not use the term for themselves, and broaches the whole question of the identity of celtic languages, it might be worth asking whether there is any evidence for the linguistic affiliation and identity of the language Caesar calls 'Celtic'. In other words, is there a systematic record from Roman times, of a dominant, indigenous non-Latin language, in Caesar's Celtic Gaul, which could be related or linked to modern insular-celtic languages?

As it turns out, there *are* numerous records of such a language, and it was spoken in just those places unambiguously identified in classical literature as Celtic or occupied by Celts from at least as early as 300 BC: France south of the Seine, northern Italy and Spain (Figures 2.1a and 2.1b). To classical authors, the two more northerly regions were known respectively as Trans-Alpine Gaul and Cis-Alpine Gaul, named from the Roman point of view: Gaul-on-the-far-side and Gaul-on-the-Roman-side of the *French* Alps.

These dead but not completely lost languages, for which there is clear archaeological evidence, are ancient Gaulish in Trans-Alpine and Cis-Alpine Gaul, and Celtiberian in east-central Spain. The evidence for Gaulish is extensive and includes bilingual inscriptions of Rosetta Stone quality and inscriptions accompanied by unmistakable depictions.[2] Gaulish inscriptions are found in southern France from the third century BC until the first century AD, initially in Greek script and later, post-Caesar, in Roman script. Some Roman inscriptions are found in central France; the latest of them, from the third century AD, was recently discovered (1997) in Châteaubleau, about 40 km south-east of Paris.

Although very similar, the Italian or Cis-Alpine version of Gaulish is usually called Lepontic. Lepontic is not an Italic language, although

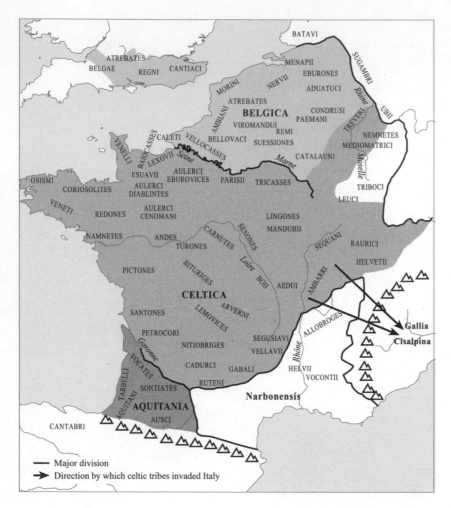

Figure 2.1a Who were the Celts in classical times? According to Caesar, Gaul was divided into three parts. The middle part, spoke celtic and was mainly south of the Seine and Marne, although there was a north-eastern extension as far as the Rhine. In the north, the Belgae were mostly 'descended' from Germani. Arrows show Celtic tribal invasions of Italy (around 600 BC, according to Livy).

the earliest inscriptions from northern Italy are dated (controversially) to the sixth century BC and are written in Etruscan script. This early celtic-linguistic date in Italy would clearly be consistent with Livy's disputed historical claim for early Celtic invasions of northern

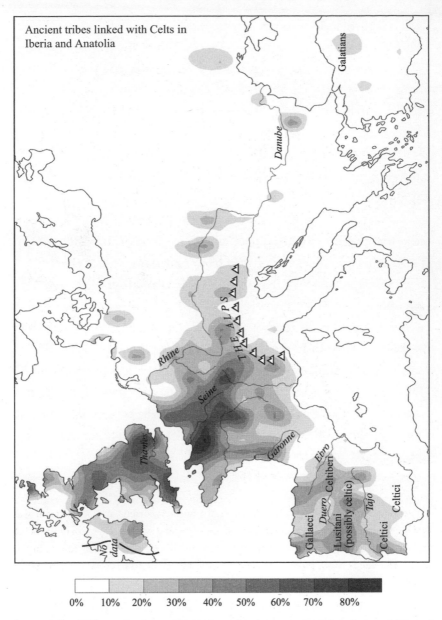

Figure 2.1b Where was the evidence for celtic language in classical times? Map of Europe and Asia Minor with the percentages of ancient place-names that were celtic shown as contours. The highest frequencies were in Celtica, Britannia and Iberia. Their absence in western Ireland is due to lack of data. Unexpectedly low rates, e.g. in northern Italy and Galatia, are due to the 10% contour cut-off point used for contrast.

Italy.[3] Gaulish, lacking its own script, would be unlikely to reveal similarly early dates for inscriptions in France before the Roman invasion.

For some reason, however, Collis seems particularly unwilling to accept Livy's early date of 600 BC for an initial Celtic invasion of northern Italy from Gaul. However, if we review Polybius and Livy for the actual origins of the first northern Italian Celts,[4] in spite of their disagreement over the exact date (respectively c.400 BC and c.600 BC), we find no suggestion that they came across the Alps from any direction other than the west – from Gaul, south of the Seine. Livy in fact gives the names of several Alpine passes, which makes it clear that these early invading Gauls came from Trans-Alpine Gaul (i.e. southern France) rather than from Germany or Austria.[5]

This early date – written in stone – is consistent with trends to move back the age of the celtic-language family using other methods (which I shall discuss further at the end of this chapter and later in the book). Also, the early break-up of celtic languages, combined with Caesar's language link, is consistent with the inference from Himilco's *Periplus* that insular Celts could have arrived in the British Isles at least as long ago as the mid-first century BC (see Chapter 1).

The quality of these extant records of an ancient Gaulish relative of the insular-celtic languages, and its clear association, in time and place, with the Celtic cultures of south-west Europe, make Pezron's and Lhuyd's suggestion of a Celtic link across the Channel less of a false leap of imagination than Simon James and John Collis imply. The critical false step seems to have been the conviction of nineteenth- and twentieth-century archaeologists that the Celtic homeland was in Central Europe. Far from being the nationalist, New Age amateurs that the Celto-sceptics paint them as, those Enlightenment bookworms Pezron and Lhuyd made the most useful theoretical advance in the past three hundred years of Celtic studies.

The Gaulish records are not the only evidence of celtic languages in south-west Europe, but they are the best, both for textual reference and for inscriptions.[6] Iberia is the other region with the clearest evidence for the coincidence of both Celtic people and celtic languages in early classical times.[7] Collis reviews a number of classical sources which refer to Celts in Iberia. These mention mainly Celtiberians, who appear in the texts as a hybrid product of invading Gauls from France and indigenous Iberians, who battled at first, then buried the hatchet and joined cultures. Other Celtic groups are referred to as being in Iberia, such as the Celtici.[8]

The record of Iberian inscriptions shows Celtiberian as a major Continental branch of celtic languages. This evidence comes from around seventy inscriptions totalling around a thousand words dating from the third to the first century BC. These were mostly written in the Iberian script but sometimes also in Roman script. The outstanding ones are three major texts on bronze tablets found at Bottorita, near Zaragoza in north-central Spain. The longest and most famous inscription is a *tessera hospitale* (a written promise of hospitality), written in Celtiberian on both sides. From the existence of four rock inscriptions in the north-west of the Iberian Peninsula, Lusitanian has also been suggested as an ancient celtic language. But this attribution has nothing like the quality of written evidence that Celtiberian has (Figure 2.1b).[9]

The archaeological evidence does indicate a specific Celtiberian cultural area coincident in time and place with the inscriptions in north-east central Spain to the west of the River Ebro. There is also extensive and overlapping textual, place-name and personal-name evidence for a Celtic presence throughout the north-westernmost two-thirds of the Iberian Peninsula (Figure 2.1b).[10] Tribes known as the Celtici on the western Atlantic coast of the peninsula seem well attested,[11] which is consistent with evidence from place-names.

The use of evidence derived from place-names to suggest a celtic-linguistic origin has been criticized. In some cases, the arguments for derivation from celtic appear circular, self-fulfilling and even deliberately misleading. More recently, stricter linguistic criteria have been applied in order to reduce dubious attributions, particularly to celtic.[12] From the distribution of names with the ending -briga ('-hill'), one celtic element that does seem to have consistency and common specificity with place-name evidence elsewhere,[13] it appears that the Celtici and other putative Celtic tribes, such as the Gallaeci and Lusitani, spread up the rivers from the coast, deep into the hinterland of northern and western Spain (Figure 2.1b).

That apart, the place-name evidence for celtic languages much east of the Rhine or in other parts of Northern Europe is not convincing. A recent workshop was convened by linguists to establish just what proportion of European place-names on Ptolemy's famous map of the second century AD could be reliably identified as celtic.[14] Several broad conclusions emerged. Numerically, the centre of gravity and greatest diversity of forms for Continental celtic place-names were in France south of the Seine, Spain and northern Italy, as predicted by the distribution of early celtic inscriptions. There were very few celtic place-names much east of the Rhine or north of the Danube. There is a similar paucity of celtic place-names in the southern Balkans, Romania and Hungary, to the south-east.[15]

Archaeologist Colin Renfrew comments on the circularity of linguistic arguments based on 'evidence' for Celtic populations in Eastern Europe: 'Very often the claims for a Celtic population in those areas are backed up by discussion of objects found there which are in the La Tène art style.'[16] Clearly, the La Tène art style is not linguistic evidence of Celticity, and its Celtic connection is based only on the Central European Celtic homeland theory, thus creating a circular loop of association. There is also a possibility that the Celtic invasions into Greece and beyond into Anatolia (in modern Turkey) took celtic languages into the region subsequently known as Galatia. This is borne out by the new systematic place-name work (Figure 2.1b).[17] But, as Renfrew points out:

> Whatever the status of Galatian Celts, they have little significance
> for the origin of the Celtic languages, since it has never been

suggested that a Celtic language was spoken in Anatolia prior to the supposed arrival of these intruders in the late third century BC.[18]

To summarize: in agreement with inscriptional evidence for early celtic-language distribution, classical historians seem to place the oldest records of the Celts in Narbonne, southern France, and also Italy and Spain, the earliest dates being around 600 BC. The simplest interpretation, and the one most consistent with the partially conflicting sources, is that the Celts originated in the south of France and then spread southwards to Spain and eastwards across the French Alps to invade parts of Central Europe, Italy and even Anatolia by at least the third century BC. The classical writers' confusing practice of conflating Gauls and Celts was common but acknowledged specifically, even by Caesar. We have to assume that he did intend the reader to understand that there were celtic speakers as far north as the Seine – as evidenced by people who called themselves celtic speakers. Their presence in the areas identified by Caesar is again supported by the distribution of place-names ending in *-briga*. However, it is not clear *when* they got there. The classical view of Celts as people speaking celtic languages and originating in south-west Europe is supported by writers such as Strabo, Diodorus Siculus and Caesar. The modern view, derived from Buchanan, Pezron and Lhuyd, that these classical celtic languages are related to modern insular-celtic languages is well supported by the finding of extant celtic inscriptions and other primary linguistic evidence confined to those areas where Celts were first attested – namely southern France, Italy and Spain.

On the other hand, it is difficult to infer a clearly Central European origin for Celts from any of the classical writers. To cite the La Tène or Hallstatt art styles as favouring such an origin is an invalid argument, based on nineteenth-century misconceptions. And for twentieth-century scholars to cite those styles as primary evidence for Celts immediately becomes a circular and invalid argument. Further, there is no linguistic evidence whatsoever that those celtic speakers originated in southern Germany or anywhere east of the Rhine. Use of the excuse 'absence of evidence is not evidence of absence' is no good either because the original assumption that Celts were associated with the very real spread of La Tène and Hallstatt art styles had inadequate textual or linguistic support in the first place.

Collis's only constructive conclusion, in his excellent deconstruction, is grudgingly to acknowledge that if there were a historical-linguistic link between Celts and celtic languages, then their origins would include central and western France rather than south-west Germany.

So, we are left with objective evidence of the presence of Celts and celtic languages in France, Italy and Spain during Roman times. By combining inferences from early Greek writers with modern linguistic analysis, we can see that the presence of Celts and celtic languages in south-west Europe, and maybe even the western parts of the British Isles, stretches back to before the middle of the first millennium BC.

In my view, this perspective vindicates Lhuyd's provisional naming of his clutch of non-English languages of the British Isles as Celtic with a big 'C', although I shall continue to use the small 'c' for language. We can now move on to look at them in more detail.

Celts and celtic languages in the British Isles

I have done my best to vindicate much of Pezron and Lhuyd's broad linguistic claims, and, unlike Simon James, I feel that these claims justify continuing the study of modern insular-celtic languages in the context of the classical Celts. But, there are several loose ends to address. The most important of these is their tacit assumption, still perpetuated, that *all* Ancient Britons were celtic-speaking (i.e. that the whole of the British Isles spoke celtic languages). While much of the western and northern British Isles was unarguably celtic-speaking, as far back as records go, the same cannot necessarily be claimed for

England. There is fairly good evidence that over the past two thousand years celtic languages were spoken almost universally in those areas – Wales, Cornwall, Cumbria, Scotland and Ireland – that we now associate with a 'Celtic heritage'. Indeed, hundreds of inscriptions on stone, made after the Romans had left, are ample evidence of how celtic languages thrived in Ireland, Scotland, Wales, Cumbria and Cornwall. However, England (apart from the West Country and Cumbria), although rich in stone inscriptions, is notably practically devoid of any in celtic dating from any period (Figures 2.2 and 7.4).[19]

As far as the subjective visual impressions are concerned, Tacitus compared the southern Welsh exclusively to the Spanish in one of the few contemporary descriptions of the people of south-west Britain:

> The dark complexion of the Silures, their usually curly hair, and the fact that Spain is the opposite shore to them, are an evidence that Iberians of a former date crossed over and occupied these parts.[20]

The Silures inhabited south Wales. Along with the Ordovices in central Wales and the Deceangli in north-west Wales and Anglesey, they put up an extremely fierce and prolonged resistance to Roman occupation – a sense of independence that has lasted. Tacitus here makes a direct migratory connection between the Silures in Wales and the people of Spain, on the other side of the Bay of Biscay. He makes this comment based on physical appearances, in a throwaway remark that is reminiscent of similar casual subjective comments still made today about the Mediterranean appearance of Welsh people.

The Spanish connection also makes one think again of the ancient Greek and Punic references to the tin trade along the Atlantic coast from Spain up to the west coast of Britain. These references support the possibility that a 'Celtic' entry to the western side of the British Isles came from the south rather than from the nearby Continent on the other side of the Channel. What other information Tacitus may have based this remark on we shall never know, but it also emphasizes

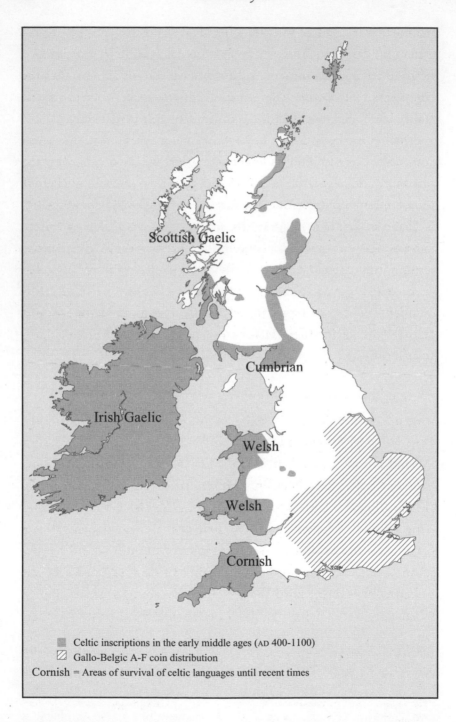

Scottish Gaelic

Cumbrian

Irish Gaelic

Welsh

Welsh

Cornish

    Celtic inscriptions in the early middle ages (AD 400-1100)
    Gallo-Belgic A-F coin distribution
Cornish = Areas of survival of celtic languages until recent times

that there was a lack of similarity between the Welsh and the English even in his day. I shall discuss the Spanish connection shortly, but for the moment such comments mean that we cannot assume that the inhabitants of Roman England had exactly the same origins as the Welsh, let alone being 'Celtic' or universally celtic-speaking.

I shall come back to the question of what other languages apart from celtic *were* spoken in Roman England in Chapter 7. For the time being, however, the patchy evidence in England tells us that we cannot assume that the Ancient Britons of England all spoke celtic, or that the English language is merely the result of complete physical and cultural replacement of Celts by the Anglo-Saxons starting from around AD 400. In those parts of the British Isles that were clearly celtic-speaking throughout (Figure 2.1b), there is abundant evidence from inscriptions, literature and modern languages to chart their history of change and movement since Roman times.

Britain: direct linguistic evidence

Most of the known recent celtic languages of the island of Britain form a group known as *Brythonic*. As far as Wales and western England are concerned, it is fairly safe to assume that Brythonic celtic languages have been in continuous use since Caesar's time, and most likely before. Apart from Welsh, Cornish is the only other living Brythonic language on the British mainland. Cornwall (or Kernow) was also known by Romans as Dumnonia and by the Saxons as the Kingdom of the West Welsh. Cornish (Kernowek) was spoken continuously until three hundred years ago. From then it declined, and just before the twentieth century it became extinct.

Figure 2.2 Gallo-Belgic coins and their British derivatives do not overlap later celtic stone inscriptions in the British Isles. That is possibly a coincidence, but it is consistent with a prior Belgic human (and/or cultural) migration to southern England, which could be inferred from Caesar and Tacitus (For detail of Gallo-Belgic coin distribution, see Figure 7.3.)

The last monoglot Cornish speaker may have died in 1676, but there were still a number of fluent native speakers available for Lhuyd's work, and he included the language in his study. As the result of work by a number of enthusiasts in the twentieth century, reconstructed Cornish has now made a dramatic revival, with an estimated 3,500 speakers today. Cornish is now officially recognized by the United Kingdom Government as a minority language under the European Charter for such.

There is good evidence that Brythonic celtic languages, probably South-west Brythonic, were also spoken elsewhere in the West Country, in Devon. After the Romans left, the Kingdom of Cornwall persisted during the Dark Ages.

Other surviving Brythonic remnants

As may already be clear, Breton, the celtic language of Brittany, was no ancient Gaulish remnant stranded on the horn of the Continent (as was mistakenly thought by Pezron) but a Brythonic tongue, closely related to Cornish, which at some point over the past two thousand years moved across the short stretch of water between Britain and Brittany. Evidence for Old Breton, dating from the eighth to the eleventh century AD, can be found in lists and glosses in documents and as names in Latin texts. There is evidence for some borrowing from Gaulish into Breton, which is either an indirect sign of its longevity in Brittany or of the presence of Gaulish so far north and so late. Although not recognized officially by the French Government, Breton has half a million speakers, thus rivalling Welsh as the most flourishing of all the modern celtic languages.

Another Brythonic language, now extinct, can be fairly safely inferred: Cumbric, which was similar to Welsh. Pictish, formerly spoken in northern Scotland, is claimed to have been Brythonic, but whether this claim covers all languages present there in the first millennium AD, apart from Scottish Gaelic, is still disputed by a few (see below).[21]

Except in Cornwall and Wales, there is little evidence from the spoken word remaining today to support the view that Brythonic languages were spoken throughout England in Roman times. However, even today, remnants can be found of a Brythonic-celtic language previously spoken in Cumbria and elsewhere in north-west England. When I was a child, the only celtic words spoken north of Wales were from dialects of a language closely related to Welsh, known generally as Cumbric. Even now, I can recall my brother being taught by a friend at school how to count from one to twenty in numbers used by North Country shepherds to count their sheep. As far as I remember it went something like this: *yan, tan, tether, mether, pimp* (5), *sether, hether, hother, dother, dick* (10), *yan dick, tan dick, tether dick, mether dick, bumfit* (15) *yana bumfit, tana bumfit, tetherer bumfit, metherer bumfit, giggot* (20). Many distinct variants of this sheep-counting system are recorded from the Lake District (Cumbria) and the Pennine Hills, north from the west coast right through to Ayrshire in the western lowlands of Scotland: in Keswick, Westmoreland, Eskdale, Millom, High Furness, Wasdale, Teesdale, Swaledale, Wensleydale and Ayrshire.

This method of counting things in four tallies of five up to twenty was known as 'scoring', hence the term 'score' for twenty. Counting in fives reflects the use of fingers, if not toes. After twenty, a scratch or score was made on a piece of stone or wood and the scoring recommenced. The word 'tally' comes from the Latin *talea*, meaning literally a stick with notches.[22] Cumbric scores have even been found in a number of places in the USA. In Cincinnati we find *een, teen, tother, feather, fib, soter, oter, poter, debber, dick*; and from Vermont there is *eeni, teni, tudheri, fedheri, fip, saidher, taidher, koadher, daidher, dik*. Cumbrian settlers presumably took these celtic counting-words to the New World. Children's counting rhymes also retain relics of Cumbric scores, although the counting is in fours, presumably to suit the rhyme: from Edinburgh, for example, 'Inty, tinty, tethery, methery; bank for over, dover, ding ...' and the universal 'Eeny, meeny, miney, moe ...'[23]

This numeric digression from Celtic remnants in England took us briefly to Welsh, which I shall come back to in slightly more detail later. Linguistic reconstruction, a well-recognized method based on knowledge of systematic sound changes, indicates that both Welsh and Cumbric are related but distinct languages sharing a common ancestor, with Kernowek, in Brythonic Celtic. This branch has descendants only in mainland Britain and in Brittany (see below), and is distinct from another group of insular-celtic languages, known collectively as Gaelic or Goidelic.

Tribal names

When we look at what Roman writers have to say about the tribes of northern England, we find that they mention one group which occupied all the areas where the Cumbrian celtic-language relicts are still found. They were known as the Carvetii, but that is about all that is known about them from the texts. Their territory included all of Cumbria (the Lake District) and parts of north Lancashire and south-west Durham (the Pennines), and south-east Dumfries and Galloway in Scotland (Figure 2.3). They were probably mainly hill farmers, with few large settlements apart from ones associated with the numerous Roman forts in the area. They did not make their own coins. For some obscure reason their entire county and its towns were left out by Claudius Ptolemy from chapter 2 ('Geography of Albion Island of Britannia') of his famous second-century *Geographia*. This omission was probably a block clerical error, since there is none of the concordant detail that should be expected, for instance cross-classification of Carvetian towns with their neighbours, the Brigantes. Their Roman

Figure 2.3 Ancient British tribal names, locations and capitals. The tribes between Dorchester and Canterbury all had connections with the Belgae; the 'Belgae' and 'Atrebates' even shared tribal names ('B' on map). In addition to the Roman defence walls (Antonine and Hadrian's), Offa's Dyke is shown, since it has been claimed to have been based partly on a previous earth defence rampart built by the Roman emperor Severus at the beginning of the third century AD.

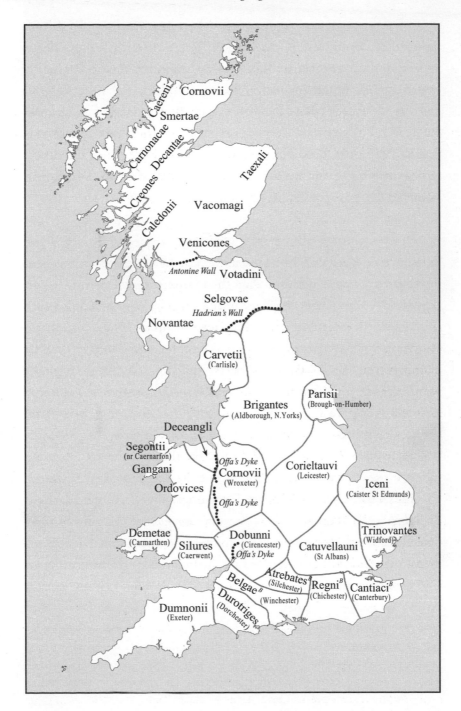

provincial capital was Luguvalium (Carlisle). Further north into western Scotland, the area occupied by the Damnonii would have corresponded geographically to where today's remnant Brythonic counting numbers are found in Ayrshire.

Although connections have been and still are made between the Carvetii and their neighbours, the Brigantes, the evidence is indirect. The Brigantes were also a putative Celtic people and much the largest tribal grouping of northern England. They occupied most of England north of the Rivers Humber and Mersey that was not occupied by either the Carvetii on the west or the Parisii in East Yorkshire.

Much is known about the Brigantes' interactions with the Romans, for whom they were generally a collaborating client tribe. They had a queen, rivalling Boudicca in fame, who feigned to shelter the fleeing rebel leader Caratacus and then promptly handed him over to the Romans in chains. They had a Roman capital at Isurium Brigantum (Aldborough, North Yorkshire). However, the Celtic claims rest mainly on their distinctive tribal name, which, having a celtic derivation, was shared with various other contemporary tribes also thought to be Celtic. The derivation of the name may be from the Celtic mother-goddess Brigit, also known as Brigantia. It is, however most likely to be celtic. Ptolemy mentions in his *Geography* three tribal locations with this name, one in northern England, another in south-east Ireland and Brigantinus Portus of the northern Gallaeci tribe. Also known as Brigantium Hispaniae, this was the ancient seaport of La Coruña in north-west Spain and the western terminus of a major trade route in tin, gold, lead and silver.[24]

There was also a clutch of similar place/tribal names around Lake Constance on today's Swiss/Austrian border. The people of Central Raetia were called the Brigantii. Their tribal capital was Brigantium Raetiae (now Bregenz), and Lake Constance itself was then called Brigantinus Lacus.[25]

John Collis has an interesting map on which he shows the relevance of such name links between the Continent and various parts of the British Isles. He assigns three levels of relevance: accepted, possible and rejected. He seems to accept the name link between the south-east Irish and English Brigantes tribes,[26] which would tend to favour the 'celtic-speaking' label as well as 'Celtic', since there is no evidence for any non-celtic ancestral languages in Ireland. However, he implies, reasonably, that such a general name may lack tribal specificity. The Roman town name Cambodunum, however, does appear in both the northern English and west-Austrian locations, implying that the sharing of the root 'Brigant-' is more than just a coincidence.

The puzzle of the British Brigantes does not stop there, since such a general label may have been applied to the Romans, as a tribal description, by celtic-speaking informants, without the tribe neces-sarily being celtic-speaking. The English historical linguist Kenneth Jackson, in his classic *Language and History in Early Britain*,[27] has a whole appendix on the problematic name used by Bede and others for the northernmost 'Anglian' tribe of Anglo-Saxon Britain, the Bernicii, who inhabited Bernicia on the east coast of North-umberland. Although he opposes the idea, Jackson shows that the name could be derived through Welsh from the tribal name Brigantes. As he points out:

> if Bede's *Bernicii* ... really represents a British tribe-name borrowed for the designation of the northernmost Anglian settlers, it can hardly have been taken over any later than the seventh century ... it is plain that it had become ... recognised by the Britons as now an English and not a British name, before c.600.[28]

The paradox of Welsh writers naming a northern 'invading Germanic-speaking tribe' by the name of the British tribe they wiped out, *and* calling them English, can be resolved if the original Brigantes of the north country were not celtic-speaking, and were not really wiped out – merely invaded by new elites from

Scandinavia. There is also the problem that the celtic stem *brigant-* (meaning 'high person' or 'high place') was used; whether it just meant highlanders or people of the goddess Brigit. The former could simply be a description of terrain, the latter of religion.

Interestingly, Collis does not seem so completely convinced of the 'possible' celtic link between the Parisii of Yorkshire with their namesakes the Parisii of northern Gaul, who gave their name to the modern French capital. The former appear to have left some very unusual cultural remains, unique in the British Isles, but with links in northern France and Belgium. The Continental Parisii, like other tribes of England twinned by name across the English Channel, such as the Atrebates and Belgae, were situated mainly north of Caesar's politico-linguistic boundary between the Belgae of northern Gaul and his 'celtic-speaking Gauls' to the south of the Seine. Admittedly, the Parisii were only just to the north of the Seine. I shall come back to the Parisii later on, but for the moment I wonder whether there is a strong reason to argue that they were, to borrow Caesar's classification, Celtic rather than Belgic.

Collis also puts the Damnonii of Ayrshire and the celtic-speaking Dumnonii of Cornwall into the category of 'possible' tribal name-links. Since there are surviving elements of Brythonic celtic language in both locations, this link could be promoted to a 'probable'.

The only other such external name-link that Collis allows for Romano-British tribes in the island of Britain is that between the Atrebates of Belgica and the Atrebates of the English south. When the Romans invaded in AD 43, the Atrebates occupied a territory which covered Sussex, Hampshire, Berkshire, west Surrey and north-east Wiltshire. On the other side of the Channel, the Gallic Atrebates were situated well into northern Gaul; in other words they were Belgic, and in spite of what Collis says, possibly non-celtic-speaking. Collis accepts the name-twinning of the Atrebates across the Channel, but strangely not that of the tribe that actually carried the name Belgae. The English Belgae were close neighbours of the Atrebates and had their capital at Winchester (Venta Belgarum, 'Market of the Belgae'). It is not clear why Collis chose to indicate the Atrebates link but not the equally general Belgic one. They are equivalent in some respects, both having general meanings in their celtic derivations: 'Atrebates' means 'settlers', and 'Belgae' 'people from Belgica'. But these names may have been labels imposed by others, so neither of these links can be taken as proof that the respective tribes were necessarily either Celtic or celtic-speaking, since even Caesar does not suggest this (see discussion on Belgae and Atrebates in Chapter 7).

There is another English tribal name-link which Collis does not mention, the Cornovii. These people lived just on the east of the northern Welsh border in Staffordshire, Shropshire and Cheshire, with a regional capital at Wroxeter (Viroconium). They shared their name with a tribe, recorded by Ptolemy, in the northern tip of Scotland. It is not clear whether this is a coincidence, since the name could simply mean, in Latin, 'people of the horn'. However, they both share one characteristic with all the western regions of Roman Britain, which were most likely to have been celtic-speaking: the absence of coinage at the time of the Roman invasion.

Inscriptions and names

Coins are useful archaeological date and culture markers, for many reasons, but for our purposes they are useful in the negative sense. At the time of the Roman invasion, *most* of southern England, excluding Cornwall and the West Country, already both made and used coins (Figures 2.2 and 7.3). Along with many other cultural features of the pre-Roman south of England, this practice – and indeed many of the imported coins – derived mainly from the rebellious northern Gallic region of the Belgae rather than from Caesar's celtic-speaking part of Middle Gaul. Unfortunately, rulers' short-form names inscribed on the coins give little more information than is already available in the contemporary literature.

In contrast to the practice in southern England, none of the other tribes I have mentioned so far – the Welsh tribes, the Cornish Dumnonii, the Carvetii, the Brigantes, the Cornovii – made or used coins at that time. This penury also applied to Ireland and Scotland, suggesting that, in the British Isles, the lack of coins in Roman times was associated with the survival of celtic language after the Romans left (Figure 2.2).

While this difference between southern England at the time of Caesar and the rest of the British Isles could simply have been a random geographical association, it is reflected in reverse by the near-total absence of celtic language to be found in stone inscriptions in the same areas of England set up either during or after the Roman occupation (Figures 2.2 and 7.4). In contrast, Cornwall, Wales and Ireland, and to a lesser extent Scotland, have a rich record of inscribed stones attesting to their celtic-language heritage from soon after the Romans left.[29]

The main positive evidence for celtic-language use in Roman England comes from linguistic research into place-names of Roman Britain and personal and tribal names cited in contemporary documents and on tablets and coins.[30] Although there is clear evidence for some celtic-derived names in Roman England from all these different sources, their relative frequency compared with equivalent parts of the rest of the Roman Empire (only 34% in England according to one study – see Figure 7.2[31]) does not give the sort of overwhelming endorsement required for confidence in a 100% celtic-speaking England. (This concern is acknowledged in recent critical studies[32]; I shall discuss this evidence in more detail in Chapter 7.)

So to summarize, the West Country, Wales and northern England were home in large part to tribes for which there is some evidence for a celtic-language link: the Brigantes and Cornovii through their names, and the Welsh tribes, the Cornish Dumnonii and the Carvetii through surviving celtic-linguistic substrata and a presumed connection with the Brigantes during Roman times. For the tribes of the south coast of England, of whom the Roman historians had most to say, there is less evidence for Celtic culture or name, or for celtic-language use.

I have mentioned living evidence for Brythonic celtic being spoken in Brittany, Wales, Cornwall, northern England and southern Scotland, but there are claims that Brythonic was previously spoken throughout Scotland (by the Picts), and even in Ireland.

Picts …

The Picts, supposedly painted, aboriginal tribes of northern Scotland, have always been a problem to place, since whatever language or languages they originally spoke have apparently disappeared except for a few scraps of evidence. They seem to have been linguistically replaced by Scottish Gaelic and ultimately by English during the first millennium AD. Scottish Gaelic was spoken by people of the Irish Dalriadic kingdom, thought to have invaded from Ireland in the fifth century AD – although, as we shall see below, there is more than one opinion on their time of arrival.

A number of theories have been put forward as to what language or languages the Picts spoke, and argument still continues, but a careful article written as long ago as 1955 by Kenneth Jackson still covers most and rejects quite a few.[33] The main problem is the lack of direct evidence of almost all kinds used in language reconstruction, such as texts or surviving linguistic remnants, which could be used to compare with place-names. Even celtic inscriptions, which are so useful in attesting to celtic language in other non-English parts of the British Isles, are absent in a large part of the areas supposedly occupied by the Picts.

The Medieval author Adamnan wrote in his *Life of St Columba* (completed around AD 692–97)[34] that Columba used an interpreter to converse with the Picts. Presumably this meant that they spoke a different tongue from Columba's, which was Irish Gaelic. Since

Scottish Gaelic is similar to Irish Gaelic, that would tend to exclude Gaelic, unless Pictish was a particularly archaic form.

Bede refers to Pictish in his *Ecclesiastical History* (AD 731). Bede lived in Jarrow in the north of England at a time when Pictish was still being spoken in Scotland. His first-hand observation should, hopefully, be more useful than his historical compilations of Gildas, who was much more concerned about religion than accuracy. Twice, Bede clearly indicates that there were four indigenous languages spoken in Britain: Gaelic, British (i.e. 'Brythonic'), English and Pictish. He describes Picts as invaders who arrived windswept in Northern Ireland in longboats from Scythia. Not being allowed to settle there, they made their home in Scotland.[35] How they reached the British Isles from Scythia, Bede does not make clear, but elsewhere in Medieval literature the region of Scythia is sometimes alluded to as the ultimate Norse homeland in the Danish and Icelandic sagas.[36] The longboats might imply the Picts were from Scandinavia (see Chapter 10), but in any case this story from Bede makes it clear that he did not think that they were either British or Irish. His linguistic skill should have been enough to work this one out for himself.[37]

Bede also refers to the Pictish name for modern Kinneil at one end of the Antonine Wall (a Roman fortified defence, stretching between Edinburgh and Glasgow, north of Hadrian's Wall) as Peanfahel. While this appears to support the view that Pictish was not the same as Gaelic, it would leave a puzzle. In a Brythonic language (also known as *P-celtic* – see p. 75), *pean* or *penn* might mean 'end', but the second half of the name, *-fahel*, would appear to be Gaelic, meaning '[of] wall'. A name meaning 'end of wall' is appropriate for the location, but the word would be a compound mixture of Gaelic and some P-celtic language. Kenneth Jackson points out that such compounds are no rarity, and gives the etymology for some other words and place-names that support the presence of a P-celtic language north-east of Edinburgh. He argues that this was 'a Gallo-Brittonic dialect not identical with the British spoken south of the Antonine Wall, different from the British-P-celtic used south of the Antonine Wall, although related to it'. According to the place-name evidence, this 'P'-celtic language would have been distributed in Scotland north-east of Edinburgh and the Forth river.[38] This distribution coincides with the main Scottish concentration of celtic-inscribed stones on the east coast (Figures 2.2 and Fig. 7.4).

However, Jackson argued, from the evidence of Ogham inscriptions (Ogham being an alphabetic script used throughout insular-celtic-speaking areas often with celtic–Latin bilingual inscriptions in the fifth century and onwards), that there was a third language in northern Scotland apart from Scottish Gaelic and the P-celtic language: 'The other was not Celtic at all, which would fit the relative absence of celtic place-names in northern Scotland, nor apparently even Indo-European, but was presumably the language of some very early set of inhabitants of Scotland' (Figure 2.1b).[39] Jackson's concept of a third language is now viewed by some as a minority view, but Colin Renfrew in his book *Archaeology and Language* chooses to take it seriously, referring to Jackson as having been 'the leading living authority'.[40]

This would leave ancient northern Scotland with three distinct languages, one of which was spoken to the west and two to the east of the Grampians for a large part of the first millennium AD. However, the western language, Scottish Gaelic, which apparently displaced the other two, is regarded as an intruder during that period. What is odd about the disappearance of Pictish is that the Picts were in the ascendant during the Dark Ages, according to both Gildas and Bede. Their attacks on England were stated by Gildas as being part of the reason for inviting the Saxons, so why should both their putative languages have disappeared so comprehensively, when Gaelic was essentially the dominant language of the Argyll west of the Grampians? I shall come back to this puzzle again in the second part of the book.

... and Scots?

The generally accepted view is that Scottish Gaelic was derived not from Scotland, or Caledonia as the Romans called it, but from Irish Gaelic, from the Irish tribe previously called the Scotti by the Romans. This may come as a surprise to some readers, but is generally felt to be incontestable. It seems that not even the name 'Scot' belongs in Scotland.

But – find an incontestable position, and there will be sure to be someone to oppose it. In an appropriately titled article 'Were the Scots Irish?' in the august archaeological journal *Antiquity*, Scottish archaeologist Ewan Campbell, of the University of Glasgow, recently did just that. He starts his polemic by outlining the conventional story of how

> the Scots founded the early kingdom of *Dál Riata* in western Scotland in the early sixth century, having migrated there from north-eastern Antrim, Ireland. In the process they displaced a native Pictish or British people from an area roughly equivalent to the modern county of Argyll. Later, in the mid-ninth century, these Scots of *Dál Riata* took over the kingdom of *Alba*, later to become known as Scotland.[41]

Campbell then claims that 'There had never been any serious archaeological justification for the supposed Scottic migration', citing a 1970 study which failed to find any archaeological evidence for cultural transplantation, into either Scotland or other parts of western Britain such as Galloway. He then goes through the inconsistencies and gaps in the archaeological and historical evidence.[42] It is Campbell's analysis of the linguistic evidence that is most likely to raise objections – from linguists. He does acknowledge that the phenomenon of Ogham inscriptions came from Ireland. However, he does not mention Patrick Sims-Williams' inference from those inscriptions that the Irish Gaelic language and names made significant inroads into Wales over the same period, although their influence subsequently faded.[43]

This evidence for an Irish linguistic intrusion during the Dark Ages does not necessarily invalidate Campbell's main alternative. He suggests that, rather than being limited by the Irish Sea until the first millennium AD, Gaelic languages and culture had extended across the North Channel between Antrim and Argyll and Galloway for much longer, perhaps back as far as the Iron Age. A glance at any map of the British Isles shows that geographically, his argument is sound. Ireland, the Isle of Man and western Scotland are very close to one another across the Northern Channel, far closer than the steps on the sea-trading links between northern Spain, Brittany, Cornwall, south-west Wales and County Wexford in Ireland. From the Bronze Age onwards, the sea route would have been far easier for trade than overland, and northern Scotland had the added geolinguistic barrier of the Druim Albin, the 'Spine of Britain', the Grampian Highlands.[44]

... and Irish?

Campbell also argues for Brythonic intrusions in the opposite direction, from Britain into Ireland. He is not alone. Ivernic, for instance, is said to be an extinct Brythonic language that was spoken in Ireland, particularly in the south-west, by a tribe called the *Érainn* (Irish) or *Iverni* (Latin). There are several independent fragments of evidence to support the notion of Brythonic languages having been spoken at some time in some parts of Ireland, either before or after the present insular-celtic Gaelic (or Goidelic) branch became dominant. David Rankin points to Ptolemy's description of eastern Ireland, which he says mentions four 'British-sounding and possibly British connected tribes, such as Brigantes'. Rankin suggests that these could have been possible refugees from the Romans in Britain.[45]

These ideas of Brythonic speakers in Ireland are also bound up with legendary concepts of multiple pre-Roman invasions of

Ireland. Not the least used of the sources is the traditional written Irish (Goidelic) record, in particular the *Lebor Gabála Érenn*,[46] drawing on the oldest traditions of all from the celtic language, which records four invasions. Such traditions record, among others, several presumedly celtic invasions, including Cruithni (Priteni, or Picts) and the Firbolgs (or *Érainn*), who comprised three groups from either Greece or Spain. These invasions are all supposed to have occurred before the final invasion of the Gaelic Milesians, also either from Greece or Spain.

David Rankin has reviewed this controversial argument, presented most strongly by Thomas O'Rahilly in 1946,[47] that there was a pre-Gaelic, Brythonic presence in Ireland. Although Rankin is equivocal, he does feel that the linguistic evidence points towards an agreement with the Irish legends that the last invasion, that of the Goidels or Gaels, was from Spain.[48] However, to make this point Rankin assumes that Gaelic was coeval with prehistoric Spanish-Celtic languages, while Brythonic was coeval with Gaulish in France.

P's and Q's

As with people and genes, trees of ancestry can be reconstructed for languages based on the degree of retention of shared characteristics. Problems arise when the retention of some characteristics do not match those of others. Rankin's attractive simplistic division for the ancestry of Welsh celtic from Gaulish in France and Irish celtic from Spain is based on the well-known shared use of a 'P' sound in Brythonic languages where Gaelic languages would use a hard 'Q' consonant, This division was used, among others, by the linguist Karl Horst Schmidt to construct a controversial formal deep-split language tree with P-celtic – which includes Brythonic, and also Gaulish and Lepontic – on one branch, and Q-celtic – which includes Gaelic and Iberian celtic – on the other. On this basis, Brythonic and Goidelic (i.e. Brittonic and Irish in Figure 2.4a) are on opposite sides of the celtic tree, and the Continental celtic languages are scattered in between.[49]

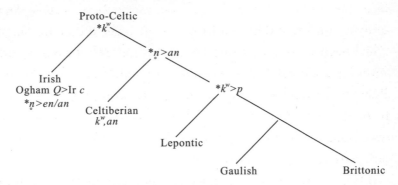

Schmidt's scheme

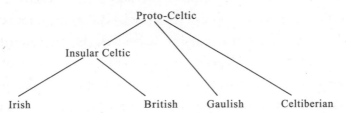

McCone's scheme

Figure 2.4 Celtic confusion. Schmidt's (a) and McCone's (b) alternative trees of insular-celtic provide radically different interpretations of origins from the Continent. In Schmidt's tree (partly based on the P/Q change), Gaelic (e.g. Irish) links with ancient Celtiberian in Spain, while Brittonic (e.g. Welsh) links with ancient Gaulish in France. In McCone's version (based more on 'areal' similarities), Brittonic and Gaelic are more closely related to each other.

The problem is that, as usual, linguists are divided down the middle. In this case, they disagree on the relative importance of the P/Q division. The deep structure of the celtic language tree would be different if P's and Q's in celtic just turned out to be a convenient description of Brythonic vs Goidelic, rather than a valid way of dividing the origins of Continental and insular-celtic languages as well.[50]

The P- and Q-celtic classification is indeed a useful rule of thumb for dividing Brythonic from Goidelic insular-celtic languages, based on the use, in various shared words, of a *p* sound in Welsh and a hard *c*, *q*, *g* or *k* in Gaelic. A simple example can be seen in Welsh numerals which use *p* in *pedair* and *pump* for 'four' and 'five' rather than the Gaelic hard *c* in *ceathair* and *cuêig*. Old Gaulish has *petor/petvar* and *pempe/pinpe* for four and five, which according to this rule would make it P-celtic.

But for some celtic linguists the existence of a consonant change between Brythonic and Goidelic is as far as it goes, and deep celtic-language 'genetic' splits based on this linguistically common shift may be unwise. Such P and Q changes have also occurred in other Indo-European languages. For example, the reconstructed root words for 'four' and 'five' at the base of the vast Indo-European language tree are *kwetwores* and *penkwe*. Like Goidelic, most Romance languages, including Latin and French, have retained the *k* sound (e.g. *quatuor*, *quatre*) but converted the *p* in *penkwe* to *qu* (e.g. *quinque*, *cinq*). However, two ancient Italic dialects, Umbrian and Oscan, went for *p* in both words, in the same way as in Brythonic. Similarly, most Greek dialects have gone the *p* way, converting the *kw* to *p* or *t* in four but retaining the *p* in five (e.g. *pessares/teseres* and *pempe/pende*). Greek is only very distantly related to most European languages including celtic. So, this *p* to *q* consonant change is common in either direction in Indo-European languages, and although it is useful, it cannot be used as a stable marker for deep language splits. In other words, if you watch your P's and Q's too closely you may get a Greek grandmother.

In spite of the apparent consistency of the P/Q split in the British Isles, Brythonic and Goidelic do have a number of similarities which make them even closer to each other than either is to the extinct celtic languages recovered from inscriptions in south-west Europe. This similarity leads to an alternative tree of celtic languages with three deep branches: the insular-celtic group, Gaulish/Lepontic and Celtiberian. We then need to ask whether the relatedness of Brythonic and Goidelic results from common 'genetic' descent, or from geographic closeness — a so-called areal phenomenon, as suggested by linguist Kim McCone of the National University of Ireland (Figure 2.4).[51]

This question of which language tree to use is not as academic as it seems, since it obviously affects any attempted reconstruction of which part of Europe (e.g. Spain vs southern France or Italy) each of the Brythonic and Goidelic branches might have come from and when. The importance of language history to anyone but non-linguists is, apart from natural curiosity, what it might contribute to the history or prehistory of peoples and their cultures. Although language is likely, on balance, to be rather more about movement of culture than about movement of people, both aspects are fascinating to the interested layperson as well as the academic. And as we shall see from the human genetics, the Irish legendary connection with Spain may not be as risible as some archaeologists make it out to be.

Dating celtic-language splits

From this thumbnail sketch of linguistic comparison we can see that, although the common ancestry of the Continental and insular-celtic languages is not in doubt, the structure of that relationship is anything but agreed by scholars. In trying to deduce the geographical origin of a group of languages in a historical context, the dating of splits and changes is as important to the historian as the reconstruction of the order of branching of the language tree is to linguists. Although in previous decades linguists made confident dates deep splits in languages, based on the evidence of word-sharing, some have had their fingers burnt, and rather than try again they have tended to avoid the practice. Others have simply never strayed from the tight confines of the last couple of thousand years, for which period the written word in documents and inscriptions is the ultimate test of time and place.

Scholars studying celtic languages have shared this reluctance. For the British Isles, they do have the advantage that the large body of extant inscriptions and other texts provides a tremendous opportunity to look in detail at sound changes over the past 1,600 years. They can cross-check their dates against those determined by the archaeologists.[52] While this makes possible a microscopic in-depth exploration using the core tool of their craft (known as the *comparative method*), the evidence on which it based needs to be rigorously determined and of high quality, which means that this approach cannot be extended back any further than the first celtic inscriptions, around 2,500 years ago. Celtic linguists are therefore content for the big questions of European Celtic homeland origins and dates, first posed so long ago, to remain on the shelf.

One of the main methods previously used to date language splits depended on measuring the degree of change in numbers of shared words (*cognates* – quite literally, words with a shared birth) between related languages. Since our vocabulary is also our dictionary or lexicon, this mathematical approach to language diversity is called *lexico-statistics*. Family trees can be constructed, based on the degree of lexical sharing between related languages. Such trees, although they may look superficially similar to some of the trees produced using the strict comparative method, are fundamentally different in concept and meaning.

The comparative method is used in a rigorous tree-building approach (as with genetics) and places groups and subgroups of related languages on the tree, with exclusively shared sound changes (called innovations) appearing in new subgroups. Not only that, but such sound changes are expected to be similar and reproducible in cognate words throughout the lexicon. For instance, *d* in German is systematically changed to *th* in English – *der* becomes *the* and *dunne* becomes *thin*. Systematic reproducibility of sound changes is a hallmark of the comparative method.

Counting words

In contrast, the lexico-statistical tree, while it ideally uses attested cognate words as the basic unit of measure, concentrates on the proportion of cognates shared between related languages, rather than structuring according to the strict rules of the comparative method. Lexico-statistics is thus more similar to *phenetic* analysis, meaning literally a comparison of phenotypes (the different varieties actually seen in a population, rather than the genes or 'genotypes' that underlie them). The phenetic approach was used in the past by population geneticists, and still is by physical anthropologists, to compare human populations around the world. Those biologists compared the frequency of common markers, such as different aspects of head shape, between populations to see how close they were to one another. The trouble with this statistical method is that for living human populations it gives very blurred trees.

On this comparison of 'genetic vs phenetic' methods, the former strict approach may sound better and more rigorous, and for most academic purposes it is, but there are some problems. It is very difficult to use the comparative method for dating language change unless, as in the case of historical documents and inscriptions, one has a rigorous, historical/archaeological method of checking dates.

Lexico-statistical analysis, on the other hand, is a quantitative technique and thus lends itself more to the use of the data to estimate dates of splits back beyond the written word. Using the comparative method, individual words, say in English and German, can be identified as sharing a common ancestor by inheritance rather than by borrowing. So, for example, the two words mentioned above (*thin* and *dunne*) can be shown ultimately to have a common Germanic ancestor which has changed in a systematic way in each language. These words are cognates. By contrast, the fact that English *beef* and French *boeuf* have a common origin and meaning results from the Norman Conquest and is an example of borrowing. The French have more recently borrowed the same word back in *biftek* (meaning 'beef steak'). Since some words drop out of use from individual languages, there is a decay process, rather as there is with radiocarbon. Counting up the proportion of remaining shared cognate words between two languages is thus some measure of the closeness of their relationship in time. This general principle can be used to reconstruct a tree, with dates on the branches. The dating method is called *glottochronology*.

The chief problem with glottochronology is that the decay appears to occur at different rates in different language groups, and that puts a fatal flaw in the method. For this reason, most linguists long ago rejected the glottochronological method of dating language splits using lexico-statistical data as inaccurate. Furthermore, since the measurement of the percentage of shared cognates, is a measure of decay of relationship, this also means that one has to assume accurate identification of cognates between more distantly related language groups if it is to be calibrated securely. Differentiating cognates from borrowed words becomes increasingly difficult the more distant the relationship.

There is a general problem which affects both the comparative and the glottochronological method: a large number of different languages on a single large landmass do not simply branch in a tree-like genetic manner from one another. Neighbourly languages co-exist in a multilingual environment and interchange their content: there may be a lot of borrowing of words and even syntax between languages. Where such neighbourly languages are largely related, as in Europe, the borrowings − although initially obvious − may become increasingly difficult to detect *as* borrowings rather than inheritance. Undetected borrowing distorts both kinds of language tree (comparative and lexico-statistical) and is probably the main underlying reason for structural differences between them.

So, what to do? There are basic differences between the disciplines of archaeology and linguistics on the one hand, and sciences such as geology and biology on the other. In their attitude to the scientific method, some linguists seem to misunderstand the meaning of, or are unable to accept, uncertainty. They interpret the scientific method as implying authority, rigour and certainty, while scientists accept that, in many situations, comparisons have to be made using measurements that have some degree of error and theories of classification with a degree of uncertainty. A statistical approach has to be used to handle such uncertainty. Unlike disagreements between academic authorities, there are standard methods of dealing with sources of observational error and of uncertainty. Archaeologists, in contrast to linguists, have learnt through experience that if a method such as carbon dating gives inaccurate results at first, it should not be thrown out of the window, but attempts should be made to sort out the problems of error and improve it.

My observer's take on all of this is, 'If at first you don't succeed, try, try and try again.' That does seem to be happening, at least amongst some linguists. A huge set of cognate data on Indo-European languages, originally published in the early 1990s by Hawaiian linguist Isidore Dyen,[53] one of the doyens of lexico-statistics, has recently been recycled in some high-profile publications.[54] Rather than just reinventing Dyen's analysis and conclusions, these publications use new tree-building methods developed to deal with similarly heterogeneous data in genetics studies.

Dyen made an observation on celtic languages which has not really been disproved or falsified in subsequent re-analyses. Although it shows some relationship with the three Indo-European branches which are dominant in Europe, Germanic, Italic-Romance and Balto-Slavic (described as 'Meso-European' by Dyen), the celtic group tends to stand on its own as a deep branch sharing fewer than 20% of cognates with them. Put simply, this suggests that celtic languages separated from the other three groups before those three split from one another. Not only that, but even the branching within insular celtic is also deep (see Figure 6.2a).

Dyen's dataset includes seven dialects from surviving insular-celtic languages, two each of Irish and Welsh and three of Breton. His analysis confirms general points from the comparative method: these three groups are internally consistent (i.e. on a regional basis their dialects are as closely related to each other as are other modern dialects, such as different types of Swedish); and that Welsh and Breton dialects group together as Brythonic, and are separate from Irish dialects (Goidelic).

But there the similarities stop, in terms of the expected degree of relationship. When Dyen analysed the two large Meso-European branches, Romance and Germanic, he found that each Romance dialect shared between 47% and 67% of cognates with each Germanic dialect. On the other hand, only 30% to 36% of cognates were shared between Brythonic dialects and Goidelic dialects, suggesting deeper splits. I should stress that this is not a false result, that might possibly follow from borrowing between Irish and Welsh[55] as a result of the geographical proximity of Ireland and Wales; rather the opposite – the older genetic relationship is apparently still strong. On this scale of percentage-shared-cognates, the deep 'celtic split' between Brythonic and Goidelic is on the same scale as that between Lithuanian and Slavic languages or between the various Indic languages.

What does this mean? In a relative sense, it is consistent with Schmidt's argument for a deep genetic split between Irish and Brythonic languages rather than McCone's (later) insular-celtic classification, based on an areal effect (Figure 2.4).[56] Schmidt postulated a deep split between Goidelic (Irish) on the one hand and all the rest, including all the Continental celtic languages (Celtiberian, Lepontic and Gaulish) and Brythonic, on the other. This implies a very different history and age of separation of the Goidelic languages from the rest.

Edinburgh professor of linguistics April McMahon and her husband geneticist Robert McMahon confirm these deep celtic relationships in a re-analysis of Dyen's data, using various different tree-building methods.[57] While the McMahons urge caution against rushing into dates,[58] Russell Gray and Quentin Atkinson of Auckland University have done just that in the journal *Nature*,[59] again using Dyen's dataset. Gray and Atkinson's estimate for the Goidelic/Brythonic split is 2,900 years ago – during the Bronze Age. This could be conservative, since their estimates for the break-up of Romance and Germanic languages are only 1,700 and 1,750 years ago respectively, and in the latter case would seem to be an underestimate (see Figure 6.2).[60]

Count everything

Just before Gray and Atkinson's paper appeared in *Nature*, the geneticist Peter Forster at the McDonald Institute for Archaeological Research, University of Cambridge, published a comparable figure (3,200 years) for the break-up of Continental and insular celtic, using a completely different dataset and method of estimation.[61] Whereas the Dyen dataset included only well-attested, living and thus data-rich insular-celtic languages, Forster bravely took information from Continental Gaulish inscriptions and compared this linguistically with insular-celtic languages, both living and extinct.[62] However, his tree explicitly acknowledges borrowing, resulting in ambiguity of some branches, thus causing the tree to become more of a 'network'. Possibly as a result of this, Forster's tree shows the two insular-celtic language branches Brythonic and Goidelic to be more closely related to each other than to Gaulish, but only just, and with a (branch ambiguity) at the break-up point.

The closest relatives to Gaulish are the insular-celtic Old Irish, Welsh, Scots Gaelic and Breton languages – in that order.[63] In Forster's tree, Gaulish is separated from the others by a very rough estimate of 5,200 years, consistent with the inference from Himilco's *Periplus* that insular celtic could have arrived in the British Isles rather early. In Forster's words:

> For the fragmentation of Gaulish, Goidelic, and Brythonic from their most recent common ancestor, the ... tree yields [an age of 5,200 years[64]], but this ... should be regarded as exploratory because it is based on only three estimators, i.e., three descendent branches. The [age of 5,200] years would represent an oldest feasible estimate for the arrival of Celtic in the British Isles, and indeed is expected to be close to the actual date if the phylogenetic split between Gaulish and Insular Celtic was caused by the migration of the Celtic language to Britain and subsequent independent development in Britain.

On a deeper scale, Forster's network tree also 'yields a date for Indo-European fragmentation in Europe, as a whole,' of 10,100 years ago,[65] which is at least comparable to the figure arrived at by Gray and Atkinson (7,300 years), who used a much larger dataset. These old estimates are consistent with Colin Renfrew's theory that the Indo-European language family originally expanded into Europe on the back of the arrival of the first agriculture there.[66] I shall return to these dating methods later (see Figure 6.2a).

Coming back to the celtic branch, these two studies – which used completely different sources and methods – seem to agree that the break-up of individual celtic languages happened perhaps twice as long ago as the separation of each of the two largest West European groups, Germanic and Romance, and roughly the same time as the separation of Lithuanian from Slavic (although these conclusions have to be acknowledged as provisional). Such datings would be more consistent with a Neolithic or Bronze Age celtic expansion than with the Iron Age one previously suggested under the southern German homeland theory (see Chapter 1). Not only that, but since Irish (Goidelic) and British (Brythonic) seem to be on opposite sides of a deep divide (Figure 2.4), the possibility of two early movements of celtic languages to different parts of the British Isles before 1000 BC is not disproved. Finally, on an even deeper timescale, Gray and Atkinson's estimate of the separation of the ancestor of the celtic branch from the common ancestor of the other two West European language groups, Romance and Germanic, is very old indeed, with a Neolithic scale estimate of 6,100 years – or even older, if we accept Forster's estimate.

Irish legends

Such a deep history of linguistic colonization, if real, might resolve an interesting apparent anomaly which lies at the core of Irish legend and culture. According to the orthodox academic view of 'Iron Age Celtic invasions' from Central Europe, 'Celtic' cultural history should start in the British Isles no earlier than 300 BC. Yet Irish legend has *all* the so-called mythological cycle of invasions from the Continent done and dusted by 1700 BC, which is the time of some of the earliest copper-mining in Ireland at Mount Gabriel in Cork. On this basis again, the first of the mythological invasions would have been at about the same time as the very earliest copper mining in Ireland, at Ross Island. These legendary estimates are based on two Medieval documents which list names and dates of Irish kings following the last Gaelic invasion, both pre- and post-Christian.[67]

The last three of the six invasions in the mythological cycle were over within the comparatively short period of 234 years.[68] The first of these last three groups of invaders were the Firbolgs, whose name is generally given a celtic derivation by linguists. The Firbolgs are said to have fled Greece, where they had been enslaved and made to carry earth in bags. They were supposed to have made ships out of these bags and sailed to Spain before arriving in Ireland, where their dynasty of nine kings lasted only thirty-seven years. Around 1900 BC the *Tuatha Dé Danann*, skilled artisans, replaced the Firbolgs and supplied another series of nine kings, who ruled Ireland for a further two hundred years, until 1700 BC.[69] The *Tuatha Dé Danann* worshipped Dana, who was synonymous with Brigit, the Celtic mother-goddess.

The last of the legendary invaders were the Milesians, who, as mentioned above, have been identified by both archaeologists and traditionalists as being Gaelic, and as coming from Spain and possibly ultimately from Asia Minor. I have already noted David Rankin's discussion on this theory, which partly relies on Thomas O'Rahilly's reinterpretation of the traditional or legendary texts; I am now coming back to it again in the context of linguistic dating. O'Rahilly accepted the orthodoxy of the Iron Age as the period for all 'Celtic' invasions and simply changed the dates of the mythological invasions to suit, by cutting out 1,600 years between the last invasion and the birth of Christ. He re-dated the Gaelic invasion at 100 BC, thus effectively killing off nearly all the numerous pre-Christian Irish kings. But

he still accepted the older mythological cycle of invasions as valid
evidence from before those king lists. O'Rahilly (and Rankin) also
accepted that there was an association between the multiple earlier
invasions, which O'Rahilly now dated over the previous 600 years,
and Brythonic languages arriving from the nearby Continent. In
arguing for these dates, and for the final Milesian invasion representing
the arrival of Goidelic or Gaelic from Spain, O'Rahilly clearly creates
a dating problem or anachronism by comparison with the much older
language splits discussed above.[70] O'Rahilly, if he were alive today,
might not accept the new linguistic dates. However, the anachronism
can be addressed by replacing the southern Germanic homeland
theory, for which there is less and less evidence, with a Spanish
homeland theory, for which – as we shall see – there is ample genetic
data and even archaeological evidence of cultural connections.[71]

Metals, beakers and celtic languages

If celtic languages moved into the British Isles not in the Iron Age
but, as suggested by linguistic dating evidence, in the Bronze Age or
earlier, then the dates of nearly 2000 BC implied by the Irish king
lists may not be so fantastic.[72] It should be remembered that the
earliest text relating to these dates[73] was written in the twelfth
century AD. However, it was based on a compilation of older texts
now lost to us, and gives considerably more credible detail and
cross-references for the last two thousand years than do any of the
chroniclers, such as Gildas,[74] whose works are still relied upon in
reconstructing the Anglo-Saxon invasion.

Barry Cunliffe gives parallel archaeological evidence for such an
early cultural spread up the Atlantic coast from Spain to Britain and
Ireland. In spite of this, in his more recent book *Facing the Ocean* he
still seems to sit on the fence when it comes to origins of 'the Celts'
and their languages. In apparent deference to the orthodoxy, he still
uses a map (Figure 1.3),[75] reproduced from his 1987 book,[76] of
Celtic tribal arrows coming out of southern Germany. But else-

where he acknowledges its evidential weakness and describes an alternative archaeological southern-origin view which argues that cultural continuity along the Atlantic coast could have fostered the development of a celtic lingua franca along the coast. He adds:

> It could be further argued that the language had developed gradually over the four millennia that maritime contacts had been maintained, perhaps reaching its distinctive form in the Late Bronze Age ... the archaeological and linguistic evidence support each other without being dependent.

Cunliffe then suggests that the famous Beaker pottery archaeological phase, which moved with early metallurgy into Ireland, might be a marker for this cultural-linguistic spread: 'In this model Beaker prospectors were the carriers of the Celtic language.'[77]

The specifics of the first metal mining dates are critical in Ireland, in view of the legendary last three 'invasions'. Cunliffe points out that, after Spain, where copper was being extracted as early as 3000 BC,

> [t]he main source of copper in the northern part of the Atlantic zone in the third and second millennia [BC] was Ireland ... Two major mining complexes have been located in the south of the country, together spanning the period from c. 2400 to 1500 BC ... The earliest so far known is at Ross Island ... Co. Kerry ... the mines were in operation from 2400 to 2000 BC ... A later more extensive series of mines was opened at Mount Gabriel ... west Cork'... [They] were in use over about the two centuries from 1700 to 1500 BC ... Finally we must turn to [north] Wales, where some thirty mining sites are now known ... [F]our ... were in operation in the first half of the second millennium.[78]

Significantly Cunliffe notes that 'the earliest [Irish] metallurgy seems to be coeval with the appearance of the earliest beakers',[79] which confirms his view of their association as cultural markers.

The distributions of so-called Maritime Bell Beakers, a distinct form of pottery traded along the Atlantic and western Mediterranean coasts of France, Britain, Spain and northern Italy in the early third millennium BC (see Chapter 5 and Figure 5.11b) are remarkably coincident with the map of inscriptional evidence for celtic languages over two thousand years later. More specifically, when tin later came to be used to make bronze, it was mined from the same key metal-rich locations of the Atlantic coast (western Spain, Brittany, Cornwall, north Wales and southern Ireland) that made up the later tin trading networks of the first millennium BC discussed earlier (see pp. 35–6).

Finally, I should like to reiterate a point about dates and the Irish legends. There is a recurring theme in Irish tradition that the ancestors of the last two invasions arrived in Spain, their previous location, by sea from an ultimate homeland in the eastern Mediterranean. The earliest linguistic dates given by Forster and Gray appear to suggest an early Neolithic date for the separation of the common celtic ancestor, well before the splitting of the other two West European language branches, Italic and Germanic.

Cunliffe has something to say on this Mediterranean trail. When discussing the theory of Beaker–celtic association (see also Chapter 5), he notes that 'Other archaeologists have taken a more radical line, in considering Celtic languages to have been carried to the west much earlier by Neolithic cultivators.'[80] 'Other archaeologists' refers to Colin Renfrew's theory of the agricultural spread of Indo-European languages and the role of celtic languages in that spread.[81] Cunliffe refers to this Mediterranean 'roots' possibility himself in terms of Neolithic trade between the Mediterranean and Atlantic coast from the fifth to the third millennium BC. For evidence of this, he cites the trade in Cardial Impressed Ware, a distinctive type of pottery which, during the early Neolithic, spread from Italy along the Mediterranean coast to southern France and through his favoured Garonne river route to the Atlantic Coast (see p. 179 and

Figure 5.1). As we shall see, there are many genetic parallels to such events during the Bronze Age, the Neolithic and even earlier.

Summary

Celts were real rather than mythological people to the Roman and Greek authors for over a thousand years. Ancient literary evidence points to their early presence in Spain, southern France and Italy, thus contradicting the nineteenth-century view of their origins in southern Germany. Inscriptional evidence indicates that languages closely related to insular celtic are found in the same southern distribution over the same period, and were the main vernacular alternative to Latin there. Given this distribution, Caesar's remark that people in central Gaul, south of the Seine and Marne rivers and west of the Garonne, referred to themselves as both Celts and celtic-speaking is consistent with Buchanan's and Lhuyd's view that there was a close link between insular celtic and the classical Continental Celts. This contradicts the Celto-sceptic view that the term 'Celtic' is based on a worthless myth best left, with its fuzzy linguistic associations, in the classical period.

There is good direct and circumstantial evidence that insular celtic was present in most parts of the British Isles at the time of the Roman invasion, although not as abundant in the one place where there should be the most evidence – England. Controversial linguistic-dating evidence may link the spread of celtic languages, not with the spread of Central European Iron Age Hallstatt and La Tène cultures as presently held in the orthodox view, but from a different region in earlier times stretching from the Bronze Age possibly back to the Neolithic.

Part 2

Colonization of the British Isles before the Roman invasion

Introduction

CAN LANGUAGE DATING TELL US WHEN THE CELTS ARRIVED?

In the last chapter we saw that any reappraisal of the scattered comments of classical historians and evidence from non-Latin inscriptions tends to challenge the nineteenth-century theory that the Celts emerged from southern Germany during the Iron Age and spilt out over Europe, invading the British Isles only a couple of hundred years before Caesar. Rather, those texts and the locations of ancient celtic inscriptions seem to point towards an earlier, south-west European linguistic origin for the celtic languages now surviving only in Brittany, Ireland, Wales, Cornwall and Scotland. This alternative south–north language arrow resonates with archaeologist Barry Cunliffe's description of an Atlantic coastal culture spreading up the coast from the direction of Spain during the Neolithic and Bronze Ages, and agrees with many accounts of Irish legends and traditional history. This perspective is clearly neither the whole story nor a secure revision of Celtic cultural origins; and the devil is in the dates.

At the more recent end of the language scale we have celtic linguists who would rather not commit themselves on the history of

celtic languages any further back than the texts, inscriptions on
stone and bronze, can take them – that is, not much earlier than the
Christian era.[1] A little further back in time, we have other linguists
disputing how deep the relative structural split is between the
Gaelic and Brythonic celtic branches, but still unwilling to commit
themselves to any absolute timescale.

When we do get dates, they are emphatically not supplied (nor
generally endorsed) by linguists, but by people experienced in
drawing genetic trees. Nonetheless, respectable linguistics has been
used to construct those trees and their derived dates. Two different
research groups recently produced results for the Indo-European
family, although they systematically differed in dates of splits by over
2,000 years.

As mentioned in the last chapter, Russell Gray and colleagues,
from the University of Auckland compared two different tree-
building methods on sets of data they obtained from two linguists
working independently. The two datasets yielded similar overall
estimates for the break-up of celtic into Irish and British branches:
one 2,500 and the other 2,900 years ago.[2] Peter Forster of
Cambridge University and Alfred Toth of Zurich University used a
rather different, so-called *network method* derived from genetic
studies and estimated 5,200 years for the fragmentation of Gaulish,
Goidelic and Brythonic from their most recent common ancestor.[3]

Now, all of these dates will have their own inherent errors, but
the implication is that we are looking at separate celtic-language and
British-culture stories starting from at least the time of the Bronze
Age, if not earlier – in any case, before the Iron Age. The earliest
evidence for metal mining in the British Isles comes from southern
Ireland and north Wales, connecting up to the pre-existing metal
industries of Brittany and the Spanish Peninsula. This finds cultural
resonance in Cunliffe's description of the simultaneous introduction of the Maritime Bell Beaker – a pottery style from the same
Atlantic regions (Figure 5.11b).

However, when we look further, at equivalent estimates for the separation of celtic languages as a whole from other Indo-European languages, a much deeper timescale appears. Dates for the separation of the ancestor of celtic languages from Italic and Germanic language roots are estimated at around 6,000 years ago by the Auckland group.[4] In other words, celtic-linguistic roots could go well back into the Neolithic period.

The Neolithic, or the New Stone Age, was also the time of the spread of agriculture into Europe, and Colin Renfrew has argued strongly that Indo-European languages spread throughout Europe on the back of agriculture. If the celtic Indo-European branch had a Neolithic date of separation, the implications for its influence on European prehistory take on a quite different perspective from that of the spread of a successful Iron Age tribe.

Languages are not the same as nations or peoples

Before opening Pandora's box and plunging back into speculative cultural, linguistic and even genetic links along the Atlantic seaboard drifting ever further back into an opaque celtic-linguistic prehistory, I would like to pause in this section and put the peoples of the European Atlantic coast in a broader time perspective. Whether celtic languages were introduced to the British Isles during the Neolithic or the Bronze or the Iron Age, those who introduced them were not the first people in Britain, and in common with Anglo-Saxons, Vikings and Normans, need not *necessarily* have made much numerical impact.

In the nineteenth and early twentieth centuries, there was a tendency to equate culture and language, and to combine them together with genetic heritage into a 'racial' package, seeing all three as moving in concerted historical and prehistoric migrations. The idea of large-scale gene flow is still with us, encapsulated in the racial use of the word 'blood'. So, for instance, we find a documentary title such as 'Blood of the Vikings', intended to imply that we could simply look for Viking genes and work out how much 'Viking blood' made its way into the British Isles, and where. While I am interested in looking for specific gene flow, I do think that it is important to get away from the idea that genes, language and culture move together in equivalent doses. France, for instance, has since Roman times spoken a Romance rather than a celtic language, but this is not to say that the bulk of French ancestors physically came from Italy. The impact of a new movement of genes into an old population depends far more on the size of the pre-existing gene pool than on any archaeologically visible cultural changes. One of the good things about the politically correct archaeological reaction to racial migrationism in the late twentieth century is that it has forced upon us the realization that culture and language can move relatively independently of gene flow; but that does not of course mean that they always do.

While it is clear that different parts of Britain do show clear genetic differences, these differences do not necessarily relate to any ethnic labels we might think of, which are mostly only a couple of thousand years old. So I shall move away from language issues to give some geographic, climatic and genetic perspectives to the recurrent patterns of colonization of the British Isles since the last Ice Age, around 15,000 years ago. The reason for this digression is not a prehistory lesson, but more a check against moving to overenthusiastic conclusions that all 'Celts' came over to the British Isles during the Bronze Age or whenever. As we shall see, there is abundant genetic evidence linking Spain with Ireland and Wales by multiple migrations. But even such parallel links do not automatically give a date, size or number for 'Celtic migrations'. Genes are a proxy for actual migrations, while language only may be. But language is, more importantly, a proxy for cultural movement. And what if there were similar recurrent migrations up the Atlantic coast before and after the Bronze Age? As it turns out there probably were – both.

3

AFTER THE ICE

The theme of this book is that the British Isles were colonized from two different parts of Western Europe. These geographic origin-trails, being reinforced by recurrent long-term contact, could have persisted independently creating cultural and genetic affiliations throughout prehistory, continuing through the historical period until the present day. This chapter is as much about climate and geography as it is about the main core of the book — people. But the former guided the latter in their movements. The initial colonization of the British Isles, after the ice melted, also set the scene and patterning of the later genetic landscape.

A clean sheet in Britain

Europe has been occupied by our kind, *Homo sapiens*, for perhaps 50,000 years,[1] with a considerable amount of mixing over those millennia. So why should we even consider it possible to get more than the vaguest story of the peopling of one corner of this long continent? It is possible, and for that we can thank the climate.

During glaciations, the sea level falls by as much as 127 metres as a result of water being locked up in the huge ice caps. At the Last Glacial Maximum (LGM) — the peak of the last Ice Age, between

22,000 and 17,000 years ago – the North Sea and the English Channel dried up, and Britain was joined with Ireland and France as part of a larger European landmass. It is almost certain that this British corner of Greater Europe was completely cleared of people at some time during the LGM. Not only was a large part of the British Isles actually covered by a thick layer of ice, extending from Scandinavia in the north-east, but those parts of the British extension of the Continent, south of the ice, were uninhabitable polar desert (Figure 3.1). This means that whoever had lived there before the ice had now gone, leaving an empty landscape and only their bones and tools – and a blank genetic sheet (Figure 3.1). So we can be sure that the colonization of Britain began afresh after the LGM.

The same bleak picture applied to much of north-west Europe. Scandinavia, the Baltic Sea and coast, and Finland were covered in ice, with sterile surrounding zones of polar desert. The neighbouring north-east European regions of Murmansk, Karelia and Archangel were polar desert, as were the coastal zones of northern Germany, Frisia, the Netherlands, Belgium and northern France down as far as Brittany (Figure 3.1).

The questions I set out to answer in this chapter are how soon the first permanent recolonization of Britain happened after the LGM, and what contribution those first settlers made as ancestors for today's British populations. I ask 'how soon' and specify 'permanent' recolonization, since another short but severe glaciation, known as the Younger Dryas Event, followed several thousand years after the main post-LGM deglaciation in Europe. It is suggested, but by no means proven archaeologically, that some of the first entrants after the LGM survived the Younger Dryas in the British Isles to stay on and form a permanent nucleus of pioneers. I aim to show that they did, and that the first settlers in that short window of opportunity immediately after the LGM, known, confusingly, as the *Late Glacial* (which coincided with a European cultural period known as the *Late Upper Palaeolithic* – see Figure 4.1), contributed a substantial

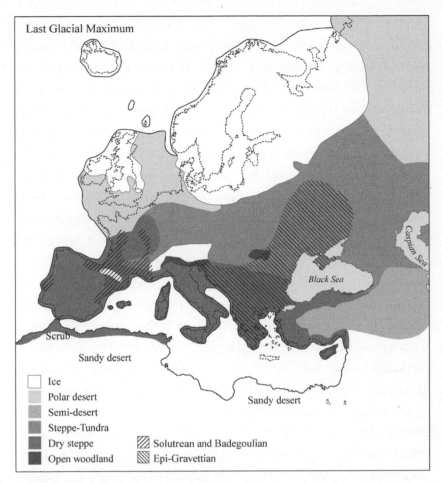

Figure 3.1 A clean sheet for the British Isles. During the last Ice Age most of the British Isles was covered by an ice cap. The rest of the extended landmass, including north-west France, was uninhabited polar desert. Northern Spain and southern France held the western refuge populations, identified respectively by Solutrean and Badegoulian cultures. Farther east in Europe, a much larger collection of refuge areas existed between the Balkans and the Ukraine. Italy was continuously occupied throughout the LGM, but at present there is no evidence for its role as a source population for recolonization.

proportion of ancestors to our modern fathers and mothers, using their advantage of first arrival.

Having argued for the sterility of much of north-west Europe during the last Ice Age, I should make it clear that the rest of

Northern Europe was not entirely devoid of people. In Eastern Europe, thriving communities extended up the tributaries of the rivers Don, Dnestr and Dnepr in the Ukraine and in Moldavia, both north of the Black Sea. Here, the expert mammoth-hunters had adapted well to the treeless Arctic landscape that now covered most of Central and Eastern Europe south of the ice. This landscape, known as steppe tundra, is similar to parts of northern Siberia today. Farther west, although the steppe tundra extended into southern parts of Germany, there is little evidence (just two sites in southern Germany[2]) for such hardy human activity north of the Alps at the LGM.

Parts of eastern France, by contrast to Germany and Britain, benefited from a slightly more benign landscape suitable for hunting, with dry grassland extending possibly as far north as the Champagne region. There are consequently a couple of LGM sites of human activity dated to around 18,000 years ago, in north-eastern France. After 16,000 years ago, such activity also reappears in Belgium, southern and eastern Germany, and the Rhineland.[3]

These few, scattered north-western LGM sites contain stone tool types matching the two Ice Age cultures,[4] found in the more populous regions farther south in France and Spain, where resident refuge populations held on throughout the freeze. So the implication is that before 17,000 years ago these heroic north-western pioneers were mounting intermittent or seasonal recolonizations, taking advantage of rapid oscillations of world temperature, rather than relict northern groups hanging on over the LGM.[5]

Recolonization from the Ice Age refuges

There is good archaeological and genetic evidence for at least four areas of southern Europe having served as refuges for Palaeolithic hunter-gatherers during the LGM.[6] I have already mentioned the extraordinary LGM human persistence in the Ukraine. This was largest and most active refuge, extending into Northern and

Eastern Europe and south-west towards Moldavia. Geographically, and for cultural (i.e. archaeological) reasons, Slovakia and parts of the Balkans can probably be included this eastern group, while Italy in the Mediterranean and south Central Europe formed a central southern refuge and remained largely occupied throughout the LGM. While these refuge regions are fairly widely separated from one another, it does not mean that they were each necessarily homogeneous refugee camps. They were probably each made up of zones with some degree of difference in genetic mix. The refuge for southwest Europe was spread either side of the Pyrenees in southern and eastern France, the Basque Country, and other northern coastal parts of Spain such as Galicia and Catalonia (Figure 3.1). I shall therefore not always stick to one label for each refuge.

In my book *Out of Eden*, in which I traced the Palaeolithic peopling of the world, I made a big deal out of a general observation deriving from the geography of gene trees, that 'once they got to their chosen new homes, the pioneers generally stayed put, at least until the Last Glacial Maximum forced some of them to move'. The recolonization of Europe after the LGM was no exception to this rule, and these pioneer hunters then chose to return to just those places they had taken refuge from. Once settled, a founding population is hard to dislodge.[7]

They achieved this just after 16,000 years ago, when Scandinavia and the Baltic were still covered in ice, by demonstrating, in both the archaeological and the genetic record,[8] possibly the highest rate of population expansion Europe would see until modern times. Archaeological records for this Late Palaeolithic period show evidence of twice as much human activity (measured in radiocarbon dates), lasting for longer (about 3,000 years) than either the Mesolithic or the Neolithic expansion, which began respectively around 6,000 and 8,000 years later (Figure 3.2).

The earliest archaeological evidence for the recolonization of north-west Europe comes from the Rhineland and southern

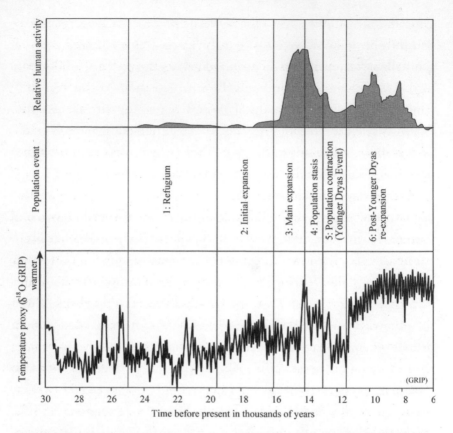

Figure 3.2 Human activity in northern Europe through the Ice Age and Younger Dryas. From zero population during the LGM (1, 2), the greatest expansion was immediately afterwards (3), falling (4) to a non-zero low during the YD (5) and rising again during the Mesolithic (6). ('Relative human activity' is a measure of human population in northern Europe, derived from the number of human-associated radiocarbon estimations (calibrated with Calpal). 'δ^{18}O' GRIP is a proxy measure of global temperature, 'up' being warmer.)

Germany, to where Magdalenian cultures (see p. 108) had spread shortly before 16,000 years ago. Belgium, the Netherlands, eastern and northern Germany, and Poland then rapidly lit up with numerous new sites of occupation as the Magdalenians spread into the North European Plain.[9]

The British north-west corner of Europe was not left out in this recolonization. Our stone tool styles have been named *Creswellian* after the cave sites at Creswell Crags (Figure 3.8). Archaeologists have recently begun to show just how early Britain was re-entered, starting only slightly later than in the Continental north-west, at 15,000 years ago. A number of sites in England, as far north as Nottinghamshire (the Creswell Crags in particular), show human activity at this time. The Creswellian culture appears to have extended through Norfolk across to the other side of the North Sea (which was dry at the time) into Belgium, the Netherlands and, in particular, Frisia (Figure 3.8).[10]

Other Late Upper Palaeolithic sites with a clear Creswellian signature are found in the West Country, in the caves of Devon and Somerset. These include the famous Cheddar Gorge and other limestone caves well known to tourists. Just what was going on farther west on the Atlantic fringe at this time is not clear. Creswellian sites are found on the south-west tip of Wales, near Haverfordwest,[11] but it has always been assumed (based on archaeological evidence from the far north, in the lower Bann Valley near present-day Colcraine, and in the south-west, in the Shannon estuary) that there was no occupation of Ireland earlier than 10,000 years ago. However, between 15,000 and 10,000 years ago Ireland was connected to the mainland in the south and had a much larger surface area, in spite of the ice cap to the north. Any earlier coastal settlements of those times would now be deep under water. The Creswellian findings in Somerset and Devon suggest that there may well have been earlier Irish habitation, and this is supported by the genetic evidence from Ireland, which I shall come to shortly.

Perhaps one of the most illuminating records of our hunting ancestors comes from their rich and varied cave art (Plate 4). Church Hole, one of the Creswell Crag caves, possesses a richly carved and engraved ceiling. Bas-reliefs are a particular and unusual feature of Church Hole,

> which is of huge importance not only because of its quantity of figures, but also their variety (at least six kinds of animals, plus two

or three species of bird, together with 'vulvas', etc). In addition, a few of the many peculiarities observed so far can be mentioned: e.g. the fact that, with the exception of the large stag and the first bison engraving detected nearby, and a possible bear bas-relief, which are complete, most figures comprise only parts of the animal, primarily the head or forequarters.[12]

First come, first served

The first colonizers of any region have the opportunity, not available to latecomers, to imprint their own particular style and determine what follows. In the genetic record there is an exact analogy to this archaeological explosion of human activity. The first colonizing gene lines often establish the genetic landscape in a way that long outlives the event. In the British Isles this imprint is still clearly with us. The genetic terms 'founder effect' and 'drift' are perhaps more prosaic than those used to describe the rock art and stone tools, discovered by archaeologists at sites such as the Creswell Crags, but they still bear the signal of first-comers.

The Leeds geneticist Martin Richards has conducted a massive study of all available evidence for migrations in Europe with thirty-seven international collaborators. They focused on mitochondrial DNA, which is passed down the maternal line and provides us with the most robust method we currently have for dating any part of our genome.[13] If we look at maternal prehistoric contributions to the modern gene pool over the whole of Europe in this study, 21% of extant lines derived from pre-glacial migrations and 51% from the Late Upper Palaeolithic just after the Ice Age (the latter from around 14,500 years ago in their study). For the rest, 11% each were contributed by the Mesolithic and Neolithic and around 4% by the Bronze Age.[14]

In north-west Europe (excluding Scandinavia), the Neolithic component was higher at 22%, but the Late Upper Palaeolithic contribution to the modern gene pool was still over 50%.[15] In other words, over half our maternal ancestors arose from southern refuges and arrived in north-west Europe just as soon as the melting ice allowed.[16] Our pioneer-mothers were tough!

Martin Richards' founder analysis of the European maternal gene pool took as its starting point the Near East as the ultimate *external* source of all genes coming in, but it is clear that the Late Upper Palaeolithic (LUP) contribution results mainly from re-expansions from Ice Age refuges *within* Europe. So, when Richards and colleagues looked at the LUP expansion in the Basque Country, the best modern representative of the south-western refuge, they found a massive 60% LUP contribution to locally extant European gene lines.[17] A similar conservative and indigenous European pattern has previously been argued for Y-chromosome data, but without the same detailed knowledge of gene line dates.[18] As we shall see, my own analysis of the Y data, separating and dating individual gene lines, supports this concept of a very large LUP colonization, although I am suggesting a larger Mesolithic component for the British Isles than in Richards' analysis of north-west European maternal lines.

The pioneers who came after the ice did not just establish first-come-first-served priority in the north, but they expanded and flourished in numbers as great as at any time in prehistory. This teeming fecundity of hunter-gatherers runs counter to the idea that humans did not achieve the sophistication and technology needed for successful population growth until the Neolithic agricultural period.[19] But it is even more stunning when we consider that most of Europe was still very cold at the time – rather like parts of northern Siberia today.

The climate itself may explain this anomaly. As anyone who has been to a wildlife park, or watched wildlife documentaries, can observe and deduce, it is easier to detect and hunt prey if you can see it. Grassland, with or without scrub, is therefore of more value for most hunters than dense forest. Likewise, most grazing beasts get more sustenance from grassland than in the forest, particularly if the forest is dense and coniferous. Immediately after the end of the LGM, a warmer period from 16,000 to 12,500 years ago transformed the dry European steppe tundra and polar desert into sweeping rich grasslands supporting a huge biomass of ruminants. But the forests had not yet returned, and, as on the Mammoth Steppe of Eastern Europe and Central Asia,[20] there were rich hunting fields if you were able to dress warmly and hunt big game.

To me, what is most interesting about the massive population expansion of the early West European pioneers is that much of the genetic landscape in the whole region was likely to have been 'set in stone' by the end of the Palaeolithic. Irrespective of the relative size of later Mesolithic and Neolithic contributions, this was long before the events that most British people now use to define their roots, such as invasions of Celts, Anglo-Saxons and Vikings. This implication of 'old blood' gives a more conservative complexion to the peopling of the region that is borne out repeatedly the more we look at the detail of genetic distributions on the landscape.

To some extent, this conservatism is predictable if one considers the founding events. As I mentioned earlier, the impact of a new movement of genes into an old population depends on the size of the pre-existing gene pool. So, the first colonization of a virgin area will tend to expand and fill up the habitable space, and the initial mix will have a greater effect on the make-up of the final population than subsequent immigrations. The Saami in Lapland, for instance still show clear signs of the very small size of their original founding female population a few thousand years ago, in that up to 52% of their population has the relatively infrequent V type of mtDNA (see below).

There is general agreement and good evidence that Western Europe was largely recolonized from the south-western refuges in southern France and Spain, which were separated from the more easterly refuges by the French Alps and some more. But before we

look at the primary genetic sources for the repeopling of the British
Isles, I would like to describe an important genetic and cultural
cross-fertilization which has only recently been brought to light and
which took place from east to west in Europe in the last few
millennia before the LGM.

A last-minute change of refuge

Between 26,000 and 22,000 years ago, in the millennia leading up
to the LGM, two new Stone Age cultures started to replace the
famous, so-called *Gravettian* styles of the Middle Upper Palaeolithic
in the south of France and in Spain. In this early glacial period, one
of these cultures predominated; it is now called the *Solutrean*, after
the village of Solutré in eastern France. Characteristically, Solutrean
stone-knappers produced finely crafted flint spear and arrow
points, which were worked on both faces using fine invasive flaking.
They also made ornamental beads and bone pins as well as creating
evocative prehistoric art.

During the build-up to the LGM, the Solutrean initially flourished on both sides of the Pyrenees, especially in Aquitaine in
south-west France, but for the period after the worst cold snap of
the LGM, 22,000 years ago, evidence for the Solutrean culture
tends to contract south of the Pyrenees to the main western Ice Age
refuge, which was northern Spain.[21] Coincident with this chilly
retreat, a new technical style became dominant in the Dordogne
refuge cultures in the south-west of France. This is known as the
Magdalenian, after the site at La Madeleine.

The Magdalenian culture, later to define human re-expansion
throughout Western Europe after the Ice Age, was previously
assumed to have arisen locally from the Solutrean in south-west
France. However, a reappraisal of the earliest Magdalenian in France
suggests that it arose from the pre-glacial, so-called Epi-Gravettian
cold-adapted culture in Eastern Europe and spread south-west to
France shortly before the LGM.[22] This new insight has, confusingly,

given rise to yet another culture name, after a place in the Dordogne, the *Badegoulian*. With this new trans-alpine insight, the Badegoulian culture can be seen to have been continuous in Europe between 25,000 and 13,000 years ago; it came to the fore in south-west France with the retreat of the Solutrean 23,000 years ago, and then thrived throughout the LGM.

The point of this digression, apart from underlining humans' extraordinary cultural capacity for adapting to extreme adversity, is that we can suggest both female and male gene-line markers to accompany this pre-glacial east–west cultural cross-fertilization.[23]

Before the LGM, two maternal gene groups dominated Europe: U and HV. The earliest of all was U (nicknamed Europa in *Out of Eden*), arriving from around 50,000 years ago and 30,000 years before the LGM. She still makes up a background of around 9% of modern European maternal lines. HV arrived in a second wave, 10,000 years before the LGM.[24]

In *Out of Eden* I described the origin of the HV group, the ultimate root of the majority of European female lines. She was born, possibly in the Trans-Caucasus, over 30,000 years ago and may have been a genetic signal running in parallel to the early Gravettian cultural spread. Early on, HV split into two branches: the smaller one known to mitochondrial geneticists as Pre-V, and the much larger branch H, both of which subsequently spread all over Europe. I shall use as a nickname for this large group the Russian name *Helina*, and the widespread Latin name *Vera* for Pre-V's daughter group, V. (Vera is the generic Latin name for 'truth', and V gave us the first clear genetic evidence for the Basque glacial refuge; I could not find a Basque name beginning with V.) Pre-V, perhaps 26,000 years old, is found infrequently in a limited distribution stretching from east to west, from the Trans-Caucasus through the north Balkans and Central Europe to southern Spain and Morocco.[25]

The female ambassadors from east of the French Alps, who could be genetic proxies or parallels for the pre-glacial Badegoulian trans-alpine culture spread, are these very two: H and Pre-V. Helina is now the most common female group in Europe, accounting for around half of gene lines in the West, including the British Isles, but this rise in matriarchal prominence and rank took place just after the LGM. The evidence for the trail of these younger female marker lines is rather good.

North-western female pioneers

Being numerically far greater, the female Helina group ought to be a better marker for the first recolonization events than its close relative Vera. However, for a long time the hope of a clear picture of Helina's movements was not fulfilled, partly because of her pan-European distribution but mainly because there appeared to be insufficient geographical pattern in the Helina subgroups.[26] Now, with the increased resolution (i.e. accuracy and number of branches) of the Helina family tree resulting from extended analysis in key parts of the mitochondrial genome, it has become possible to trace at least eight specific Helina subgroups expanding in Europe after the LGM. The geographical picture for Vera, with her smaller numbers and more limited distribution, had already emerged, in 1998,[27] so I shall present the data for Vera first.

The Vera group, Pre-V's genetic daughter, tells us rather clearly about the next stage. Today in Europe, Vera is found most frequently in western and northern regions, suggesting that she was one of those gene groups who expanded out of the south-western refuge. She almost certainly began her life in one of the southern refuges, in either south-west France or Spain, and has been in Western Europe now for 16,300 years.[28] Soon after the LGM she spread out from the refuge, heading north and west along the Atlantic coast (much later forming a large component of the Saami ancestors), but also south to Morocco and east to Central Europe. Scandinavia and Lapland have the highest Vera rates, but because of the presence of a large residual ice sheet they were not colonized until the last 10,000 years (discussed in detail in Chapter 4).[29]

After the Saami and Basque, in Western Europe Vera is found at highest frequency in south-west France (over 10%) and at a progressively declining frequency up the Atlantic coast (Figure 3.4). Most of northern France and the British Isles achieve around 5% Vera, for instance. So, if we look at the continental shelf, including the British Isles but excluding Scandinavia, it can be seen that Vera's frequency falls off north of the 53rd parallel and both north and east of the Rhine.[30] This distribution may well reflect the distribution of the ice sheet and environmental conditions 16,000 years ago, just after the end of the LGM, and the limit of the initial human spread (Figures 3.1 and 3.3).

At least a dozen specific subgroups of the Helina group have dispersed in Europe since the LGM. One of them – H1 – seems, like her aunt Vera, to have originated in the west Iberian refuge (the Basque Country) and expanded early into north-west Europe after the LGM. She now accounts for 66% of Helina lines in Spain and 45% of Helina in the rest of north-west Europe (overall, including the British Isles).[31] Smaller subgroups, H2, H3, H4 and H5a, may have had a similar history in the Iberian refuge, but expanding during the Mesolithic, and account for a further 16.5% of Helina in the United Kingdom, bringing the proportion of *identified* subgroups of the Helina contingent there to over 60% of British Helina types. Clearly it might eventually be possible to trace and date more British Helina subgroups back to Iberia, but on these figures the specific early expansions of Helina and Vera from the south-western Ice Age refuge before the first farmers account collectively for around 60% of all modern Helina and Vera groups in the British Isles.

Of course, this is not the whole picture since we are looking only at specific dated Helina and Vera subgroups, and not at other, less well-resolved subgroups or other gene groups apart from Helina and Vera. These two gene groups as a whole contribute about 42% of the total extant British gene pool, so the specific individual earlier subgroups H1 to H5 and Vera would together account for only around a *quarter*

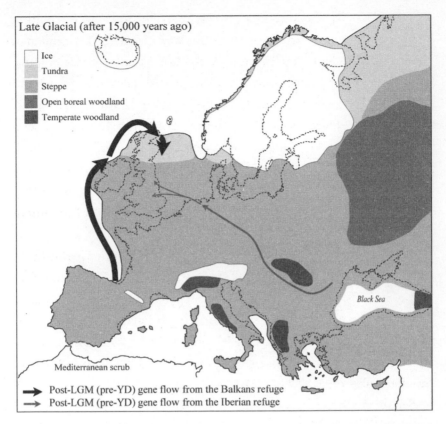

Figure 3.3 Colonizing 'Greater Britain' after the Ice: a summary map of early recol-
onizing gene flow into northern Europe and the British Isles 15,000–13,000 years ago.
The bulk came from the Iberian refuge, which contributed perhaps a third of maternal
ancestors for the British Isles during this cool period. The reason for this apparently high
gene flow is only partly that they were founders; the other reason is that most of
northern Europe was grassland and rich in big game.

of today's maternal lines. However, since most of Helina expanded in
Europe during the Late Upper Palaeolithic anyway, 42% may not be a
gross overestimate for this period (Figure 3.5).[32]

One reason for trying to resolve the Helina tree better is that it
makes it much more personal and specific if we can show when, and
from where, more than half our British maternal ancestors arrived.
For a historian of migrations, measuring the size of the older migrations is useful in estimating what is left over for the more recent

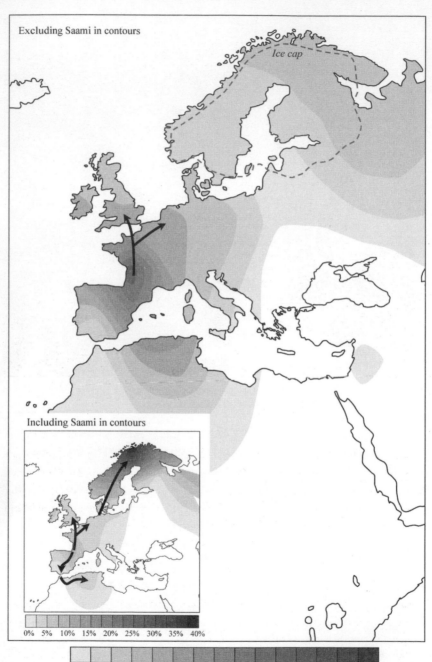

Excluding Saami in contours

Ice cap

Including Saami in contours

0% 5% 10% 15% 20% 25% 30% 35% 40%

0% 1% 2% 3% 4% 5% 6% 7% 8% 9% 10% 11% 12% 13% 14%
Gene frequency

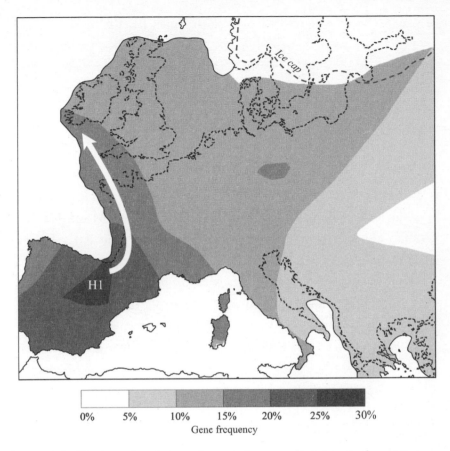

Gene frequency

Figure 3.5   The immediate impact of maternal re-expansions into north-west Europe from Iberia after the Ice Age from 15,000 years ago. Contour map of H1 (Helina main sub-group) gene frequency – arrows indicate direction of gene flow based on the gene tree and geography. Contours follow greater land area resulting from low sea level and avoid the ice cap (Scandinavia excluded from analysis).

Figure 3.4   Ancestor Vera, who gave the first clear genetic evidence of Ice Age refuges and early re-expansion from Iberia into the north-west. Contour map of Vera gene frequency – arrows indicate direction of gene flow based on the gene tree and geography. The lower (inset) image shows a Mesolithic founding event in Lapland – which is removed from the analysis in the main figure to avoid distortion and because it happened after the Scandinavian ice cap melted. Vera is a smaller group than Helina, but is more specific to the colonization of Western Europe from the Iberian refuge.

invasions. As to who and what were the main British ancestors, we can say they were largely Ice Age hunting families from Spain, Portugal and the south of France. Obviously, the Basque refuge area has since received intrusions of its own, particularly from the Mediterranean and North Africa, but these still constitute only a small percentage of that region's present-day gene pool.

Male pioneers

Can the Y chromosome, which is carried only by men, tell us a similar, different or better-resolved story? The answer is very similar, although as usual the geographic patterning is far better resolved and more informative than the female picture.

The large Y gene group R (nicknamed *Ruslan*) is the closest male equivalent of the maternal Helina group, accounting for ancestors of half of Europe's men; he moved into Central from Eastern Europe 30,000 years ago and thence to south-west Europe before the LGM.[33] As a result, presumably, of genetic drift, only one discrete R gene group emerged from the western refuge after the LGM – this is the sub-group labelled R1b.[34] In the Basque Country, R1b achieves his highest frequency at around 90% of the male population, confirming his claim to be the main, and probably the *only surviving*, south-west European gene group as a result of severe genetic drift during the Ice Age. The *overall* R1b gene group I shall refer to as *Ruisko*. This is a Basque male name, chosen for its coincidental similarity to *Russkiye* ('Russian') and intended to imply an ultimate preglacial origin with Ruslan as ancestors in Eastern Europe.[35] Ruisko's frequency declines the farther north and east along the mainland European coast one goes, from 90% in Basques, 70% in the Dutch, 60% in the French and 50% in Germans, to 36% in Danes, around 25% in Norwegians and Swedes, 21% in northern Russians, and 18%, 16%, 7% and 0%, respectively, in Estonians, Poles, Latvians and Finns around the Baltic coast (Figure 3.6a).[36] The sharp fall-off in frequency of Ruisko going across the mouth of the Baltic to

3.6 Male re-expansions into north-west Europe from Iberia immediately after the Ice Age. Contour maps of gene frequency – arrows indicate direction of gene flow based on the gene tree and geography. Contours follow greater land area resulting from low sea level.

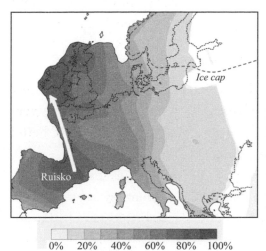

Figure 3.6a Ruisko gene group (R1b). This map shows the impact of all male re-expansions from the south-west European refuge; Ruisko (R1b) represents nearly the entire source gene pool from there. The densest gene flow follows the Atlantic façade, thus favouring Ireland. Since Ruisko also covers later periods, contours are shown in Scandinavia beyond the ice line.

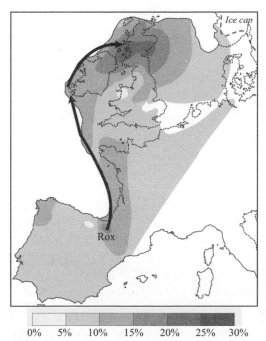

Figure 3.6b Rox gene cluster (R1b-9). This map shows the impact of the earliest male re-expansion from the south-west European refuge, 15,000–13,000 years ago; Rox is the main source gene cluster for that period. The gene flow follows the ancient extended coastline, favouring Ireland and Scotland.

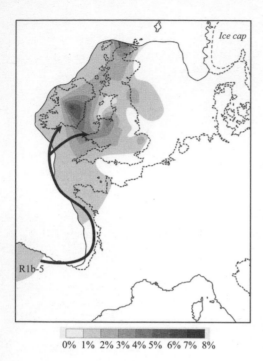

Figure 3.6c R1b-5 gene cluster. This map shows another of the earliest male re-expansions from the south-west European refuge 15,000–13,000 years ago. R1b-5 derives from Rox and also concentrates on the Atlantic façade, featuring in Ireland, Wales and northern Scotland.

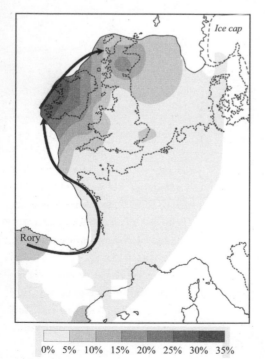

Figure 3.6d Rory gene cluster (R1b-14). This map shows Rory, one of the larger early male clusters to re-expand from the south-west European refuge 15,000–13,000 years ago. Rory derives from Rox, also concentrating on the Atlantic façade, featuring particularly in Ireland and less so in Scotland. He is strongly associated with Irish men with Gaelic names – but this does not mean that Gaelic arrived so early!

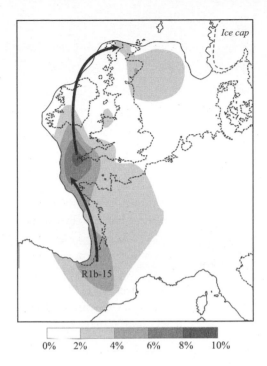

Figure 3.6e R1b-15c gene cluster. This map shows R1b-15, another sub-cluster of Rox which re-expanded from the south-west European refuge 15,000–13,000 years ago and concentrates on the Atlantic façade, featuring in this instance in Cornwall.

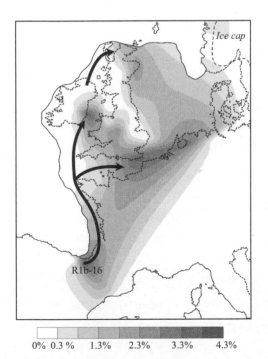

Figure 3.6f R1b-16 gene cluster: this map shows R1b-16 another early male re-expansion from the south-west European refuge (15,000–13,000 years ago), that derives from Rox and also concentrates on the Atlantic façade, featuring in Ireland and Wales. Unlike the others, this cluster also moved up the region of the English Channel along the ancient river valley of the Seine to Kent and the Continental Low Countries.

Sweden and thence east to Russia ties in with an early recolonization of mainland north-west Europe from the south-west, while the bulk of Scandinavia was still covered in ice. However, even the 25% rates in the far north, in north-west Scandinavia, indicate the ultimate geographical extent of spread from the Spanish refuge.

In north-west Europe, Ruisko is replaced partly by R1a1,[37] a distantly related pre-glacial R1 gene group coming from Eastern Europe, the Balkans and ultimately Asia. Not only is R1a1 a minority lineage, but it appears to have arrived in the west via Scandinavia and Poland well after the Younger Dryas.[38] I call him *Rostov*, after cities in the Ukraine and Russia where some of the highest rates are found (not forgetting to mention Tolstoy's famous character of the same name). There is also another exclusively European gene group, I, deriving from the Balkans and found at high rates in Scandinavia and northern Germany[39] (30–40%, for map of total I distribution see Figure 3.7). His Balkan origin earns him the pan-Slavic name *Ivan*. Rostov and Ivan are nearly absent from the Iberian refuge, thus making it easier to measure discrete north-eastern and south-western sources of gene flow in Western Europe and the British Isles, an approach I use throughout this book. I shall come back shortly to these interesting eastern refuge lines, which are important after the Younger Dryas freeze-up.

Re-analysis of the Y-chromosome data

Just comparing the relative frequency of the Ruisko gene group in different countries to determine their relatedness is a crude and somewhat circular approach since it simply lumps together all members of the same group (Ruisko) that dominates Western Europe and fails to give any perspective in time or space. Unfortunately, the eleven or so clearly defined gene groups (such as Ruisko, Rostov and Ivan) which feature in published West European datasets have yet to have their finer branch structure resolved.[40] From my perspective, this is a problem in particular for the Ruisko

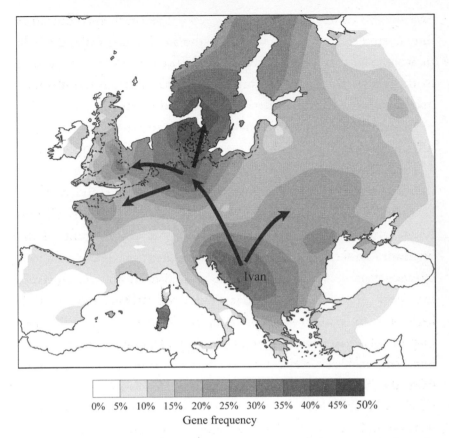

Figure 3.7 Ivan (I gene group): male re-expansions into north-west Europe from the Balkans and Ukraine after the Ice Age. Ivan is the second largest male gene group in Europe. This map shows his likely Ice Age refuge in the Balkans and his main regions of dispersal – Slavic countries, north-west Europe and Sardinia. While these dispersals most likely occurred during the Mesolithic and Neolithic, they may have started even before the Younger Dryas freeze-up 13,000 years ago.

gene group, but also more generally for all the other main Y gene groups represented in the British Isles.

A number of researchers have used detailed Y-chromosome markers known as *STRs* (*single tandem repeat sequences*) in an attempt to overcome this problem, and over the past few years have accumulated a huge British dataset of STR gene types within the established Y gene groups in order to break them up into more detail.[41] STR

testing adds much more information for each individual than just
the group markers because STR markers are more rapidly mutating
than the group markers, but the information is less easy to use in
constructing unambiguous gene trees. Lacking this last step, of
breaking the gene groups into smaller branches, the data, especially
for British males, are under-analysed with respect to defining more
genetic branches.[42]

Early on in writing this book, I decided to address this problem
and use these STR gene-type data to map the detail of British colonization, by defining more geographically useful clusters of gene
types, each closely related to identifiable founding gene types in
Western Europe within each of the main gene groups.[43] I have reanalysed a huge Y dataset of 3,084 individuals, collected mainly
from British and continental European populations this century.
The Y-chromosome results have been published, as comparable but
separate datasets in a number of different papers which are in the
public domain, covering Spain, the Basque Country, Turkey, north-
west Europe, Scandinavia, Iceland and the British Isles. I pooled all
these datasets for re-analysis. My aim was to chart the post-glacial
re-expansion from the Iberian and East European refuges. My
methods for this analysis and for other European male lines entering
the British Isles are described in the Appendices, where the naming
and definition of groups and clusters are clarified, and there are
dated trees both for maternal and paternal lines.[44]

For the British Isles, I have collected all the available published
data on the Ruisko, Rostov and Ivan gene groups and compared
them with mainland Europe.[45] This reveals an even more dramatic
British dominance of Ruisko than on much of the neighbouring
mainland: he accounts for around 90% of all males in Ireland, north
and south Wales, and Cornwall (Figure 3.6a).[46]

Time and the massive expansion of Ruisko into Western Europe
from the Iberian Peninsula over the past 15,000 years have resulted
in a complex network of interrelated gene lines whose origin,

ancestry and temporal relationships have to be sorted out before we can draw more specific conclusions about the male colonization of the British Isles.[47] When all the individual Ruisko gene types are assessed for the similarity of their STR markers, they initially split up neatly into sixteen gene clusters, all having very closely related gene types, and each with a central or modal (most common) gene type. These clusters each have a very clear geographical localization, so it is likely that their modal types represent ancestral founders. I have labelled the clusters R1b-1 to R1b-16 (see Appendix C). All clusters are descended from one of the two main gene types from Spain, the oldest of which I shall discuss next.[48] (I should emphasize at this stage that I have extended this cluster approach to the entire dataset, i.e. breaking up Rostov and the Ivan subgroups as well.)

The first male expansion from the south-western refuge after the ice

Further analysis revealed that, ultimately, all clusters and their constituent gene types can be traced back to one root type found in cluster R1b-9 which expanded northwards from the Basque refuge around 16,500 years ago[49]; for this reason I label it the *Basque Haplotype* (Figure 3.6b).[50] This root type differs by one STR mutational from its nearest relative, known as the *Atlantic Modal Haplotype*, which resides in cluster R1b-10, and did not expand until much later (see Chapter 4).

In fact, the major early expanding male clusters can be divided fairly clearly on the Y tree into those dating in the British Isles from between 16,000 and 13,000 years ago, which derive directly from the Basque Haplotype in R1b-9, and those expanding from Iberia after 11,500 years, which all derive from the Atlantic Modal Haplotype in R1b-10 (see Figure A4 in Appendix C, and Figures 3.6b and 4.5). The root male gene cluster R1b-9 can be called *Rox*, and his son, R1b-10, *Ruy*; both are Basque names. Rox and his descendants show significant expansions from their refuge at least twice if not three times since the last Ice Age. The first such expansion seems to have kicked off smartly from the southern refuge around 15,600 years ago, since a number of Rox's sons found in Europe or Britain (or both) date to that time (see Figure A4 in Appendix C, and Figures 3.6c–3.6f).[51] Six of the sixteen British male founding clusters (R1b-4 to R1b-6 and R1b-14 to R1b-16) descend directly from Rox and date from before 13,000 years ago. In total, around 27%[52] of modern British men can claim descent through their fathers from the seven clusters arriving in this early post-LGM period. This is certainly within the bracket of the 25–42% I estimated for maternal descent, but obviously nearer the lower limit of 25%. But even a 30% contribution of Basque Late Upper Palaeolithic male and female ancestors for modern British imposes a completely different balance on our 'roots' perspectives.

I shall return to the events taking place after the 13,000-year threshold, but it is likely that this genetic watershed between the initial Late Upper Palaeolithic recolonization period and what came later, during the Mesolithic, is not just a genetic accident. The watershed may reflect the profound climatic reversal that occurred 13,000 years ago, known as the Younger Dryas Event, a short worldwide freeze-up which ended abruptly around 11,500 years ago with another equally dramatic warm-up (see below).

A Danubian line from the Balkan refuge?

During the LGM, Ivan (group I) may have diversified in isolated local refuges in the Balkan region, into at least four descendent gene groups,[53] followed by several post-LGM dispersals into Northern Europe. Ivan arrived much later than Rox in north-west Europe and the British Isles. But one minority branch of the Ivan gene group, I1c (named here *Ingert*, a German name to fit his location of highest frequency), appears to have placed at least one of his three clusters in Britain as long ago as did Ruy, around the time of the Younger Dryas. This raises a question whether Ingert arrived in Western

Europe and Britain before or after the Younger Dryas, or a bit of both[54]; I shall return to this later.

The first British Atlantic settlers: still on location

As the ice receded in the north, the earliest Ruisko clusters began to move northwards up the Atlantic coast. As beachcombers, they would have met no English Channel, and no Irish Sea – they could have walked in a straight line from the tip of Brittany across to the south-west coast of Ireland. To the east, the continental shelf was dry land (Figure 3.3), and the Channel Islands, Land's End and the Isles of Scilly were all well inland. The majority of our Atlantic coastal beachcombers would then have been able to walk onwards, round the west coast of Ireland, where the coastline has not changed much since. From there they could have spread straight on towards the Western Isles, Scotland's north coast and Orkney, which at that time all formed one landmass, whose west-facing coastlines are also still mostly unchanged. After the initial deglaciation 15,000 years ago, there was still a small ice cap in the Scottish Grampians spreading just across into Antrim in Northern Ireland.

The lower sea level also exposed the bed of the North Sea continental shelf in a straight line across from Scotland to Denmark. The Scottish ice cleared over the next couple of thousand years, and the North Sea coast receded somewhat, allowing access to Aberdeenshire. Although it is possible that both Ireland and Scotland were colonized as early as 15,000 years ago, there is no archaeological evidence for this. The ultimate spread of these gene lines north along the Atlantic fringe to Scotland from Ireland, as evidenced in the genetic record, could have been somewhat delayed.

In spite of this delay, the distributions of the eight earliest Ruisko clusters, in particular the main founder Rox, still actually reflect the extended beachscape and coastline of the Greater British Atlantic coast before the sea level rose to create the English Channel and the Irish Sea split off Ireland from Wales (Figures 3.3 and 3.6). Largely shunning Wales, and southern and eastern England (except for pockets on the south coast), these early founders distributed themselves round the less changed parts of the old western coastline, including western Ireland and the Western Isles, right up to Scotland and Orkney. They also feature in northern England, for instance in Nottinghamshire, near the Late Upper Palaeolithic site of Creswell Crags (Figure 3.8).

The highest rates for Rox, of up to 27% of all gene lines, are found in Scotland, with a flow round to the Scottish east coast to Stonehaven, where 23% are Rox. This distribution perhaps reflects the coastline at that time (Figure 3.6b), while the high rates most likely reflect genetic 'founding events' along a beachcombing route – which is not entirely surprising if Scotland was their terminus. Rox is notable by his low rates on the nearby Continental coast (2–5%), which would have been joined to Britain, but well inland at that time and inaccessible to Scottish beachcombers because of the Scandinavian ice cap.

The westerly distribution of Rox is reflected in that of his *derived* branches: cluster R1b-5 features in Ireland, Wales and northern Scotland (Figure 3.6c); cluster R1b-16 characterizes north-east Scotland, Ireland and south-west Britain (Figure 3.6f). There was some later regional re-expansion and spread during the Mesolithic in several other Rox clusters (see also Chapter 4); so cluster R1b-4 features in Scotland and north Wales, being absent from Ireland, while R1b-6 expanded in eastern England and R1b-15 re-expanded in Cornwall, central Wales and, to a lesser extent, in Ireland, Scotland and the Channel Islands. Three R1b-15 sub-clusters focus on the Atlantic fringe regions, each with slightly different emphasis. However, only one of these three actually expanded on arrival (Figure 3.6e); one more re-expanded immediately after the Younger Dryas, while a third re-expanded during the Neolithic.[55]

Cluster R1b-14 is perhaps the most interesting migrant gene line of this early post-LGM period. He has recently been strongly associated with Irishmen with Gaelic surnames, for which reason, and that of his location, he has been given the nickname *Rory* (Gaelic for 'little king'). Rory's distribution correlates with that of ancestral Rox,[56] missing out on Wales and southern England. But while featuring in Scotland and Orkney in the far north, Rory is charac-teristic of Ireland, being found at rates of up to 30% there (Figure 3.6d). This distribution again fits with the dry Irish Sea and English Channel at that time.

As I have included the same dataset that gave rise to this dramatic claim in my analysis, it is natural that I should make the same gen-etic association with the 'Gaelic' sub-sample, and that I should find the frequency of Rory to be higher than in other Irish samples unselected for surname. However, my re-dating of Rory's arrival in the British Isles to around 16,000–15,000 years ago is quite differ-ent from the date arrived at in the original report in *Nature* by Emmeline Hill and colleagues at Trinity College Dublin, which suggested a Neolithic age. Before my new dating is taken as a suggestion that Gaelic arrived in Ireland 15,000 years ago, I should point out that there are much simpler explanations for the associ-ation.[57] I shall come back to this and to the resurgence of three descendent clusters of Rory during the Mesolithic and Neolithic, but for the time being it should be clear that their male ancestors arrived in the west long before then.

Apart from Rory, several other derived Rox clusters (R1b-4, 5, 6 and 15) re-expanded and split up again thousands of years later, and will be revisited in Chapters 4 and 5, when I come to discuss the Mesolithic and Neolithic periods. The fact that their ancestors arrived so early obviously reduces estimates of the size of later immigration.

Early eastern British male gene lines on the North Sea Plain

As I have shown, while Rox dominated the early post-LGM scene, and ultimately spread throughout the whole of the British Isles, there is a strong bias in his relative frequency towards the greater Atlantic coastal region and away from England, and particularly from regions in the south-east such as the south coast and Norfolk. This bias relates to their initial accessibility to beachcombers at that time, since the coastline still extended hundreds of miles out and they were way inland; but the geographical implications of this ancient coastline were long-lasting. After the Younger Dryas, the sea level rose and these less densely populated eastern and south coastal parts became more accessible to Ruy (as we shall see in the next chapter).

So, in the Late Upper Palaeolithic, after the LGM, the eastern side of Britain is rather quiet with respect to input from the Basque refuge. Instead, from the European mainland came another male gene group, Ingert (group I1c – see p. 124), who originated in the Balkan Ice Age refuge. Ingert is characteristic of north-west Europe, but particularly northern Germany and eastern England, although found at rather low rates (maximum 13%) even there.[58] This early intrusion from the east has a bearing on one of the issues that is central to our story of British origins: the apparent similarities between England and the parts of the Continent that face us across the North Sea, such as Holland and Frisia. Is the North Sea genetic fellowship we see today a reflection of Anglo-Saxon invasions or of something older? Several studies have shown that measurements of genetic distance (these are based on similarity of specific gene frequencies – see Chapters 6 and 11), depending on the relative frequencies of the broad male gene groups, always tend to place Frisia (on the coast of northern Holland) very close to East Anglia and Norfolk (see Figures 11.2b and 11.4b).

My re-analysis confirms the general trend of similarity across the North Sea. However, there are older reasons – and evidence – for this genetic neighbourliness than a massive swamping of England by Anglo-Saxon invaders from north-west Europe during the Dark Ages. The drowned North Sea Plain is one of the oldest geographical indicators of the beginning of an eastern British identity, and there is genetic evidence to support this.

At the time of the great post-LGM European expansion of 15,000 years ago, there was no North Sea. Instead, there was a flat grassy plain stretching all the way from Poland and the southern Baltic, through southern Sweden, Denmark, Germany, Frisia and Holland across the North Sea and into eastern England (Figure 3.3). In fact, had they wished, our forebears could have walked in a straight line all the way from Berlin to Belfast, although in practice they seemed to prefer wandering along beaches.

If it still existed today, the North Sea Plain would be in the centre of the Ingert distribution (Figure 3.8). Ingert dates overall in Europe to 21,000 years and may have originated in a Balkan Ice Age refuge (see below).[59] Three British founding clusters from Ingert (I1c-1, 2 and 3) date to around 13,000, 14,000 and 12,000 years ago, respectively.[60] This suggests a pre-Younger Dryas (i.e. Late Upper Palaeolithic) spread for at least part of the Ingert branch. While Ingert is present at a low rate of about 3.3% throughout the British Isles, this figure rises to over 10% on parts of the English north-east coastal region, in particular York and Norfolk. Given this distribution, the age of Ingert in the British Isles,[61] and the fact that he is no more common on the neighbouring Continent, the chances are that this represents the echo of an ancient intrusion. To me this is the first of a series of specific, dated early British genetic intrusions from the Continent which tend to mitigate claims of a later Anglo-Saxon genocide.

There is in fact some archaeological evidence for cultural links between the North Sea Plain Continental cultures and the Creswellian sites of Norfolk and Kent in the final stages of the Late Upper Palaeolithic, before the Younger Dryas. This would be consistent with the distribution and age of Ingert there (Figure 3.8).

Apart from the Ingert story, much of the apparent ancient similarity between Frisia and Norfolk relates to shared Ruisko gene lines, derived in parallel from Iberia. This linkage can also be explained from the perspective of the drowned North Sea Plain. As we have seen, the first hunters to return up north from Iberia, along

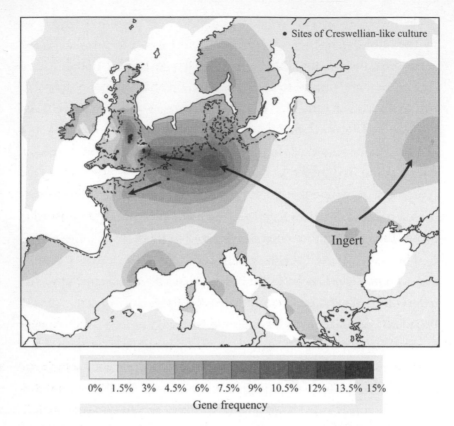

Figure 3.8 Ingert (I1c gene group), the earliest Balkan males to reach Britain? This map shows the distribution of Ingert, a sub-group of Ivan, who most likely originated near the Black Sea and started spreading to north-west Europe just before the Younger Dryas freeze-up 13,000 years ago and continued during the Mesolithic 11,500 years ago. Black dots indicate finds of Creswellian style artefacts, dating to just before the YD both sides of the North Sea.

the Atlantic coast, would have found themselves entering a hugely extended continental shelf. Arriving at a point halfway between the tips of what are now the peninsulas of Cornwall and Brittany, they would have split at the mouth of the River Seine, which then flowed south-west out of the dry bed of the English Channel.

Those crossing the Seine would have continued along the Atlantic coast towards a vast flat region which has now shrunk back to the more raised regions consisting of Cornwall, southern Ireland and

Wales, while the others continued up the river basin now occupied by the English Channel. The genes of those who turned west would form the main stock of today's Atlantic fringe 'insular Celts'.

The descendants of those who went up the old Seine–Channel river would become the prehistoric East Anglians, Dutch and Frisians, while their vanguard would continue into Germany and Poland. In this Late Upper Palaeolithic recolonization scenario, which is implied by the Creswellian archaeological links across the North Sea, it would not be surprising to see some genetic similarities across the North Sea, but they would reflect ancient cousinship rather than any more recent west-to-east gene flow.

This scenario does seem to be borne out in the pattern of Ruisko gene types found in Frisia and England. The apparent similarity between Frisia and eastern England results, largely, from a similar broad mix of shared gene groups and Ruisko gene types which were derived independently from Spain. After removing these matches from the analysis, there are no other Ruisko gene types which are shared between Norfolk and Frisia, and only a couple of matches elsewhere in Britain, implying a common Ice Age refuge origin rather than similarity resulting from later invasions. As we shall see in Chapter 11, the nature of this parallel relationship with Frisia contrasts with the situation for genes shared between eastern England and the Continental 'Anglo-Saxon' homelands, for which there is clear evidence for a small recent local east-to-west gene flow across the North Sea, in addition to the common southern refuge influence.

Time blunts the genetic reality of Celtic and Anglo-Saxon origins

Theoretically, it should be possible to determine for every individual living in the British Isles approximately when each of their paternal or maternal ancestors arrived. What I am trying to do in this book is to put dates of arrival and numbers (i.e. relative frequency) to those ancestral founding gene lines, as represented in today's population. Clearly, the deep time perspective and the size of ancient male and female recolonizations of north-west Europe rather blunts simplistic claims, based on similarity of genetic markers, that the ancestral insular-celtic languages must have come from Spain, or that English results from a replacement of Celts by Anglo-Saxons based solely on the extraordinary genetic similarity between people of the respective regions. This is partly because of the huge time gap since those recolonization events after the Ice Age, and partly because there is absolutely no evidence of what languages those early hunters and gatherers may have spoken. I shall come back to this issue later in the story, after we have looked at the Neolithic and the Bronze Age.

Apart from confusing the language story, the substantial size of the Iberian contribution to the initial colonization of the Atlantic coast makes it important to assess the numerical contribution of each new arrival to the British Isles from the south of other related gene groups over the next 15,000 years. Clearly, we dilute any discussion of British roots if we do not attempt to get a better time resolution for each Ruisko cluster. There are several ways of looking at this aspect, but before doing that we should go back in the story to the rapidly changing climate following the initial recolonization.

Rapid climate change in the Late Upper Palaeolithic: the Younger Dryas

As average world temperatures worsened in a ratchet fashion from an immediate post-glacial high 14,500 years ago to a series of lows starting 13,000 years ago,[62] the grasslands of Western Europe were gradually replaced by steppe woodland. Central and Southern Europe were mainly open forest, while in the north and west steppe forest predominated, with sparse birch and pine (Figure 3.9). Paradoxically, with this increase in tree cover, archaeological traces of human activity in Northern Europe and Britain tend to decline slightly. Whether this resulted from the increasing variation in the

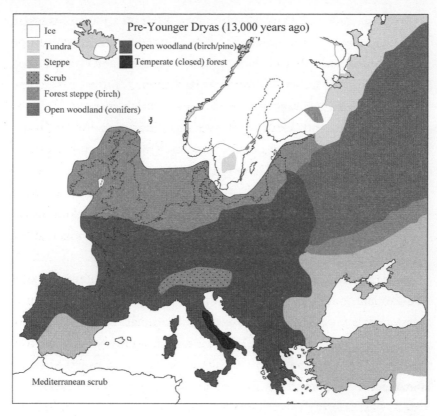

Figure 3.9 Pre-Younger Dryas climate map. This map of the climate just before the YD shows that, in spite of worsening weather leading up to the YD, open woodland cover had been increasing in Southern Europe. However, north-west Europe, including 'Greater Britain', still had some grassland (steppe – more suitable for hunting) with scattered birch forest.

weather caused by frequent wide temperature swings,[63] or just from the decline in open grassland for hunting, is a matter of speculation. One piece of evidence from Britain and Europe which could support the latter view is the later shift in the incidence of archaeological sites from the plains into low hilly country, where there were fewer trees.[64] The North Sea Plain probably remained as steppe, which may explain the expansion of Ingert into Norfolk apparently only shortly before the Younger Dryas cold event.

Over the centuries leading up to 12,500 years ago, the weather became more and more erratic; it grew colder, and human activity declined. Around 12,300 years ago Europe plunged into another severe glaciation, known as the Younger Dryas – 'Younger' because there had been a couple of other chills in the preceding few thousand years, and 'Dryas' because the hardy polar wild flower *Dryas octopetala* flourished during these cold spells, and is detected in deposits by its pollen.[65] The Younger Dryas was extremely cold and arid, and lasted about 1,500 years. Ice caps re-expanded over Scandinavia, with a resulting fall in sea levels, and even reformed on the Scottish Grampians and the Pennines.

It is tempting to wonder whether this cold snap put the recolonization of Britain back to the Ice Age, with Northern Europe depopulating and our ancestors scurrying back south to their refuges. This is an important question. However, there are several pieces of evidence to suggest that our hunters were just as hardy as the polar wild flowers and hung on in there, albeit in lower numbers, in all the places they had reoccupied. There is archaeological evidence for continued human activity in Northern Europe and Britain during the Younger Dryas.[66] In Britain there are bone, antler and ivory artefacts from the Younger Dryas, although there is a gap in the record of more elusive evidence for actual human remains and butchered animal bones.[67] In any case, the vegetation in north-west Europe did not revert to polar desert, remaining as steppe tundra. This is an environment modern humans tolerated in a number of other places such as the northern Ukraine all the way through the LGM.[68] Situated on the edge of the continental shelf, Britain may even have had more temperate and less extreme weather – as is the case today. There is also the dated and quantified genetic evidence, already discussed and based on today's populations, which suggests that a major north-westward European expansion of mtDNA- and Y-gene groups preceded the Younger Dryas.[69] Several of the founding Y clusters dating to *before* the

Younger Dryas, such as Rory, are unique to the British Isles and so could not have 'recolonized' from somewhere else *after* it.

The Younger Dryas finished even faster than it started, perhaps over a period of just fifty years. Effectively this was the end of the last ice age; it would not get as cold again, except for a minor freeze-up around 8,500 years ago. [70] In 'uncalibrated' radiocarbon years, the warm-up comes out as 10,000 years ago. The last 10,000 years of our modern post-Ice-Age (also known as Post-Glacial) era is also conventionally known as the *Holocene* epoch (meaning roughly 'completely new') to differentiate it from the preceding nearly two million years of the *Pleistocene* (or Great Ice Age). However, in spite of this apparent agreement of radiocarbon dating and geological era, the dramatic warm up after the Younger Dryas actually happened around 11,500 *calendar* or *corrected* years ago, not 10,000.

The Holocene was indeed something completely different. Its onset changed the hunting landscape for ever and took European human culture into the Mesolithic. The vast Subarctic steppe habitat known as the Palaearctic Biome simply disappeared from the northern hemisphere, and large, cold-adapted beasts such as the mammoth, woolly rhinoceros and giant deer vanished eventually with it. Much controversy rests on whether human hunters or the changing weather (or perhaps both) were responsible for the loss of mammoths in Europe. [71] However, an increase in the incidence of mammoth remains from the time of maximum hunting by humans, just before the Younger Dryas, rather goes against our ancestors having had a primary role in the extinction, at least in Europe. [72] After the Younger Dryas, most of the Palaeolithic hunters' rich, chilly grasslands disappeared, along with the mammoth's habitat, to be replaced by woodland. Other herds, of elk and reindeer for example, moved farther north, to be replaced in the south by wild pig, red and roe deer, aurochs and a variety of smaller mammals.

There was much ice left to melt down, apart from the vast sheet remaining over Scandinavia, but it did not melt gradually. Immediately after the Younger Dryas, sea levels rose at an initially dramatic rate, which eventually slowed until another warm-up hit and produced the highest post-LGM temperatures, causing the final over-topping 'flood' of the Black Sea around 7,500 years ago.[73] In the next chapter we move on to find out what our ancestors were doing in this post-glacial springtime, also known as the Mesolithic.

4

ULTIMATE HUNTERS AND GATHERERS: THE MESOLITHIC

The cultural period following the Younger Dryas Event, the Mesolithic, saw the final and most sophisticated flourishing of the hunter-gatherer lifestyle in Europe. It was the golden age of coastal hunter-gatherers. The Mesolithic was a time of rapid innovation in stone tools and increasing use of microliths – very small, multi-purpose stone tools which had already been in use in Africa and India for 20,000 years – and preceded the Neolithic agricultural revolution. Quite a bit is known about how Mesolithic hunter-gatherers lived, and one major feature in the evolution of their lifestyle in north-west Europe was a reduction in big game hunting and an increasing reliance on the beach and sea. In making this change, our Mesolithic ancestors resembled their African forebears, who were the first humans to see the great advantages of seafood,[1] but the reason for the change in Europe was the encroachment of the forest, not the desert.

In terms of stone tool technology in the Early Mesolithic, some has been found in Uxbridge, Middlesex, and in Suffolk, where there was a short-term carry-over of tool types similar to those from before the Younger Dryas, and to those found in Germany. This would be consistent with continuous occupation of Britain throughout that cold period.[2]

Can't see the game for the trees

Perhaps the best-preserved and researched record of Mesolithic life on the north-west European plain comes from Denmark, which in spite of rapidly rising sea levels (Figure 4.1),[3] was then still connected to England across the southern North Sea Plain (Figure 4.2). Archaeologists recognize three phases in the Danish Mesolithic, recording a relative decrease in the availability of larger game as the forest became denser, and an increase in coastal settlements and reliance on fishing and beachcombing. Conditions for the preservation of vegetable matter vary from one site to another. But the best-preserved artefacts are beautifully fashioned bone fish-hooks, carved painted wooden paddles, neatly made wickerwork fish-traps, spears, bows and arrows, ropes, woven textiles and even log-boats. These give us a unique window onto the more perishable aspects of the lifestyle of fishing folk and hunter-gatherers of the north-west European plain.[4]

North of the rapidly shrinking plain, east coastal Scotland shows similar Mesolithic settlements around the Tweed and Forth river estuaries.[5] On the west coast of Scotland, archaeologists have reconstructed a detailed picture of Mesolithic folk angling for fish such as saithe (a species of pollock) all year round on Oronsay Island, off Jura, and raising huge shell middens with at least seven species of shellfish. In other places, such as the Island of Risga at the mouth of Loch Sunart in Argyllshire, the pickings were even richer. In addition to the usual shellfish, nine fish species of varying size, eleven species of water birds – including the great auk, gannets and geese – two seal species, not to mention red deer and wild pig, all came to the hunter's table, along with more than just two vegetables.[6]

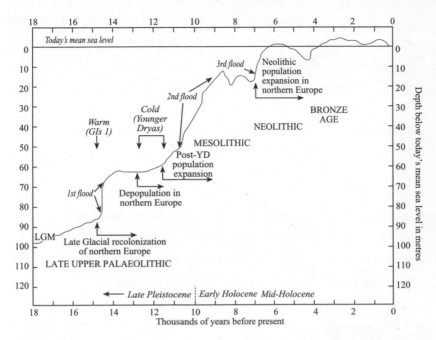

Figure 4.1   Sea-level rise after the last Ice Age. This graph shows mean world sea-level rises since the last Ice Age. These took place in a series of three dramatic warming and flooding episodes during the Late Glacial, Post-Glacial, post-Younger Dryas and Early Neolithic periods, separated by cold periods, such as the Younger Dryas Event. The three warm, flooding periods correspond to three recognized periods of human cultural and geographical expansion into northern Europe: Late Upper Palaeolithic, Mesolithic and Neolithic.

Ireland was no exception to the trend towards coastal and river-based subsistence, with Early Mesolithic settlements along the coast in the north-west and up the estuary of the River Bann. Large wooden buildings up to six metres in diameter imply an increase in communal settlement. Later in the Mesolithic, as Ireland separated from Britain, there was a change in stone tools from microliths to flint blades, and more reliance was placed on beachcombing, as reflected by huge shell middens, for instance in Sutton and in Dalkey Island, in Co. Dublin. On the Dingle Peninsula in the south-west, this lifestyle continued in pockets long into the Neolithic period.

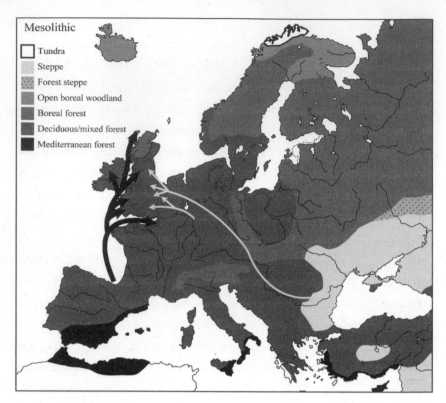

Figure 4.2   Mesolithic colonization of north-west Europe. Extensive forest cover during the Mesolithic reduced grassland hunting and forced increased coastal exploitation. The Balkan and Iberian refuges both acted as gene pools for re-expansion after the Younger Dryas, leading to two different sources of gene flow into Central Europe, while the British Isles still received the bulk from the south-west.

However, the greatest density of settlements in the extended Mesolithic continental shelf of 'Greatest Britain' was in the south of England, where the emergence of two distinct territories with different resources was already apparent. In the south-west (Cornwall, Devon and Dorset) there were mainly stable coastal settlements, who sustained themselves by both beachcombing and hunting small mammals in the hinterland, much the same as in north coastal Ireland. The south-eastern territory was around the Isle of

Wight (which was no island then) and its hinterland. Trade in prized stone materials thrived between the two regions (Figure 4.3).

Somewhat similar coastal Mesolithic settlements can be traced down through Brittany to Spain and Portugal. One would like to know what regional connections these Atlantic coastal cultures may have had with one another, either of a cultural nature or in terms of northerly migration. Barry Cunliffe examined cultural connections in his excellent book on this region and found some evidence of unity in their culturally and technologically innovative lifestyles. But these links were nothing in comparison to the trans-continental cultural trails of the Neolithic period and later. For instance, Cunliffe mentions the similarities between harpoon points found in Spain and on the western Scottish coast – only to dismiss this as serious evidence of migration. Although there is clear evidence of trade along the south coast of England, technical styles of tool-making vary from location to location in north-west Europe, suggesting that communities were beginning to settle down and create individual cultural identities. This would be the first stage in a process that would exclusively and visibly link such identities to territories.[7]

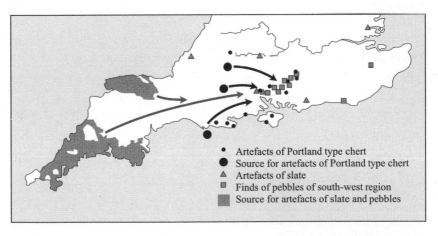

Figure 4.3   Mesolithic settlements on the south coast of England. There is evidence for long-distance movement of raw materials across southern England and Cornwall.

British Columbia in the British Isles

In its striking and rich 'novelty' the Mesolithic of the Atlantic coast is analogous to the unique and much more recent and famous non-agricultural, fishing cultures of the wet temperate coasts of the Pacific coast of North America (Washington State, British Columbia and Alaska) with their advanced maritime skills and complex artistic-spiritual life.

I recently had the privilege of visiting and staying with some of the surviving communities of this culture, on the coast of British Columbia in Canada. Although ravaged and nearly annihilated by smallpox after first European contact and then having had their ritual culture objects and practices – totem poles, potlatch meetings, and so on – comprehensively burnt and outlawed by colonists and missionaries, sufficient stragglers had remained to pull together an existence under their animal totem clans, such as the whale, the bear, the eagle and the raven. The Haida of Queen Charlotte Island lost less of their cultural inheritance than most and are now a famous tourist attraction.

The people I visited farther south, accessible only by ferry and then by aluminium speedboat, had lost everything, culture and sacred objects together, in the missionary-imposed iconoclastic holocaust. Today they are still squatters in their own territory, since no treaties to establish reserves were ever made. What they had not lost, however, was their ability to survive on a thin rim of coast sandwiched between the forest and the sea, without becoming farmers. The hinterland of the coastline is dense, impenetrable temperate rainforest with a range of wonderful flora and fauna, including bears. But they still manage to live off the beach and the sea, and, if given the chance, might be better custodians of the vanishing salmon stock than the joint interests of the Canadian Government and the fishing companies.

The Atlantic Mesolithic coastal foragers had careful burial rituals – often in close association with their large shell middens. The quality and richness of grave goods, from small personal ornaments to elaborately carved antlers and the use of ochre, could indicate variation in status, but their presence all along the coast also suggests common belief systems spread over the whole region.[8]

One intriguing, presumably coincidental echo of the Pacific coast of British Columbia was found at Stonehenge during an excavation to extend the car park there. A set of post-holes, dating to over 9,000 years ago, had previously held three very large pine trunks each nearly a metre in diameter. They predate the famous stone circles by some millennia:

> These three timbers (and there may be more) represent the first truly monumental structure of the Mesolithic period known to us. What form they took (carved totem poles perhaps?) and the reason for their erection we will never know, but the Stonehenge timber alignment is unlikely to be unique.[9]

The thaw after the Younger Dryas (YD) and its cultural counterpart, the European Mesolithic, was a springtime of environmental rebirth and cultural efflorescence after the long shadow of the last Ice Age. Archaeological clues give evidence for a dramatic increase in human activity in Northern Europe heralded by the thaw, although the peaks of archaeological activity did not quite match those seen in the brief and lesser thaw before the YD. It is reasonable, however, to suppose that populations in Northern Europe expanded during the Mesolithic. In addressing the genetic record, I shall be asking how much of the expansion was among indigenous lines already present from before the YD, how much was from immigration from elsewhere in Europe, from where it came, and to where, in the British Isles, it went.

Evidence for female gene flow into the British Isles in the Mesolithic

Since the archaeological/cultural message is ambivalent on the extent of south-to-north coastal migrations during the Mesolithic,

we might expect a relatively stable genetic picture for this climati-
cally welcoming period in north-west Europe. This ambivalence has
been shared by mtDNA specialists. When we look at the potential
for a post-YD Mesolithic intrusion into north-west Europe, and
more particularly the British Isles, we do however find fresh gene
lines, again deriving from the Iberian Peninsula. Dealing first with
maternally transmitted genes (i.e. mtDNA), I have already
mentioned Martin Richards' founder analysis of identifiable Near
Eastern contributions to the modern maternal gene pool in Europe
(see p. 106). Richards showed that about 11% of modern types
arrived in the Mesolithic. Although this is a larger proportion than is
accounted for by later immigrations, it would still have been only
about a fifth of that attributable to the Late Upper Palaeolithic
(LUP), just before the YD.

Clearly, each era brought people who were ancestors for the
next, and only a minority of 'farmers' ancestors' actually arrived
during the Neolithic. Instead, the earlier Upper Palaeolithic and
Mesolithic colonizations of the British Isles brought many of the
ancestors of people who would become farmers later during the
Neolithic. Ireland has a much lower component of putative
Neolithic gene lines than other parts of Europe, suggesting even less
dilution during the Neolithic and a relatively greater contribution of
LUP and Mesolithic gene lines to modern peoples.[10]

I think that the figure of 11% for Near Eastern contributions to
the modern European maternal gene pool during the Mesolithic
underestimates Basque gene flow. Also it depended on factors,
unknown in 2000, when Richards and his colleagues published their
analysis, which affected the apportionment of founding lines to
before or after the YD.[11] A subsequent reanalysis has described a
number of new subgroups of the Helina (H) lineage. Their ages and
distribution suggest that some of these subgroups may have spread
along the Atlantic coast during the Mesolithic period immediately
after the YD. In total, four of these possible Mesolithic Helina

subgroups (H2, H3 and H5a) account for 11% of Irish and 11% of maternal lines in the UK in general. Subgroup H3 is notable here, since it dates to around 9,000 years ago and by itself contributes 7% to the Irish gene pool – a figure two-thirds higher than in the UK.[12]

There are other specific gene groups, T, T2 and K, whose estimated age brackets straddle the period of the Younger Dryas and are found in both Iberia and western Britain, including Cornwall.[13] Overall, the reassignment of these small gene groups to the Early Mesolithic could increase the maternal Mesolithic immigrant component by a further 10% at the expense of lines coming in during the Late Upper Palaeolithic, making a total maternal Mesolithic intrusion of around 22% (Figure 4.4). But however you decide to allocate these arrivals among the YD, the LUP and the Mesolithic, the results is that these early Atlantic coastal re-expansions would still between them contribute more than half of today's maternal lines.

Iberian male migrations of the Mesolithic

When I started looking at the Y-chromosome tree, I could see clear evidence for a large pre-Neolithic intrusion of new Ruisko (R1b) clusters to the British Isles from Iberia after the YD. Certain Ruisko clusters predominate in different regions of the British Isles, but it is important to try to date specific ones to a particular period (e.g. pre- or post-YD, or even later).

The post-YD element is made up of nine Ruisko clusters (R1b-1 to 3, R1b-7, R1b-8 and R1b-10 to 13), which include, and all derive from, the new major Mesolithic gene cluster, *Ruy* (R1b-10). These clusters expanded at various times from the Mesolithic onwards. Ruy is a Basque name; the cluster derives from Iberia, and dates overall to around 11,500 years ago (Figure A4 in Appendix C). Although Ruy's ancestor Rox (R1b-9) arrived in the very Early Mesolithic, not all nine descendent clusters expanded simultane-ously at that time. Some (such as R1b-7, R1b-8, R1b-11 and R1b-12) re-expanded from common Mesolithic ancestors, but did so

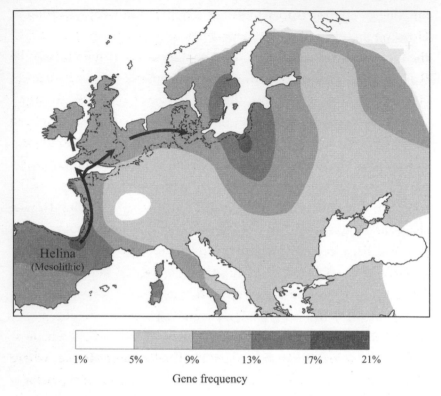

1%   5%   9%   13%   17%   21%

Gene frequency

Figure 4.4   Maternal gene flow into the British Isles during the Mesolithic. Gene flow follows the European coastlines, and a founder effect is seen at the end of the trail in the Baltic. (Combined gene frequency of Helina Mesolithic sub-groups H2, H3, H4 and H5a in Europe – arrows indicate direction of gene flow based on the gene tree and geography. Contours follow greater land area resulting from low sea level.)

during the Neolithic and later (see the next chapter). The Ruy cluster arises as a one-step derivative of Rox (i.e. it differs by one mutation). Ruy's core or root gene type is also known as the *Atlantic Modal Haplotype* (AMH), since it is by far the most common single Y gene type in Western Europe, accounting for 18% of British and 19% of Basques in the dataset I am using.[14]

Ruy's distribution is subtly different from that of his closely related forefather Rox from before the Younger Dryas. Not surprisingly, the ubiquitous Ruy is now found in all thirty-one British and all but one of the thirteen Continental population samples studied. However, in terms of relative frequency there is a bias to the parts of the Western Atlantic façade and the south coast of Britain missed by Rox in the first migration before the Younger Dryas freeze-up (Figure 4.5). By Ruy's time the Irish Sea was open, St George's Channel had already split southern Ireland from St David's Head and Haverfordwest in south-west Wales, and the Channel, although not quite open at the start of the Holocene, was occluded by just a narrow strip and soon opened up. All the now familiar features of the southern coastline of the British Isles today were beginning to take shape, and the coast became as accessible for settlement as it is today.

Local Ruy frequency favours the Welsh coast at 28–29%, and Cornwall, the Channel Islands, the south coast and Norfolk at 17–27% in order of increasing frequency (Figure 4.5).[15] In Llangefni in north Wales, for instance, 26% of males belong just to the single core gene type AMH.[16] This contrasts with Rush, near Dublin just across the Irish Sea, and the Gaelic-name sample, where males carry rather less of this type, at 10.5% and 18%, respectively.[17] Even parts of Scotland and Ireland, which had missed out on Rox the first time round, had a higher dose of Ruy. So, Castlerea in the centre of Ireland and the Western Isles of Scotland received 26% and 22% Ruy, respectively.[18]

So we find Ruy dominating the parts that Rox had not, on a first come, first served basis; so much so, that the sample sites with higher rates of the first-comer Rox in the British Isles are associated with lower rates of Ruy, and vice versa. The earlier lack of access to Northern Europe, immediately after the LGM, for the beachcombing Rox would no longer have been such a problem for Ruy with the new coastline changes. Consistent with this selective sea level access hypothesis, rates for Ruy across the North Sea, although lower than in Britain, are considerably higher at 7–13% than for Rox (2–5%).

The eight closely related Ruisko clusters, derived from Ruy, also seem to follow the same pattern of filling in the parts of the modern

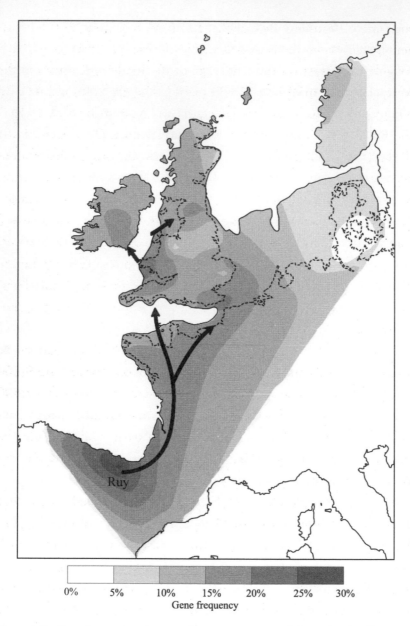

Figure 4.5   Ruy, the main male gene cluster moving into the British Isles during the Mesolithic from 11,500 years ago. Ruy (R1b-10) is the largest Ruisko cluster and the main source of gene flow from the south-western refuge for that period. The gene flow follows the ancient extended Atlantic coastline, which had by now opened up to show the southern part of the English Channel (arrows indicate direction of gene flow based on the gene tree and geography.)

coastline of southern Britain that had been obscured by the continental shelf when the sea level was lower. We find that they divide roughly east–west on the landscape of the British Isles: two in the west, especially in Wales (R1b-11 and 12 – Figure 5.5b, and R1b-13 – Figure 4.6); two more on the south and east British coasts (R1b-7 and 8 – Figure 5.5c); and three small ones diffusely in the middle (R1b–2a, R1b–2b and R1b–3). These three sets of Ruy descendent clusters account respectively for 10%, 15% and 2% of extant British male lines. Collectively, these derived clusters account for 27%, which, when added to Ruy, make a total 45% Iberian ancestral contribution to modern British lines from these Mesolithic arrivals. However, clusters R1b-7, 8, 11 and 12 did not actually generate their re-expansions until the Neolithic several thousand years later, so 21% of the 45% ancestral contribution was effectively deferred to that period.[19]

Three closely related clusters, R1b-11, 12 and 13, started to break up from a common Mesolithic British ancestor around 9,500 years ago. The first to split was R1b-13, which now dates to 7,800 years ago (Figure 4.6). It accounts for about 4% of British males today,[20] and has a similar distribution to Ruy, mainly along the western Atlantic fringe of Britain, with highest frequencies in north Wales (Llangefni) and Cumbria (Penrith).[21]

This British triple cluster R1b-11-to-13 re-expanded much later, during the Neolithic, into R1b-11 and R1b-12 (Figure 5.5b and Figure A4, Appendix C). The largest of these clusters was R1b-11, which characterizes the west or Atlantic side of Britain, from the Channel Islands through Cornwall, Ireland and Wales right up to Scotland and the Shetlands, and is uncommon on the east coast of England. R1b-11 accounts overall for about 6% of modern British males.[22] But, compared with R1b-13, there is a more northerly centre of gravity, in north Wales and Scotland (see Chapter 5, in particular Figure 5.5b).

So, although the ancestor of the two clusters R1b-11 and R1b-12 appears to have arrived during the Mesolithic, and they jointly

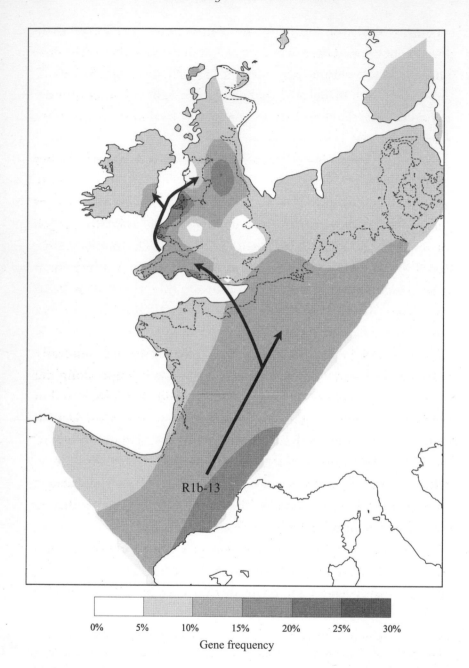

R1b-13

0%   5%   10%   15%   20%   25%   30%
Gene frequency

account overall for 7% of extant British lines,[23] most of their diversity developed later with Late Neolithic population expansions. Since both these clusters expanded during the Late Neolithic period, they effectively made up for the relative lack of contemporary incoming Neolithic lines on the Atlantic coast compared with elsewhere in Europe.

Another larger cluster descended from Ruy is an interesting puzzle for the European Mesolithic. This is R1b-8, which now constitutes 13%[24] of British male lines and is the second largest in our whole dataset, after Ruy. With apologies to the clan MacGregor for imperfect location, I am giving R1b-8 the nickname *Rob*, after Red Rob ('Rob Roy') because of his highest frequency in the north of Scotland (Durness 29%). Rob is defined by the root *Frisian Modal Haplotype* (FMH), which constitutes 7.6% of the full dataset and 8.6% of the British Isles.[25] The FMH is well represented in Denmark and Frisia, hence the name. But it is also generally common throughout the British Isles, where it is found along the south coast, including Cornwall, Midhurst and Dorchester, and in northern Scotland (Figure 5.5c). In each of the main British regions, at least one different Rob gene type is present in addition to the FMH: for example, the south-coast gene type is unique to Britain, with the exception of one instance in Frisia. The greatest diversity of Rob is thus found in the British Isles, suggesting that he arrived in Britain at least as early as in Frisia.

Some geneticists argue for a recent Frisian invasion of England during the Dark Ages (see Chapter 11), based on the 'evidence' of Frisian/English genetic and linguistic similarity. In so doing they fail

Figure 4.6 Welsh colonists of the Mesolithic. This map demonstrates the impact of another common cluster derived from Ruy, R1b-13, which expanded northwards from the south-west European refuge during the Mesolithic. The gene flow follows the ancient extended coastline, which had now opened up to show the southern part of the English Channel. In this instance south-coastal England, Cornwall, Cumbria and Wales were targeted (arrows indicate direction of gene flow based on the gene tree and geography.)

to recognize this alternative perspective of similar, parallel 'ancient history'. The geographical distribution, diversity and splitting of these closely related twigs of Rob either side of the North Sea are much more likely to reflect trail choices made by Mesolithic colonizers than any recent invasions by Anglo-Saxons or Frisians.

Although Rob is aged around 10,000 years in Western Europe as a whole (Figure A4, Appendix C) and joined the other Mesolithic clusters with Ruy from Iberia, he later broke up into much younger clusters in Northern Europe and Britain. The main expansions of Rob and his descendant cluster R1b-7 within the British Isles are local and later than the Mesolithic (Figure 5.5c) but none of them can be connected to the recent 'Anglo-Saxon' invasions.[26] I shall come back to Rob and his descendants, later during the Neolithic discussion in Chapter 5.

Overall, then, the Y-chromosomal evidence suggests that new Mesolithic immigrants from Iberia went mainly to the western and southern British Isles, contributing initially about 24% of modern lines, which is rather similar to the maternal figure. However, other Rob lines later evolved into more new clusters so the Mesolithic immigration of these lines *ultimately* contributed an overall 45% of today's British male lines, which would be larger than the previous Late Upper Palaeolithic contribution.

Additionally other clusters, already in the Western British Isles from before the Younger Dryas, split and expanded in the Mesolithic period (Figure 4.7). They included Rory (R1b-14a & b), R1b-4 and R1b-15a. These internal British re-expansions contributed a further 5.4% of Mesolithic lines surviving until today.[27]

Another line on the western side of Britain dates locally from the Mesolithic period. This is I1b-2, which is found at rates of 3% and less in the Channel Islands, south-west England, Wales and Ireland. I1b-2 almost certainly came in during the Late Mesolithic from the western Mediterranean, where it is characteristically found, via the Iberian region, having originated in the Balkans (see discussion of

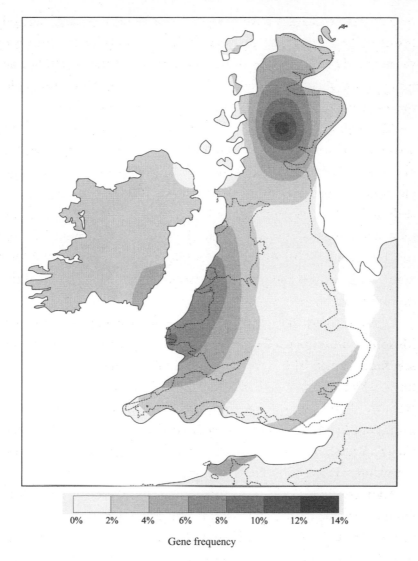

Figure 4.7 Mesolithic indigenous re-expansions in the west. During the Mesolithic much population expansion occurred among pre-existing populations in the west from before the Younger Dryas. In Wales and Scotland this included R1b-14b&c, R1b-4 and R1b-15a.

the Ivan (I) group later in this chapter).[28] The other part of I1b, denoted by I1b*, is very uncommon in Western Europe but has a similar age and distribution to I1b-2 in the western British Isles.

A quiet eastern Britain during the Mesolithic?

So, what else was happening in eastern Britain at the onset of the Mesolithic if we are seeing rather less Ruy offspring appearing on that side during the Mesolithic? The answer is not much. There was one descendant of Rox surviving there from before the Younger Dryas but re-expanding immediately after the YD.[29] This was R1b-6, the core type of which is unique to Britain. It is found in only 1% of the British sample population. The R1b-6 core type is a one-step minority derivative of Rox (i.e. one mutation different). Two-thirds of R1b-6 representatives are found in eastern England, the Channel Islands and Dorchester, while 82% are restricted to eastern England: Norfolk, Southwell, Bourne and York (Figure 4.8). The distribution of the founder type suggests R1b-6 may have mutated from Rox somewhere in the English Channel area (when it was mostly dry), and from there moved up into Norfolk and north-east England.

The decision of whether to go north-east, right up the Seine/Channel Basin, and explore the 'European' side of Britain or just continue west along the Atlantic façade marks the first step in a pattern of recurrent geographically determined relationships that still divides England genetically and culturally from the rest of the British Isles.

Norfolk and East Anglia were still linked to the Continent at this stage. We saw in Chapter 3 how Ingert (group I1c) may have moved up from the Balkans into north-west Europe and eastern England just before the YD (Figure 3.8). I1c-3 dates to just after the YD and features on both sides of the North Sea, in Norfolk and the Fen country and in Frisia. Several other Ivan lines found in eastern England date to the Mesolithic. These also link to the nearby Continent.[30] This first evidence of gene flow into Britain from the Low Countries the other side of the North Sea (Saxony, Frisia, the Netherlands and Belgium) leads me next to examine the relevance of what was going on in the northern part of north-west Europe during the Mesolithic.

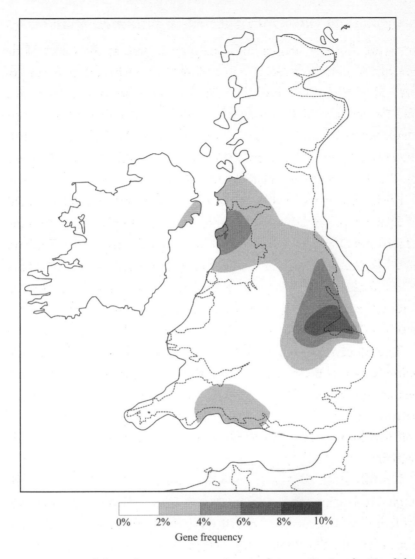

Figure 4.8 Mesolithic indigenous re-expansions in the east. During the Mesolithic population expansion occurred among pre-existing populations in the east from before the Younger Dryas. This was rather less than in the west and involved just one sub-cluster, R1b-6, also found on the Isle of Man.

Scandinavia after the melt

Moving away from the British Isles briefly, I should first like to paint in a bit of the picture of possible genetic events in northern

Scandinavia during the Mesolithic. This is not out of any obsession to give the fullest possible West European picture, but because the rather special Scandinavian genetic make-up, with its mixture of East and West European inputs, can be used to understand and pinpoint Scandinavian intrusions during the later British prehistoric and historical periods. It seems from their maternal lineages that although northern Scandinavians – Finns and Saami – subsequently received much of their male genetic heritage from Eastern Europe, the very first Mesolithic immigrants may have belonged to the Atlantic coastal set from southern France and Spain.

During the Younger Dryas freeze-up between 13,000 and 11,500 years ago,[31] although the polar caps started to re-form, the whole northern rim of the Scandinavian coastline was free of ice. This thin coastal strip was, however, uninhabitable polar desert, jammed up against the still massive Fenno-Scandian ice sheet (Figure 4.2).[32] All that changed after the end of the YD, 11,500 years ago. The ice sheets began to retreat dramatically, though a lot remained on the hills and in the high Norwegian cordillera. The last major deglaciation did not take place until around 8,000 years ago.[33] But before this time, during the Mesolithic, the Scandinavian lowlands, including the coastal strip of Norway facing Britain, had become more habitable and much less densely forested than the rest of Europe, as also were modern-day Lapland, most of Eastern Europe and Russia.[34]

The archaeological record tells us that the period after the YD saw two pioneering cultural movements moving like pincers into the Scandinavian Peninsula from south and north (compare with genetic map in Figure 4.9). The first of these, known as the Mesolithic Komsa and Fosna-Hensbacka cultures,[35] appeared as early as 8,600 years ago,[36] and most likely derived from the Ahrensburgian culture farther south in northern Germany and Denmark. These cultures collectively extended northwards through Scandinavia, and as far east as Finland and the Kola Peninsula. It has also been suggested that they might have been the ancestors of some of the present day Saami.[37]

The second of the two Mesolithic pincer movements into the Scandinavian Peninsula came from the north-east, with carriers of cultures known as post-Swiderian, via Russian Karelia and Finland. These two Mesolithic cultural movements met up in Norway in the Scandinavian Early Neolithic, 6,500–6,000 years ago.[38]

The entry of gene lines into Scandinavia during the Mesolithic

Is there other support for this archaeological perspective on the actual origins of Norwegians and Saami? Perhaps the genetic record contains analogues for the two pincer movements. For instance, could some gene lines from the south-west European ice age refuge have arrived in the far north during the Mesolithic, or even during the Neolithic? The former seems most likely. Scandinavia and Lapland still hold a significant proportion of gene lines derived from the Franco-Spanish refuge in the south-west,[39] with the same general mix as in the British Isles but with less diversity and a relatively higher proportion of the main founding types.[40]

At present, about 60,000 Saami live in the northern regions of Norway, Sweden and Finland, as well as in the Russian Kola Peninsula.[41] Because of the ice cap, these pioneers from the south-western refuge simply could not have come here before or during the YD, and so must have arrived as founders either during the Mesolithic or the Neolithic. The Saami, traditionally semi-nomadic reindeer herders, effectively still preserve elements of a Mesolithic lifestyle in parts of Lapland, while the Norwegians, for their 'Mesolithic part', are one of the world's most skilled fishing nations.

Females joining the Saami from Iberia

If we start by examining the mtDNA evidence, we find that Norwegians have a high combined overall frequency (50%) of the two most common West European, post-LGM mtDNA founding

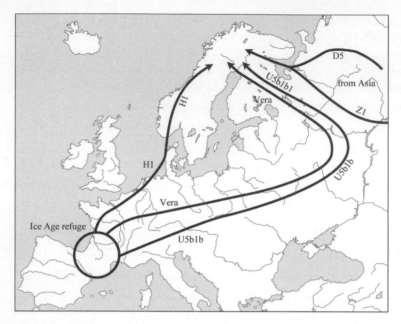

Figure 4.9a Maternal gene lines in Lapland and Scandinavia derive predominantly from Iberia via both the east and west routes.

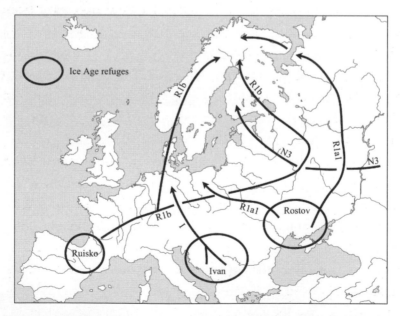

Figure 4.9b Male gene lines in Lapland and Scandinavia derive predominantly from the Balkans (Ian) and Eastern Europe (Rostov and N3) rather than Iberia (Ruisko).

4.9 Pincer colonization of Scandinavia. Lapland and Scandinavia received two colonizations during the Mesolithic, one via Finland and Archangel, the other via Denmark. The western component derived from Iberia; the eastern pincer came both from Iberia and from farther east in Europe. This mix makes it possible to distinguish Norwegian and Danish inputs to the British Isles. Norway received over 70% of its male lines via Lapland, while Denmark received most via northern Germany.

lines H1 and H3 (for Helina's daughters H1 and H3 in the British Isles). H1 dates to between 16,000 and 14,000 years ago in Western Europe as a whole, but, as mentioned, H1 clearly could not have arrived in Norway before the Younger Dryas, because of the ice. The unusually high frequency of H1 in Norway is likely to be an Early Mesolithic founder effect in western Scandinavia (shown in a recent re-analysis),[42] via the Danish landbridge. There is no evidence of any subsequent Neolithic spread of H1 into Norway from Spain.

H1 is found only in Norwegian Saami, not in more eastern Saami groups. This again suggests that the respective northern Mesolithic pincer founding events in Norway and Lapland may have been discrete and separate. H3, although not over-represented in Norway, has a similar frequency there (5.6%) to that found in the UK (4.3%). H3 is aged between 11,000 and 9,000 years in Western Europe.[43] The fact that H3 has a younger West European age than H1 in general suggests that the former moved north after the YD during the Mesolithic, rather than before.[44]

The Saami also show a clear founder event for another post-LGM maternal founding line, the previously mentioned Vera (V), which clearly arose in the south-western refuge, 16,300 years ago.[45] Like H1 and H3, Vera could have arrived in northern Scandinavia only after the YD, in other words during the Mesolithic or Neolithic. Vera reaches frequencies in the Saami of anywhere up to 70% of the population (Figure 3.4).[46] Because of the evidence for extreme genetic drift among the Saami, it is difficult to estimate the age of Vera directly there. If Vera did not reach the Arctic Circle at the same time as elsewhere in Western Europe, we have to ask when she did get there and by which route.

These timing and route questions in Scandinavian colonization were recently investigated more fully by Estonian geneticist Kristiina Tambets and colleagues (Figure 4.9).[47] Tambets looked in detail at the distribution of V, H1 and other founding lines of northern Scandinavia. While concluding that H1 had moved into Norway directly via a landbridge which is now partly represented by Denmark, they argue that Vera had more likely moved through north-east Europe, through Eastern Europe, round the eastern side of the Baltic, to Lapland in the Arctic Circle, thus making an eastern pincer arm (Figure 4.9a). If Vera did move in via such an eastern loop, it may be possible to estimate when this happened, since Italian geneticist Antonio Torroni, who has pioneered studies of maternal post-glacial re-expansions, dated Vera to the Mesolithic at around 8,500 years ago in Eastern Europe.[48] To parallel this estimate, the oldest archaeological evidence of settlement in Finland dates back approximately 9,000 years.[49]

Perhaps the simplest answer to whether Vera came into Lapland as a Neolithic or as an earlier Mesolithic intrusion lies with the surviving traditional Saami culture, which is non-agricultural and is indeed the only surviving example in Europe. While it is likely that the Saami did receive some cultural input from farther east in Europe, for instance their language, it would be illogical to infer that, as settling agriculturalists they should then have changed their farming lifestyle *back* to being Mesolithic nomads. There are no convincing examples of this kind of cultural devolution happening elsewhere.[50]

The Saami also show further genetic proofs of their ultimate origin in south-west Europe – an uncommon gene group called U5b1b accounts for the other main part of their maternal lines, apart from Vera, and again indicates a very strong founding effect.[51] The U5b1b group has been dated to 9,000 years old, in the

Mesolithic. U5b1b has also been shown to have had a strong founder effect on the Berbers of North Africa, confirming that Northern Europe was repopulated from the Franco-Spanish refuge.[52] To complete the picture for the Saami (though it has had no relevance to the British population), a later and smaller (3%) gene flow of Asian super-group M (Manju) has been suggested to have come from the East.[53]

Male Lapps

The picture for the Saami Y chromosomes also supports this pincer mix, but shows the opposite balance of eastern and western founders and adds further tantalizing detail. While the Saami maternal picture is overwhelmingly West European, the expected Y representative of Basque refuge lines, Ruisko, is not common, reaching only 9% among the Saami of the Kola Peninsula, 6% among Swedish Saami and less than 2% among Finnish Saami. Instead, his East European sibling Rostov (R1a1), common throughout Central and Eastern Europe and Siberia, and accounting for 54% of males in the refuge zone of the Ukraine, represents the Ruslan group better than Ruisko among the Saami, at rates of up to 22%.[54]

As I have mentioned, Rostov is definitely not derived from the western refuge, being absent there, but is very common elsewhere throughout Central and Eastern Europe and in Siberia. Rostov is very common in the Ukraine, suggesting that region as his Ice Age refuge. This gives us two specific male refuge markers: an eastern Rostov (Figure 4.10) and a western Ruisko (Figure 3.6a). Their ages, based on European diversity, are each consistent with an expansion from Ice Age refuges, although Rostov and the ancestral Ruslan (R1) are both ultimately much older in Asia, probably arising in India during the Early Upper Palaeolithic.

Clearly this east–west Saami mix makes any search for a single Mesolithic expansion source for Lapland problematic. Comprehensive genetic studies of the Saami suggest that Rostov entered the

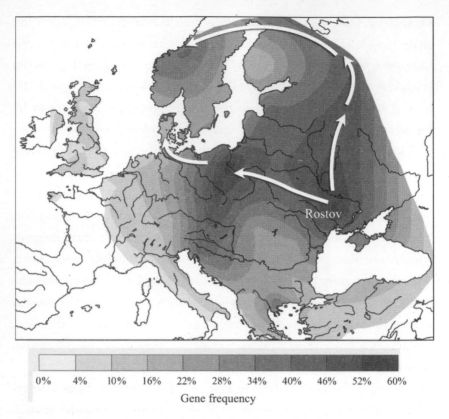

0%   4%   10%   16%   22%   28%   34%   40%   46%   52%   60%

Gene frequency

Figure 4.10   The men from the Ukrainian refuge. Rostov (R1a1) is the commonest male gene group in Eastern Europe and has three foci of highest frequency: Ukraine, Poland and Norway. Rostov most likely dispersed from the Ukrainian glacial refuge via the routes shown by arrows – as based on the gene tree and geography. Poland could have been colonized by Rostov soon after the LGM, but Norway not until the Late Mesolithic. Britain received Rostov from the Neolithic onwards.

eastern Fenno-Scandian Peninsula during the Mesolithic directly from Eastern Europe.[55] Supporting evidence that Rostov did not enter northern Scandinavia via Denmark comes from my analysis of the actual Rostov gene types. Denmark has not only less Rostov than Norway, but also a different spectrum of individual types.

Geneticist Zoë Rosser of Leicester University came up with a possible reason for a more recent arrival of Rostov into Scandinavia through central Eastern Europe, suggesting horse-riding Kurgan nomads arriving from north of the Caspian Sea (Figure 4.9).[56] The main Rostov cluster that my analysis identifies in Norway (R1a1-2b) dates to 5,700 years ago (Figure 5.6a),[57] which would be consistent with this view. More simply, however, this cluster may signal the arrival of the Mesolithic down the Scandinavian Peninsula, and its later expansion during the Scandinavian Early Neolithic.

Either way, since Scandinavia is the likely source of British Rostov types, this east–west direction is of interest to us. The highest Rostov rates among Saami groups are in the Swedish Saami (20%) and Kola Saami (22%), possibly suggesting gene flow direct from Archangel in the European Russian Arctic into the Kola Peninsula and Lapland. This route would have more or less bypassed Finland altogether, which makes sense, since much of Finland would have been uninhabitable during the earliest part of the Mesolithic (Figure 4.9b). Such a route may also be inferred from Rostov's unexpectedly low frequency among Finnish Saami (less than 3%) and among Finns in general (7%) (Figure 4.10).[58]

A somewhat similar east–west pattern is seen for N3 (also called TAT), a Y-chromosome line which is absent from the British Isles but characteristic of north-east Europe and accounts for 47% of Saami male lines. However, in this instance the distribution and frequency gradients of N3 in Northern Europe are rather different from those of Rostov, implying another source and route of entry to Lapland. Unlike Rostov, who is most common among Slavic groups and most likely arrived from Archangel, N3 is uncommon among Slavs, and is commonest in Finns (63%) and Finnish Saami (55%), together with a number of other similar groups, including other Finno-Ugric speakers, in north-east Europe (Finns, Saami, Karelians and Estonians), the Volga–Urals area and Siberia. Whether this north-westerly expansion of N3 occurred along with the spread and expansion of Finno-Ugric languages – *and* whether this happened parallel with or independently of Rostov during the Mesolithic, Neolithic or later 'Kurgan' periods – is a matter of speculation.

I prefer the view that N3 expanded separately and later than Rostov, in which case it would be a better candidate for the much more recent Kurgan expansion from the region north of the Caspian.[59] N3 is uncommon in Scandinavia and never made it to the British Isles at all, while Rostov did[60] – and in sufficient numbers to help us determine the degree and locations of northern Scandinavian influence there.

Rostov is common among Norwegians. Also consistent with the hypothesis of its introduction via Lapland to the east, the frequency of Rostov declines north-to-south through Scandinavia from 37% in Trondheim up in the north, to 26% in Oslo and then down to 13% across the water in Denmark (Figure 4.10).[61] It consists of four related clusters, three of which can be dated as local founder events. The largest, cluster R1a1-2b, is 5,700 years old (Figure 5.6a)[62] and concentrates in northern Norway, being uncommon to the south in Oslo or in Denmark. This is consistent with the archaeological picture of some delay before the supposed Mesolithic migration from Lapland established itself there.[63]

The distribution of Rostov could reflect Denmark having been at the end of a founding event which spread from the south-east. More simply, however, Denmark and Schleswig-Holstein in northern Germany may have received their modest component of Rostov directly from Eastern Europe south of the Baltic (Figure 4.9b): the more southerly Rostov clusters in Scandinavia are much younger, with a modest Bronze Age date of 3,000 years ago. Whatever the prehistoric reasons for these differences, they are helpful in establishing which parts of Scandinavia acted as sources for migration to Britain and for dating those events.

In summary, on the basis of the putative male and female Mesolithic founder lines I have discussed so far, the ancestors of the Saami would seem to have arrived in Lapland via a tortuous East European route. In the process they suffered severe population bottlenecks and eastern male intrusions, keeping their West

European female lines and swapping their Western male lines for mainly East European ones.

We can use this scheme of western and eastern Mesolithic genetic inputs to the Fenno-Scandian region to examine the make-up of Norwegians on the west coast of Scandinavia. While it is clear that there is a preponderance of eastern male lines among these populations, later to be counted as Vikings (see Chapter 12), this effect is less extreme than in the case of the Saami.[64]

However, in the context of male lines found in the British Isles, new appearances of Rostov tend to imply a northern Scandinavian intrusion rather than a German or Frisian source (see Chapters 11 and 12). Specific Rostov gene type matches confirm this trend across the North Sea, which are more between Norway and Britain (and Iceland to the north-west) than between Britain and Denmark or Schleswig-Holstein in northern Germany. What is more, because of the previous ice sheet over Scandinavia, such an intrusive event would have to be Neolithic or later. As we shall see, Rostov's Norwegian intrusion into Britain started early, during the Neolithic.

Ivan and the Mesolithic colonization of eastern Britain

The Ivan (I) clan is nearly entirely confined to Europe, although he may have originated ultimately in the Trans-Caucasus around 50,000 years ago and initially spread into Europe before the Last Glacial Maximum, 24,000 years ago.[65] Since the age of the main Ivan gene group does not tell us when his sub-branches arrived in their target regions, the question remains as to exactly when the migrations did occur – soon after the LGM, in the Mesolithic (i.e. after the YD) or in the Neolithic. One problem is in determining where each of Ivan's four main subgroups spent the last Ice Age. That would help tell us from where they each re-expanded after the Last Glacial Maximum – and when and where they each went after that. Whereas the Ivan gene group as a whole is pre-glacial in age, dates for the origins of his four main subgroups, although old, are all post-LGM,[66] suggesting that they may have re-expanded from one or more of the European glacial refuges. The Balkans and the Ukraine both carry all four branches at sufficiently high rates and diversity, in my view, to make that whole region the most likely *composite* East European Ice Age refuge for Ivan.[67]

The most comprehensive, up-to-date work on the male group Ivan in Europe comes from Estonian geneticist Siiri Rootsi and colleagues based mainly at the University of Tartu in Estonia.[68] All Ivan subgroups are present at appreciable frequencies in south-east Europe, an area including the Balkans and the Ukraine and encompassing the East European Ice Age refuges (Figure 3.7).[69] The highest frequencies of Ivan and his subgroups I1a, I1b*, I1b-2 and I1c (Ingert) occur in the Balkans (overall, Ivan's frequency is around 40% in Bosnia, Croatia and Slovenia), and around the north coast of the Black Sea (Moldavia at 28%, Ukraine and north Caucasus at 22–24%).

The only other regions with such high Ivan frequencies are north-west Germany for Ingert and southern Scandinavia, for I1a, which I therefore name *Ian* (pronounced 'Yan', a Danish version of 'John') (Figure 4.11a).[70] Ingert could have come from the Moldavian refuge before the Younger Dryas, up the Dnestr into Poland and then into Germany (Figure 3.8). Ian came rather later from slightly farther west in the Balkans, nearer to Slovenia, and then down the Elbe/Oder rivers towards Denmark, where he is now found at his highest rate, 37%.

Ian also encroaches on France and Britain, but at generally lower rates. One theory, based on diversity, is that this subgroup evolved separately in a Franco-Spanish LGM refuge, but this does not explain his absence from the Basque Country; neither does it explain his co-distribution with other Ivan subgroups, which suggests a birthplace in the Balkans–Ukraine region. My view is that during the LGM Ivan diversified in isolated local Balkan and Ukrainian refuges into his four main subgroups,[71] followed by a post-LGM dispersal of Ian and Ingert into Northern Europe.

Ian and Ingert make up respectively 11% and 3% of the British population, mainly focusing in England and Scotland, and arrived much later than Rox in north-west Europe and the British Isles. Ingert, the minority subgroup of the pair, appears at least as old there as Ruy, dating to around the time of the Younger Dryas Event. This raises the question of how much of Ingert arrived in Western Europe and Britain *before* the YD and how much *after* it.

4.11 The men from the Balkans: Ivan's sons in north-west Europe during the Mesolithic.

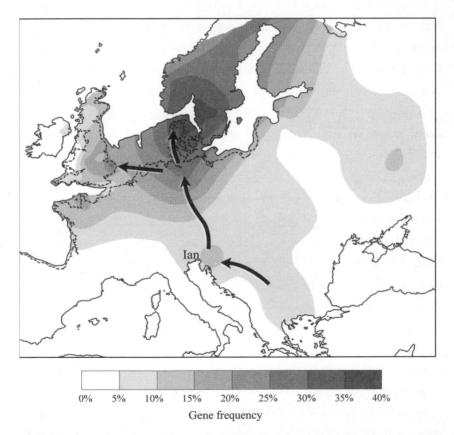

Figure 4.11a From the Mesolithic onwards, Ian (I1a) expanded in southern Scandinavia, where he is now the dominant group, and also in northern Germany. He moved to Britain starting from the Late Mesolithic, and expanded there particularly during the Neolithic.

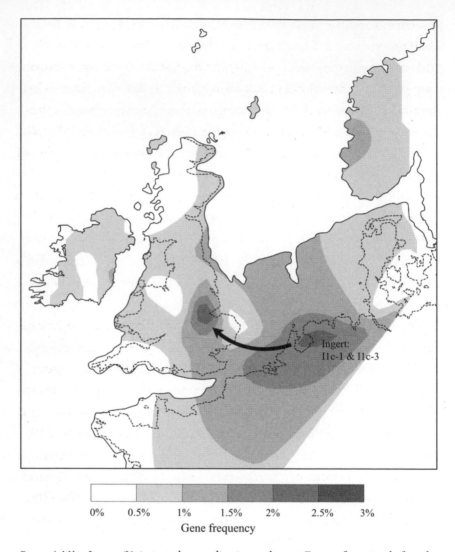

Figure 4.11b   Ingert (I1c) started expanding in north-west Europe from just before the Younger Dryas but continued during the Mesolithic particularly in Frisia, the Netherlands, eastern England and the North Sea Plain, as shown here (composite contours for I1c-1 and I1c-3 – arrow shows direction of gene flow based on tree).

Norfolk and East Anglia were still linked to the Continent during the Early Mesolithic. I have already introduced Ingert in Chapter 3, entering Britain from the east before the YD when the North Sea was a grassy plain. The main representation of Ingert is in northern

Germany and the areas bordering the North Sea, including eastern Britain, southern Scandinavia, Holland and Frisia. The degree of diversity emerging with new Ingert clusters in north-west Europe suggests an initial pre-YD migration into the north-west,[72] which logically fits with its main representation, centred in northern Germany, and relative under-representation in Scandinavia, which was ice-bound at the time (Figure 3.8). A post-LGM origin for Ingert, near the Black Sea, probably in Moldova), with a pre-Mesolithic migration and expansion north-west up the Danube then looks likely. Two of the three Ingert clusters then show further Mesolithic expansions into Britain just after the YD event.[73] I1c-3, for instance, dates in Britain to just after the YD and features both sides of the North Sea: in Norfolk and the Fen country, and in Frisia (Figure 4.11b).[74]

By contrast, the highest frequency of Ian in Northern Europe centres around Denmark and Oslo in southern Scandinavia (Figure 4.11a). Relevant genetic dates are younger than for Ingert,[75] suggesting that both his migration into and expansions within Britain occurred during the Early Neolithic and onwards, after the Scandinavian coastline opened up in the south. With Ian's Neolithic age, he could have combined with Ruisko to provide the input for the western Danish Neolithic entry into Norway, and ultimately Sweden, while Rostov was the East European Mesolithic/Neolithic founder, coming in the other way down through northern Scandinavia as explained above.

Arrivals in Britain during Mesolithic and Neolithic times

The overall British contribution of Ian and Ingert during the Mesolithic was only 3%, of which Ingert was the greatest component.[76] So, what real relevance does this Scandinavian and German Mesolithic digression have to the peopling of the British Isles? Well, it may help us to disentangle the various northern Continental inputs to the British Isles throughout the Mesolithic

4 (*top inset*): A 13,500-year-old rib engraved with a hunted horse, from Robin Hood Cave, Creswell Crags, a site of Britain's earliest re-occupation after the ice retreated. 5: The remarkable intact Neolithic settlement of Skara Brae (*top main*) in Shetland, uncovered during a storm, dating to over 5,000 years ago. 6: Megalithic tombs proliferated along the Atlantic fringe, including the truly monumental Newgrange passage grave (*above*) in Ireland from 5,300 years ago.

7: The tradition of chamber tombs spread widely in the west during the early Neolithic, such as here at West Kennet (*above main*). 8: Neolithic influences from the east included the beaker burials, like the famous Amesbury Archer (*above inset*) found near Stonehenge. 9: Henges – including the recently exposed Seahenge (*below*) on the sands of Norfolk – favoured the eastern half of the British Isles, and had counterparts in Germany.

(see Chapter 11, especially Figure 11.5).[81] And finally, in spite of its geographical proximity and the Dark Age stories of Frisian merce-naries in England, there are virtually no specific Frisian matches.

In summary, we have fairly clear maternal evidence for Norway and northern Scandinavia that the first settlers there were Mesolithic and came predominantly from Iberian Atlantic-coastal stock rather than farther east in Europe. The Y-chromosomal evidence suggests a common Balkan refuge source of the 'Ivan' group, and Britain may have shared in the spread of Ingert during the Mesolithic.

However, it is a little more complicated to distinguish what proportion of the British and Irish hunter-gatherer ancestral male gene lines arrived after the Younger Dryas, during the Mesolithic. If we take overall ancestral contribution, the figure would be around 47%,[82] of which by far the largest component (96%) came from the south-western refuge. As we shall see in the next chapter, the general figure would be much lower if the true apportionment was made to the Neolithic, since the same lines were to re-expand then. Whichever figure is used, my estimate for male Mesolithic gene flow is considerably higher than the 11% derived from studying maternal Helina lines.

Before leaving the Mesolithic, it is worth mentioning that there are several possible reasons for the high degree of genetic carry-over from this period into the Neolithic along the Atlantic fringe. Because of the British Isles' position at the far west of the huge European 'peninsula' and the rich marine resources available, the hunter-gatherer Mesolithic lifestyle lasted rather longer than in Central and Eastern Europe. This persistence of the Mesolithic is best seen in Ireland.[83]

The introduction of agriculture, and possibly the language that went with it, to Ireland and other places, could have resulted as much in an expansion of pre-existing Mesolithic lineages as of new introduced Neolithic ones.[84] This does indeed seem to be the case, as we shall see in discussion of the Neolithic in the next chapter.

5

INVASION OF THE FARMERS:
THE NEOLITHIC AND THE METAL AGE

The Neolithic (literally the 'new stone age') was a package of cultural innovations including systematic domestication of plants and animals. Large parts of the world, such as South-east Asia, still practise some form of labour-intensive agriculture, many elements of which would be familiar to their Neolithic ancestors, apart from the change to rather more effective metal implements such as iron ploughshares and iron hand-held rice-cropping knives. The Neolithic is therefore rightly seen as a culturally more dramatic threshold than our more recent Agricultural Revolution (immediately preceding the Industrial Revolution), although the size of its effect on population growth in Europe was controversially less than previously thought.[1] The Agricultural Age, as the Neolithic is sometimes called, is certainly the most researched period of European prehistory.

The Neolithic Revolution in the western part of Eurasia started at least 10,000 years ago, but did not impinge on Britain and Ireland until 6,500–5,500 years ago.[2] Not only were the British Isles at the very end of the long trail of agricultural influence that began in the Near East, but they seem to have received their

Neolithic inputs, agricultural and otherwise, via two completely different routes: north and south through and round Europe from the Balkans and the Near East. One of these routes to Britain was from the nearby regions of north-west Europe to the east, while the other was up the Atlantic coast via Iberia, thus again following previous patterns of influence.

A later Neolithic in the British isles

Following some forest clearance around 6,500 years ago, the earliest clear evidence for a Neolithic agricultural lifestyle was in eastern Britain, at Shippey Hill in Cambridge and at Broome Hill in Norfolk, about 6,300 and 6,200 years ago, respectively.[3] The cultural trend rapidly spread north over the next two to three hundred years, reaching Cross Mere in Shropshire, then Northumberland and Grampian, and right up to Orkney, where today we have the remarkable treasure of the intact Neolithic settlement of Skara Brae, first uncovered in a storm in 1850 (Plate 5).

It is logical to suppose, from the sequence of events, that the impetus for these changes in eastern Britain came into East Anglia from north-west Europe across the Channel and the North Sea. However, at the same time, things were happening and ideas were moving up on the Atlantic side, which were likely to have been introduced via seaways from farther south down the French coast. An isolated group of hunter-gatherers on the Dingle Peninsula in south-west Ireland had acquired polished Neolithic stone axes and cattle by 6,100 years ago.[4] Neolithic sites appear in Ireland in Ballynagilly and Carrowmore just over 6,000 years ago. In connection with Carrowmore, Barry Cunliffe suggests that the tradition of building large stone monuments may have actually moved from Brittany to Ireland earlier, as part of an established maritime cultural relationship, long before the Neolithic agricultural package itself arrived. This is a surprise, since passage graves in Ireland, such as those at Carrowmore, are generally associated with the full Neolithic:

It has been claimed, on the basis of radiocarbon dates, that some of the tombs in the cemetery of Carrowmore, Co. Sligo, date back to the early fifth millennium [BC, i.e. nearly 7,000 years ago]. If so they would be broadly contemporary with the earliest Breton passage graves and would predate by more than 500 years the appearance of the Earliest Neolithic in Ireland.[5]

These anachronisms of Neolithic axes, cattle-ranching and stone graves arriving before agriculture, far from depicting the Atlantic-coast Mesolithic Irish as backward and ripe for colonization, paints them as sophisticated traders with long-term, long-distance connections and an eclectic choosiness for the Neolithic fashions hailing from Brittany. As we shall see, this fits with the overall undiluted genetic antiquity, or conservatism as it seems to me, of Irish male and female markers. As I mentioned, my analysis shows that the most important Irish R1b-14 founder arrived in Britain and Ireland long before the Neolithic. R1b-14 then re-expanded as three clusters dating from the Mesolithic into the Irish Early Neolithic period.[6]

Elements of the 'Neolithic package' did not necessarily travel together through Europe, from their homelands in the Near East, at the same time. This is one of the many reasons for wondering how much of the movement was of culture passing through a network, or being copied by example, and how much was a real movement of people.

The ability of most domesticated plants and animals (domesticates) to move great distances without riding on the back of great migrations was already apparent at this early stage of European farming. Not all the early European domesticates came from the Near East. Broomcorn millet (*Panicum miliaceum*) appears early in the Neolithic of south-east Hungary, yet it was apparently first domesticated in China or Central Asia, rather than the Near East.[7]

One of the elements of the package, which can be traced most specifically, was pottery. Pottery styles are a distinctive marker of cultural spread, and fragments of pottery are sufficiently durable to survive in most sites.

Danube Neolithic?

One particular style of pottery is most frequently associated with the spread of the Neolithic in Central and Western Europe. It has a distinctive decoration, with straight or linear bands, and is usually known by its German name, *Linearbandkeramik* (LBK), which simply means 'ceramics (pots) with linear bands'. Much has been made of the rapid spread of the LBK style, apparently up the Danube from its homeland in Hungary about 7,500 years ago. Reassessment of the sequence has only highlighted the rapidity: LBK reached Austria, and then Frankfurt in Germany, almost before it left Hungary, covering 800 km within 100 years.[8] However, the land use of these early western LBK-associated settlements was still what might be described as transitional Mesolithic, with foraging and only small-scale animal husbandry and horticulture.[9] An aversion to full-on agriculture might suggest that they were unable to clear forest, but this does not seem to be the case, so perhaps it was the pots, domestic animals and ideas which were moving and being adopted within a pre-existing network of Mesolithic tribes in Central Europe, rather than new people migrating and replacing them or clearing spare forest.

After reaching the Central and North European plains up the Dnestr and Danube Rivers, LBK then spread rapidly north down the Vistula, Oder and Elbe. It did not, however, move right up to the Baltic or Atlantic coasts (so avoiding the settled coastal Mesolithic communities), but instead swung west and south-west through the Netherlands and Belgium, arriving in northern France by 7,000 years ago and finally reaching Normandy and its coast 500 years later.[10]

An eastward movement of LBK went round the Carpathians to Poland and on to the Ukraine at the same time as the westward spread, but pots had already appeared among Mesolithic hunter-gatherers in

Poland and western Russia and the lower Volga long before, around 9,000 years ago.[11]

Mediterranean coastal Neolithic?

At the same time as the Early Mesolithic (i.e. pre-agricultural) spreads of pottery were taking place in north-east Europe, another type of ceramic flow sailed west along the north coast of the Mediterranean from the Italian and Sicilian coastlines to the Mediterranean coasts of France and Spain nearly 9,000 years ago.[12] The ceramics in this instance had attractive dotted and dashed lines created by making impressions on the wet clay with the edge of a cockle (*Cardium*) shell. Appropriately, this pottery is called Cardial Impressed Ware or just Cardial Ware.

Significantly, none of the early Cardial Ware sites show evidence of a full farming lifestyle (e.g. cereals and animal domestication) apart, possibly, from some sheep.[13] Cereals did not make their appearance at these sites until after 7,500 years ago. From then, and over the next thousand years, Cardial Ware settlements appeared on the Atlantic coasts of France and the Iberian Peninsula. Their distribution (Figure 5.1) indicates that they probably took both of the same two routes that would be used by Phoenician and Greek tin traders much later on, namely through the Straits of Gibraltar and inland via Carcassonne across the Aude–Garonne corridor (see Figure 5.1 and Chapter 1).[14]

La Hoguette is yet another style of early pottery, found mainly in France from 7,500 years ago which, with its associated material culture, seems to bridge the transition from the Late Mesolithic to the full Neolithic with ceramics, small-scale pastoralism and horticulture with cereal agriculture. The La Hoguette also fills in the geographical gap between the growing pincer movements of Cardial Ware in the south of France and LBK in the north. La Hoguette style was first found in Limburg, in the Netherlands, but is defined by bone-tempered pottery found at a site near La Hoguette in central

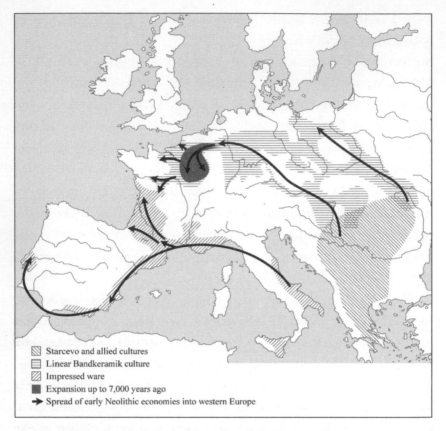

Figure 5.1   The early spread of the Neolithic into Europe, as told by pots. Two types of pottery spread through Europe with the Early Neolithic from the Near East: Linearbandkeramik (LBK) spread up the Danube from the Balkans to the north-west, while Cardial Impressed Ware spread along the Mediterranean coast and up through France.

Normandy. Although mainly found in the upper Rhone Valley, northern France, Switzerland and south-west Germany, in style La Hoguette resembles Cardial Ware farther south. Consistent with this south–north direction, a species of poppy, *Papaver segiterum*, originally from the south of France, was carried north as well. La Hoguette probably antedates LBK around the middle Rhine region where the styles overlap and even hybridize.[15]

In La Hoguette, and its association with elements of hunter-gathering, forest clearance, horticulture and limited animal husbandry of sheep and goats, we see yet another example of Mesolithic networking and opportunistic cultural acquisition (an example of *acculturation* in action). The direction of movement of La Hoguette from south-west to north-east towards the Rhine, suggests that the importance of the LBK to the West European Neolithic to the south-west of the Rhine may have been overstated and is only part of the story. What is more, the two directions of influence on the British Isles, one from the south-west and the other from central Northern Europe, seem at this point to be forming a recurrent pincer pattern, setting the scene for the next 7,000 years.

So, it appears that the earliest movement of pottery among Mesolithic groups, both in the north and the south of Europe, was a harbinger rather than a trademark of the spread of farming and farmers. It may be that the view of a mixed hunting, foraging, pastoral and horticultural lifestyle of the Late Mesolithic as only transitional is a biased perception. This pattern of 'mixed rural economy' is still the norm for so-called farming cultures in the interior jungles of South-east Asia and New Guinea today, where a tiny minority are nomadic specialist hunter-foragers, but the majority of sedentary farming folk also hunt and gather. I shall return shortly to the implications of mixed lifestyles for the question of population replacement vs acculturation in the Neolithic.

The year-by-year spread of two branches of the Neolithic cultural package from the Near East via different routes towards the Atlantic coast and finally the British Isles has been studied in minute detail by archaeologists. We have a dated progress report of how farming and pots gradually impinged on, if not necessarily replaced, Mesolithic hunter-gatherer lifestyles. Yet there are still several big questions which are very relevant to the peopling of the British Isles. One is how many people were involved in the revolution: in

other words, did the newcomers replace the old established populations? Another is what languages did they speak? A third is whether
the British Isles had two separate Neolithic revolutions. A fourth –
and this has only recently been asked – is whether the Neolithic
agricultural lifestyle was necessarily more nutritious, healthy and
successful in generating population expansion in rich environments
in Europe, than was huntin', fishin', shootin' and gatherin'.

Does farming provide better nutrition?

Rather than starting with the traditional questions, I shall go for the
last first and work backwards. In my previous book on earlier
humans (*Out of Eden*),[16] I pointed out that until recently most
modern human populations have been considerably shorter than
their hunter-gatherer forebears since the start of the Neolithic. We
have to remember that today's hunter-gatherers are obliged to eke
out an existence in semi-deserts or deep in closed, inhospitable
rainforests and are not representative of previous hunter-gatherers.
Before the arrival of farming, they had a choice of richer environments. The main reason for this Palaeolithic/Neolithic height
difference and for the variation between the heights of modern
regional populations is nutritional rather than genetic. Today, some
of the shortest peoples are those who consume a high proportion of
grain products, or other bulk carbohydrate, as their main source of
calories. Some of the tallest are those who have a high proportion of
animal protein in their diets. Economically rich populations all over
the world have increased protein as a proportion of their calories
over the past century and have shown quite dramatic increases in
average adult stature.

For the sake of brevity, I have rather oversimplified the picture,
because the physiological reasons for these effects are actually quite
complicated. What seems to be particularly important for adult
stature is diet during the weaning period and whether weaning starts
early with the use of grain-based porridges. Early infant weaning also

has a profound effect on the breast-feeding mother in that it stops lactation, restarts her monthly periods earlier, allows her to become pregnant sooner, and thus ultimately leads to larger families.

The point of this preamble is that there is no good evidence that the agriculturalist's diet was actually more nutritious than that of the hunter-gatherer in a rich environment, although it may have resulted in larger families.[17] Arguments for the success of agriculture based simply on improving individuals' quality of nutrition are therefore likely to be specious.

There are of course other advantages that farmers had over hunter-gatherers. One is visible, fixed ownership of land, which in itself tends to move on the hunter-gatherers without recourse to violent conflict. This was probably a very important factor behind the apparent Neolithic takeover of Central Europe. However, we should remember that the Mesolithic fishers and foragers of the Atlantic coast had already settled and made their identity and territorial ownership very visible.

Another often-stated advantage is that a given plot of land is likely to have a higher per-hectare yield, more particularly in the form of vegetable calories and protein, in the form of vegetable than for animal produce. This is true today, but may have had less relevance for labour-intensive Neolithic farming with the lower population densities of the past. It is also a somewhat artificial division, since both hunter-gatherers and farmers have mixed diets and lifestyles that depend on the characteristics and productivity of the landscape they live in. In the jungles of Borneo, where I have spent some time trekking and working, farmers bring a considerable proportion of their diet to the table straight from the bush. Rice is consumed in these jungle longhouses more for status than as a staple.

Clearly there was – and is – an economic advantage to farming, otherwise we wouldn't depend on farmers for food now, but early European Neolithic communities were not sufficiently densely populated to reap this bonus. However, in Mesolithic Europe there may have been a special territorial edge for anyone using the new culture, immigrants and indigenous peoples alike. This was the clearing and opening-up of the huge, dense forests that covered Europe during the Mesolithic (Figure 5.2). As we saw, during the

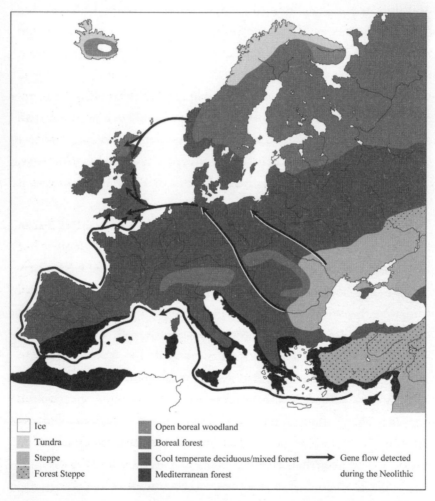

☐ Ice		▨ Open boreal woodland		
Tundra		Boreal forest		
Steppe		Cool temperate deciduous/mixed forest	⟶	Gene flow detected
Forest Steppe		Mediterranean forest		during the Neolithic

Figure 5.2 Invading the European forests and coasts. A summary of gene flow into western Europe and the British Isles as described in this chapter, also showing natural vegetation during the Neolithic. Two routes from the south-east had already been established during the Mesolithic. One went north-east of the Alps along central European rivers such as the Danube, while the other followed the coast of the Mediterranean, ending up on the Atlantic coast.

Mesolithic the hunters progressively lost their huge cold prairies rich in large visible game and had to move to the hills, where the forest cover was thinner, or to the coast, where they could rely on supplementing their diet with seafood.

Closed forest is not the paradise implied by the Robin Hood legends. Game is less plentiful and less visible; favoured fruit-bearing shrubs and trees are less accessible. Some academics studying the prehistory of peoples of South-east Asia even suggest that an independent nomadic life in the tropical rainforest without interaction with the outside world is an impossible myth.[18] This is an exaggeration of reality, but present-day forest peoples, such as the Penan and Semang in Malaysia, do need to be super-specialized hunter-gatherers to be able to survive independently in their environment. Given the hardships of the closed forest, it is not surprising that population densities of hunter-gatherers there are low.

There still are a few truly non-agricultural forest-dwellers left in the South-east Asian, African and Amazonian tropical jungles, but their independent existence is vanishing rapidly as their formerly protective forests are logged and torched. What replaces tropical forests after felling or torching is extremely degraded in terms of biological productivity, because of the poor soil, but any post-Neolithic agriculturalist can scratch a subsistence crop on it, raise cattle or grow cash crops. Although the overall environment is impoverished, forest clearance does allow potentially higher population densities. The herding of displaced forest nomads into resettlement camps before they are 'assimilated' into the mainstream population may be accomplished with minimal violence – but often is not.

In contrast to the true forest nomads, who number very few, there are other traditional forest-dwellers throughout the world who maintain relatively higher local population densities by mixing hunter-gathering with a rotation of crops and food-trees on ever-changing small forest clearings called swiddens. What is more, marauding pigs and deer provide better targets for keen hunters in these opened-up areas. In the tropics, governments and their under-the-counter partners, the logging companies, are often keen to vilify and pass the blame to these modern mixed-economy groups as forest-destroying 'slash and burn' agriculturalists. As both sides know well, this is a lie. The boot is on the other foot. In fact, the swidden farmers have managed and preserved the forest for thousands of years.

The soil situation in Europe is different, because of the rich, thick loess deposited in large areas by the wind during the ice ages, so farming is and was more rewarding, given sufficient rain; but the practical message is the same as today: opening up closed forest can dramatically increase population density.

How could all these these comparisons between Mesolithic and Neolithic lifestyles help to explain the delayed Neolithic changeover in the British Isles? First, taking an overview, it could help us to understand why the whole north-western Atlantic coastal region received less immigration during the Neolithic than did the rest of Europe (Figure 5.4c). It could be because the long, tortuous Atlantic coastline was already occupied by well-fed, settled Mesolithic communities who had diverse food sources, identified with their own particular part of the landscape, and already had traditional trade links with south-west France and Iberia and the western Mediterranean.

For most of the heavily forested Central, Northern and Eastern Europe, on the other hand, the main water bodies were rivers, which acted as conduits for immigrant farmers from the south-east penetrating back to the Balkan and Anatolian agricultural homeland. These were rivers such as the Danube and Dnestr, which discharge to the south-east into the Black Sea and would have provided a route for the Anatolian and Balkan farmers up to the north and west. The source of the Dnestr points towards Poland through the western Ukraine. Travelling upstream, the great Danube, Europe's aorta, creates a corridor penetrating between Romania and Bulgaria through Hungary, then on via Austria to southern Germany (Figure 5.2). Once across the Central European watershed, migrants would

have found other rivers, such as the Vistula, Oder and Elbe, beck-
oning politely northwards and downstream towards the Baltic, and
the Rhine and Seine inviting them towards northern France,
Belgium and the Netherlands.

As Barry Cunliffe says:

> Neolithic communities were well established in the middle Danube
> Valley and the Hungarian Plain by the middle of the sixth
> millennium [7,500 years ago] … The speed with which these
> pioneer horticultural groups were able to spread through the
> forested loess lands of middle Europe was remarkable. It may, in no
> small part, have been aided by the sparseness of the foraging popu-
> lation. The forests of the loess were dense, and over very large areas
> supported in sufficient biodiversity to attract hunter-gatherers. For
> horticulturists ready to ring-bark ancient trees and to burn under-
> growth, allowing the ash to fertilize the soil, the old forest was a
> congenial zone to colonize. By 5000 BC huge tracts of Europe from
> the Vistula to the Seine had been settled …[19]

As this passage implies, the idea that the Neolithic takeover of
Europe resulted from forest clearance rather than the superiority of
farming as a way of life is not new, but the problem has been in
assessing population densities archaeologically and in determining
how many people were real newcomers.

The European Neolithic: maternal genetic evidence

If we are to get any genetic handle on new migrations into the British
Isles during the Neolithic, as opposed to acculturation and expansions
of pre-existing Mesolithic populations, it is important to understand
the evidence for so-called Near-Eastern Neolithic gene markers. What
we need to establish is which of the two routes each gene line may have
taken from south-east Europe – up the Danube to Germany, or south
of the Alps along the Mediterranean coast to Iberia. While there is
clear evidence for both routes having been taken by maternal lines that

ended up in Britain, the southern route via Iberia and southern France seems to have flowed round the Basque Country, leaving it as a Mesolithic island with only 7% intrusion from the Neolithic.[20]

Three gene groups of Near Eastern origin have been fairly clearly identified as contributing 'Neolithic' entrants into Europe: J, T1 and U3, in that order of importance.[21] These were not the only contributors to Neolithic immigration, but together with others, they amount to around 20% of the extant European mtDNA pool.[22] While this proportion is much less than some enthusiasts had previously postulated, it still represents a significant event in European prehistory.

However, the entry of these maternal immigrant lines into Europe has been dated to around 9,000 years ago,[23] which, although appropriate for the Neolithic in Asia Minor and southern Turkey, would still fall during the European Mesolithic even for the Balkans in the south-east. The uncertainties in the dates are so large that these lines could have entered Europe either when most of the region was still enjoying its Mesolithic or after the start of the Balkan Neolithic, or even both. This should be borne in mind when considering how much of the expansion into Europe was a result of the new technologies and how much was anticipated by northward population spread in the Mesolithic, simply resulting from the dramatic climatic improvement after the Younger Dryas freeze ended.[24]

The J branch contributes the largest proportion of Near Eastern Neolithic maternal lines overall to Europe. While most of the J component is not sufficiently well defined to allow us to trace exact routes within Europe, there are three J sub-branches which clearly had separate trails and can be used for this purpose. When Martin Richards and colleagues broke down the 'Neolithic immigration' into regional contributions, north-west Europe (including northern France, the Netherlands, Frisia and Lower Saxony) had the highest proportion of maternal immigrant lines, totalling about 22%. Closer examination shows that a much lower proportion of such lines actually arrived in the British Isles (Figures 5.3a and 5.3b).

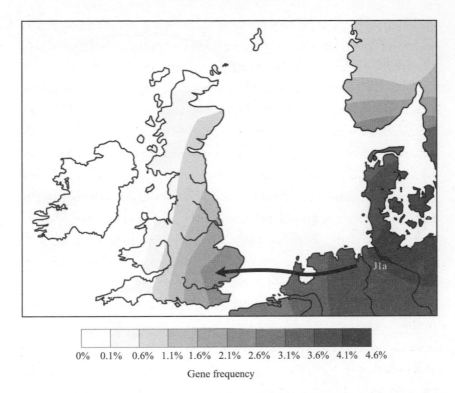

Figure 5.3a Maternal representatives of the 'LBK line' reach England: this map gives maternal evidence of the final movement of the north-western European Neolithic into England in the form of the J1a gene line (originally from the Near East). (Arrows indicate direction of gene flow based on the gene tree and geography.)

One particular J gene group, J1a, is strongly associated with north-west Europe, and has been labelled the LBK line, and even the Germanic line,[25] since it also follows the distribution of the Germanic group of languages in that area, including rates of up to 2.5% in eastern Britain. Dated in Europe to 5,000 and 7,000 years ago in two separate reports,[26] it would certainly fit a Neolithic expansion, although the uncertainties in these dates do not rule out the Late Mesolithic. J1a is also present in traditionally English-speaking areas of England, although at rather lower rates (2% vs 4%) than in northern Germany (Figure 5.3a).[27] The question, then, is when did J1a arrive in England – with the Anglo-Saxons, or

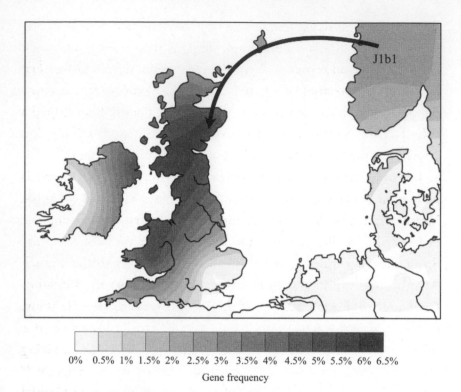

Figure 5.3b Scandinavian mothers in northern and western Britain during the Neolithic? Maternal genetic evidence for migration carrying the Near Eastern line J1b1 into northern Britain during the Neolithic probably from Norway. (Arrows indicate direction of gene flow based on the gene tree and geography.)

earlier? There is supporting evidence, as we shall see later, for at least some intrusions into England during the Dark Ages from the putative Anglo-Saxon homelands. But the latter may have been a limited, males-only, elite invasion which left at home another maternal gene line, the female 'Saxon' marker that we would expect to have been on the longboats had colonization been the aim.[28]

The geneticist Peter Forster, an Anglo-German, suggests an alternative view to the idea of J1a arriving with the Anglo-Saxons in the Dark Ages: since, logically, J1a could not have come from north-west Germany without her sister, the Saxon marker, she may have derived from more westerly Germanic-speaking Dark Age invaders such as

the Dutch, Frisians or Caesar's Belgae. My simpler default explanation, which makes fewer assumptions, is that J1a arrived much
earlier in England as part of the final movement of the north-western
Neolithic, and so would not have included the Saxon marker anyway.
Supporting such an early move is the fact that J1a itself developed a
sub-clade in Northern Europe dating to about 5,000 years ago
which, like the female 'Saxon-H', does not feature in Britain.[29]

In competition with the northern J1a gene group, there is
geographical evidence that a second main J branch, J2, known as the
'Mediterranean-celtic branch', took the southern route along the
coast of the Mediterranean from the Near East via Italy, Sardinia,
Spain and Portugal, bypassing the Basque Country to the French
Atlantic coast. In its distribution thus far, J2 mirrors the spread of
Cardial Ware moving west along the Mediterranean. After Brittany,
however, J2 jumps across the Channel to the British Isles, where it is
now found in particular association with Goidelic-celtic-speaking
areas. J2 dates very approximately to 7,000 years in Europe,[30]
which within the margin of error, would fit the spread of Cardial
Ware. In other words, J2 could be a population marker for the Early
Neolithic spread of Cardial Ware pottery along the Mediterranean
coast. Whether this is in fact the case, and whether J2 is further associated with the spread of celtic languages, are matters of tantalizing
speculation, although at an overall frequency of less than 1.5% in
the British Isles, J2 could hardly be called a major founder.

A third J subgroup, J1b1-16192,[31] characterizes British populations, especially of Scotland and other non-English areas. The distribution of the immediate J1b1 ancestor (without the 16192
mutation) in the rest of Europe is diffuse, although it is rare in non-
Scandinavian Germanic-speaking areas. J1b1-16192 is called 'British
Celtic'[32] by Forster on the basis of its British distribution. The
problem is that this label does not really explain its significant
presence in Norway and Iceland (2.6% and 3%). The highest
frequency of J1b1 including the British 16192 marker is in mainland

Scotland at 5.6% (5.4% in Wales and 3% in Cornwall). The simplest interpretation is that this subgroup came into Scotland from Norway (Figure 5.3b),[33] and its presence in Iceland is presumably derived either from Britain or, more likely, Norway (see Chapter 12).[34]

Any thoughts that this British J1b1 Neolithic maternal type might just be a more recent Viking female Valkyrie intruder can be discounted, since her even distribution in Wales and Cornwall is not consistent with a Viking origin. In any case, Forster dates the 16192 mutation expansion in Britain to at least 4,000 years ago. That is the age it would be if the 16192 mutation originated in Britain and then spread to Norway and Iceland; but if, as seems geographically more logical, she originated in Norway, then the calculation is different and entry was possibly as long ago as 6,000 years. So, the default explanation based on the Norwegian match is that this 'British' J1b1 type may have arrived in Britain from northern Scandinavia, either during the Late Mesolithic or when the Neolithic finally reached Norway around 6,000 years ago.

The J1b1 distribution in northern Britain is similar to those of male clusters from Scandinavia: I1a-5 and the much more common R1a1-2, dating to the Neolithic.[35] The latter are Neolithic entrants, also deriving from Norway rather than the Anglo-Saxon homeland, and are present throughout Britain, reaching rates in northern Scotland and its offshore islands of 5–14% (Figure 5.6a).[36] The alternative suggestion that J1b1-16192 is a Celtic marker is unsupported (even to the most orthodox adherent of the Iron Age Central European Celtic homeland),[37] based on its distribution and the Norwegian connection.

Here is a partial solution to the puzzling recurrent hints of a pre-Roman, pre-Viking, Neolithic Norse connection favouring northern and eastern Britain. As we saw in Chapter 1, Tacitus made an ethnic comment about Scottish and Scandinavian redheads, a feature which is still with us today and now has a genetic explanation.[38] Then there is the issue of the origins of the Picts and of

Bede's tale of their putative origins in Scythia, distinctive prehistoric culture and Scandinavian ponies, which could find a solution with this third J branch. There was also Avenius' confusing précis of Himilco's sixth-century BC exploration of Britain (see Chapter 1), in which he tells us darkly that the original inhabitants of the freezing north had previously been chased off the coast (presumably in this case the west coast of Scotland) by the Celts and had taken refuge in the hills.

The rest of the immigrant maternal lines from the Near East into Europe belong to branches T1, U3, a few sub-clusters of Helina (H) and W (another minority West Eurasian subgroup). Except for T1, these do not help us much further with specific routes since they are found on both the southern and northern trails, and U3 does not seem to have made it to England virtually at all, although present at low rates in Scotland and Ireland. Group T and subgroup T1 are more common in Northern and Eastern Europe generally. T1 dates to about 6,500 years ago among Germanic-speaking populations[39] and, like J1b1, also features among Scots, but not farther south in eastern Britain (e.g. not in East Anglia).[40] Like the British J1b1 Neolithic maternal type, T1 may have arrived directly across the North Sea from Scandinavia during the Neolithic, rather than later.

The landmark analysis of spread of maternal lines into Europe carried out by Martin Richards and his colleagues and completed in 2000 looked specifically at immigration across the Bosporus threshold from the Near East. While this was the best overall way of plotting the prehistory of various Neolithic immigrations into the European 'peninsula' from the Near East, it could not identify all Neolithic expansions and migrations of indigenous European lines moving *within* Europe.

This analytical problem specifically affects the measurement of how much of the Neolithic migrations were from the Near East and how much was acculturation of indigenous Mesolithic peoples and of re-expansions from the old Ice Age refuges retracing well-worn routes.[41] Richards pointed out that the Near Eastern Neolithic gene lines bypassed the old Basque Ice Age refuge.[42] While the evidence from invading Near Eastern maternal gene lines gives some indication of the relative lack of effect on the north-western Atlantic fringe, it does help to highlight the large flow of indigenous maternal gene lines north up the Atlantic coast to the British Isles during the Late Mesolithic (Figure 4.4).[43]

The problem of differentiating immigrant farming lines from the Near East from Neolithic expansion and migration of local European populations across Europe may actually be greater for the LBK-marked expansion from south-east to north-west, since the cultural (i.e. as revealed archaeologically) expansion of LBK clearly took place within Central Europe as a secondary event after the start of the Balkan Neolithic.

In summary, if there were better resolution of maternal lines we might be able to add more endogenous European lines to that 22% of Neolithic immigration to north-west Europe. A more recent figure puts the Neolithic immigration of gene groups J, T1 and U3 to Ireland at only 13%, 'a value that is toward the lower end of the range, found in Europe and similar to areas such as Scandinavia and the western Mediterranean (Iberia).'[44] The lower Irish figure of 13% is consistent with my even lower male Neolithic estimates for Ireland (7%, see Figure 5.4c), but the figure of 22% for north-west Europe is somewhat low, by this measure (see below). But these percentages relate only to the maternal genes that specifically came from the Near East. What about those lines that arose primarily in the Balkans and moved north and north-west through Germanic countries to the Baltic? Hopefully, other internal maternal Mesolithic migrations will be detected in the future.

The current status of geographic differentiation of maternal lines in Europe may be insufficient to answer all these questions, but in my view there is a potential answer to be found within the Y chromosomal

Figure 5.4   How much male Neolithic gene flow was there into the British Isles via the southern route, and how much via the northern route?

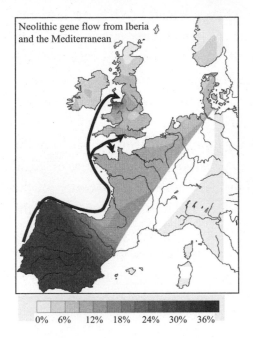

Figure 5.4a Neolithic Y-chromosome gene flow into the British Isles, from the south via Iberia, concentrated on the English south coast and western Britain, with a hotspot around Ormes Head copper mine in north Wales, but avoiding Ireland (composite of E3b, I1b*, I1b2 and J).

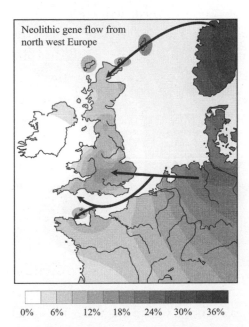

Figure 5.4b Neolithic Y-chromosome gene flow from the north into the British Isles came from Scandinavia and had two main routes: (1) arriving from Norway to the east coast of Scotland and offshore isles; (2) arriving from northern Germany/Denmark to eastern England (composite of I1a-4, I1a-5, I1a-2'1b, I1a-7b, I1a-6b, R1a1–1 and R1a1–2a&b).

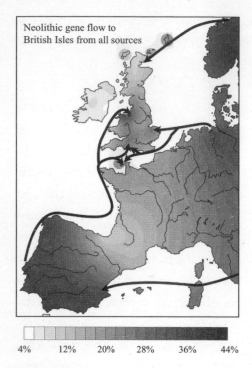

Figure 5.4c All Neolithic Y-chromosome gene flow into the British Isles. Net Neolithic gene flow into England and north Wales matched the nearby Continent, ranging from 16 to 40%. The lowest rates were in Ireland, peninsular Wales, Cornwall and western Scotland, ranging from 5 to 10% (composite of Figures 5.4a and 5.4b).

lines. Martin Richards wonders why, when 'The mtDNA picture ... suggests a value of about 20% for Near Eastern lineages in the northwest ... By contrast, both classical markers and the Y chromosome indicate few or no Neolithic markers in the British Isles.'[45] A solution to this mismatch problem – and more – can, as I have suggested, be supplied by the European Y branch Ivan, which accounts for 20% of English male lines[46] and takes us on conveniently to look at the male signals for the Neolithic spread in Europe.

Male lines of the European Neolithic

Part of the reason why there appear to be relatively fewer male than female candidates for Near Eastern migration to Europe during the Neolithic may be lack of published Y dating, which I aim to address here (as elsewhere in this book). However, the other reason is likely to be disregarded internal European migrations and expansions. The reality may be that there were as many if not more males than females moving during the Neolithic, but mainly within Europe.

Local British re-expansions

In my discussion of Late Upper Palaeolithic and Mesolithic Y-chromosome movements into the British Isles, I mentioned several lines which were likely to have arrived during those earlier periods, but only re-expanded significantly during the Neolithic. These include representatives from both the Iberian and more easterly LGM refuges.

Dealing first with old founders from Iberia, one of the most interesting male clusters of all is Rory (R1b-14). As mentioned in Chapter 4, he has a limited distribution, mainly in Ireland and the western British Isles (8% in the British Isles overall[47]). Rory has a particularly Irish flavour and has very small representation in the Iberian refuge populations and mainland Europe. He has an inter-mediate rate in Britain, being more common in south-west Wales, the Western Isles of Scotland and Orkney. Rory is, however, very common in Ireland, being found in 18% of the east coast and 23% of the west coast samples (Figure 3.6d), while his highest rate of 33% was found in a large sample of Irish men with Gaelic names, making him the most common cluster in that sample. The most common single Rory type, which could be called the 'Gaelic Modal Haplotype', constitutes 13% of Gaelic-named men.[48]

This specific celtic-surname association was originally though to suggest a later entry of Rory, possibly during the Neolithic.[49] But that is not consistent with the diversity and my date for the founding of the whole Rory cluster, which arrived before the Younger Dryas. After the YD, Rory broke up into three clusters, two expanding in the Late Mesolithic (see Chapter 4) and one in the Irish Early Neolithic (Figure 5.5a).[50] This alternative scenario, that the Irish Mesolithic and Neolithic were associated with population re-expansions of pre-existing lines, suggests several possible reasons for Rory's 'celtic associ-ation'. Either celtic languages arrived during the Palaeolithic, or there

Figure 5.5 Indigenous re-expanding clusters of the Neolithic.

Figure 5.5a Almost the entire Neolithic expansion of Irish male gene lines was indigenous, coming from a sub-cluster of Rory (R1b-14a) and spreading across to Argyll. Another indigenous sub-cluster (R1b-15b) expanded nearby, in north Wales (composite of R1b-14a and R1b-15b).

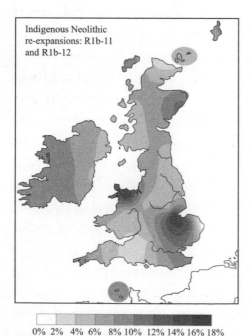

Figure 5.5b Re-expansions, in the Fens and north Wales, of the related clusters R1b-11 and R1b-12, whose ancestor arrived in the Mesolithic (composite of R1b-11 and R1b-12).

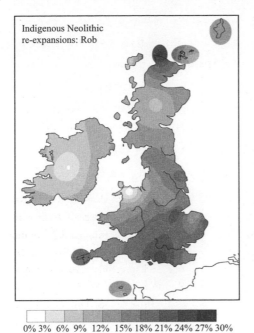

Indigenous Neolithic
re-expansions: Rob

0% 3% 6% 9% 12% 15% 18% 21% 24% 27% 30%

Figure 5.5c The main indigenous re-expansion in Britain occurred in the large Rob cluster (R1b-8) and his small brother (R1b-7), whose common ancestor arrived in the Mesolithic. Although their highest frequencies are in eastern and southern Britain and Scotland, they cover nearly the whole of the British Isles (composite of R1b-7 and R1b-8).

was a simple regional association within Ireland, or the chief cause of linkage was cultural continuity, irrespective of language. The first possibility is unlikely, and I suspect a combination of the other two.

Moving on to other re-expanding 'indigenous' lines, R1b-13 arrived during the Mesolithic (Figure 4.6) and subsequently gave rise to two Neolithic clusters, R1b-11 and R1b-12. While both are strongly represented in Wales and Ireland, the Fen country, and along the Atlantic coast of Britain, R1b-11 also characterizes Scotland and the Western Isles, and around Norfolk (Figure 5.5b). They jointly account overall for 7% of extant British lines,[51] but most of their diversity developed with Late Neolithic population expansions. Their re-expansion partly explains why the rate of contemporary incoming Neolithic lines on the Atlantic coast of Britain is lower than for eastern Britain and elsewhere in Europe.

I mentioned in Chapter 4 the large British cluster Rob (R1b-8), who is another Mesolithic entrant with later Neolithic expansion.

Rob is defined by one root type, which is the third most frequent founder type from Spain in Europe. Found in 16% of men of Valencia, this is the Frisian Modal haplotype (FMH).[52] While along most of the Continental Atlantic coast the Atlantic Modal Haplotype founder is generally the most common single Ruisko (R1b) type, Frisia has a rather higher frequency (17%) of FMH. In the British Isles the Rob cluster as a whole is only slightly less common (13%), but much more diverse than in Frisia (20%) (Figure 5.5c).

While the FMH and a common derivative with a single mutation difference[53] characterize Frisia in particular, they are nearly as common in eastern and southern Britain, where there is more diversity. There are two other closely related gene types in the Rob cluster, which are more specifically limited to the British Isles. The Rob cluster occurs generally throughout the British Isles, being more common in eastern Britain (Figure 5.5c), including England and north-eastern parts of Scotland. It has its highest frequency, however, in Durness (37% of Ruisko, 27% of total).[54] Rob, along with his small descendant cluster R1b-7,[55] both re-expanded during the Neolithic (respectively 4,700 and 4,000 years ago). These two Neolithic lines, derived from an Iberian founder with its highest frequency in Valencia, numerically contribute a large part of the apparent similarities between Frisia and England, but the distribution and diversity patterns are more consistent with parallel founding events than with a Frisian invasion, recent or otherwise.

Overall, these five indigenous Ruisko lines (R1b-7, 8, 11, 12 and 14) contribute, by their Neolithic re-expansions, 19% of extant British male lines.[56]

Neolithic re-expansion of I within Europe from a Balkan refuge

There are two Ivan (I) gene groups, which I mentioned when discussing the Mesolithic, that are good local candidates for post-glacial recolonizations of Northern Europe. These are Ingert (I1c) and Ian (I1a). Before focusing on their contribution to the

British Neolithic, I suggest that the Ivan gene group as a whole represented Northern Europe's main internal migrating male Mesolithic–Neolithic component. Ivan makes the largest non-Iberian contribution to the British Isles (16% of all males), in particular in England, where it is most common (Figure 3.7). The ancestral Ivan root is generally accepted to have originated in Europe before the LGM.[57] My estimate puts it back in the very Early Upper Palaeolithic[58]; subsequently, Ivan spread over Europe. But that is not the whole story, nor is the trail of the Ivan subgroups straightforward.

The main three post-glacial gene groups of Ivan found in Western Europe are Ian, I1b and Ingert. In my analysis, they all ultimately root to the period around the LGM.[59] As I explained in Chapter 4, the expansion of Ivan subgroups into Western Europe took place most likely at various times after the LGM, with Ingert and Ian moving up the Danube to the north-west and I1b-2 along the north coast of the Mediterranean.[60] Ingert and Ian expand most strongly in the north-west during the Mesolithic and Neolithic, respectively (Figures 3.8 and 4.11a). I1b split during the Mesolithic into two clusters, I1b* and I1b-2, most probably in the Balkans.[61]

On this basis, Ivan would have been confined mainly to south-east Europe, sheltering in the Balkans refuge throughout the LGM.[62] The restricted immediate post-LGM Balkans location would place the source and highest frequency of Ivan subgroups at centre of the oldest European Neolithic cultural zone, including the Starcevo, and allied Balkan Neolithic cultures, Kőrős, Cris and Karanovo.[63]

As mentioned above, Ingert may have been the first Ivan subgroup to move up to north-west Europe, just before the YD, later re-expanding during the Mesolithic (Figure 3.8).[64] Several other Ivan subgroups date to around 8,000 years ago, which corresponds to estimates of population expansions within each of the Ian and I1b subgroups.[65] This in turn provides a Balkan geographical origin and an explanation for the westward movement of I1b-2 along the Mediterranean coast 8,000 years ago, for Ian up the Danube to Germany 8,800–6,800 years ago, and for I1b* north-east to Russia 7,000 years ago.[66]

Not only does a Balkan homeland for Ivan fit the oldest Neolithic location in Europe, but it also, ties up geographically with the early archaeological origins of the LBK cultural expansion from its homeland in the adjacent region of Hungary 7,500 years ago.[67] However, in this case Ingert would have been in the vanguard, moving up the Danube towards Germany mainly during the Mesolithic, followed by Ian mainly during the Neolithic. These two branches, however, are numerically much more important than the south-western branch I1b2, accounting for 20–40% of male lines in north-west Europe and more than filling in the Neolithic 'Y gap' posed by Martin Richards.

I1b* seems to have stayed localized to south-east and Eastern Europe, concentrating in the Balkans, where its expansion dates to over 7,000 years ago, consistent with the Balkan Neolithic. From there, I1b* extended east into Moldavia but also strayed north of the Black Sea towards the Ukraine and Russia. The smallest branch of all, I1b-2, appears to have moved west along the north Mediterranean coast around 8,000 years ago,[68] about the same time as Cardial Ware pottery and the Late Mesolithic.

Gene group I1b-2 is found at its highest diversity and frequency (41%) in Sardinia, presumably as the result of an early founding settlement, but also extends along the Riviera into the Basque Country (6%) and Bearnais (south-west France) at 7.7%, again matching the spread of Cardial Ware. I1b-2 is rare elsewhere, although low rates are found in the Pyrenees, and Spain and notably in Normandy (2%), the Channel Islands (3%) and southern Britain and Ireland (1%), for where my dates suggest a Late Mesolithic presence, though Early Neolithic is also possible.[69]

Ian: the northern Neolithic line?

Given the Neolithic expansion ages for Ian and his sub-clusters in Western Europe and Britain, when we look in more detail we find that he could adequately fulfil the role of the main north-western expanding Early Neolithic line, in both distribution and dates. A trail up the Danube from Hungary can be discerned in the distribution of Ian in the Balkans and Central Europe (Figure 4.11a).[70] As mentioned, however, although common in north-west Germany, by far Ian's highest frequency is in Denmark and the adjacent areas of southern Scandinavia. The latter regions are rather farther north than is usually associated with the spread of LBK pottery, which is thought to be the trademark of the northern Early Neolithic. In spite of this, early LBK does have cultural parallels and there is evidence of interaction between LBK settlements and the northern region, including Denmark.[71] The very high frequency of Ian in southern Scandinavia could indicate a founder event.

In north-west Europe, Ian did not confine himself to Germany and Denmark. Like Ingert, he spread to occupy roughly the present distribution of Germanic languages – that is, southern Scandinavia (e.g. Denmark 37%), Germany (25%), Holland (16.7%), Switzerland (5.6%) and England (10–32%). The British distribution is particularly interesting, since it excludes most of Wales – and largely misses Ireland. In addition, Ian is also found in France, although favouring the north, particularly Caesar's Belgic Gaul (23%) and Normandy (at 11.9%), rather neatly fitting the ultimate spread of LBK pottery (Figures 4.11a and 5.1).[72]

With the notable exceptions of north and south Wales, Ian is found throughout Britain, even in the heart of Scotland at rates of 9% and above. The highest rates are found on the east coast of England and around the Wash (over 30%), followed by 20% in the Channel Islands and 15% in central southern England, although the south coast sample sites from Penzance to Kent only have around 10% Ian.

In the same way as with the Ruslan lines, Ian can be split into a small tree of clusters (Figure A5, Appendix C). There are seven of these, of which three large ones, I1a-2, 3 and 4, dominate, accounting together for three-quarters of British Ian types. One of the two largest of these clusters, and by far the youngest in Britain at approximately 1,200 years, is I1a-3 (30% of British I1a). I1a-3 derives from southern Scandinavia, centring in the Bergen/Oslo region, where it dates to the Bronze Age.[73] By evidence of this date and its coastal distribution round Britain, I1a-3 is a good candidate for a genetic component for more recent Norwegian invasions of the British Isles. I shall discuss later whether these migrations were entirely defined by the historical Viking raids (see Chapter 12).[74]

A much smaller cluster, I1a-5 (5% of British Ian),[75] like R1a1-2 and the maternal British J1b1 gene line, derives from farther north in Norway, and dates to the Neolithic in northern Britain, being found particularly in the Western Isles of Scotland (6%) (Figure 5.6a).

The largest British cluster is I1a-2 (32% of British Ian) (Figure 5.6b).[76] Although this cluster is found throughout Scandinavia, it centres more on Schleswig-Holstein and north-west Germany (part of the putative Anglo-Saxon homeland) at 14%, and to a lesser extent in Denmark south of the Baltic mouth (9%), spreading also to Frisia (6%).[77] This cluster dates to the Neolithic in Britain, with a wide distribution but concentrating mostly in England including the northern part. So, although I1a-2 features in the so-called Anglo-Saxon homeland, its age, distribution and unique diversity in England suggest that much of the movement had occurred during the Neolithic. This was *long* before Gildas' 'Three long-boats' (see p. 262) arrived to herald the 'Anglo-Saxon' advent.

Cluster I1a-4 also characterizes southern Scandinavia. But like I1a-3 – and unlike I1a-2 – it favours Oslo (9%) and Bergen (8%) rather more than Denmark and Frisia (4% each). Like I1a-2, I1a-4

5.6 Scandinavian male migrations to Britain during the Neolithic.

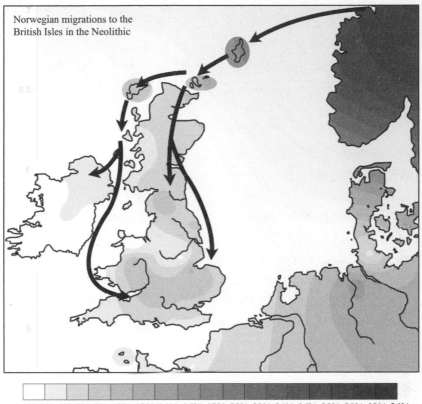

Norwegian migrations to the British Isles in the Neolithic

0% 2% 4% 6% 8% 10% 12% 14% 16% 18% 20% 22% 24% 26% 28% 30% 32% 34%

Gene frequency

Figure 5.6a Norwegian Neolithic arrivals in Britain did not just enter Shetland, Orkney, the Western Isles, eastern Scotland and Cumbria in the north, but penetrated as far south as Norfolk, south Wales and the Megalithic centre of Wessex (composite of I1a-5, I1a1–4 and R1a1–2a&b).

has a scattered distribution in eastern England and around the British coast, and an age suggesting a Neolithic or Late Mesolithic entry (Figure 5.6a).[78]

I1a-7 is the most southerly of the Ian clusters, having a centre of gravity well below Norway. However, it lacks a clear Danish identity, being found equally in Denmark, north-west Germany

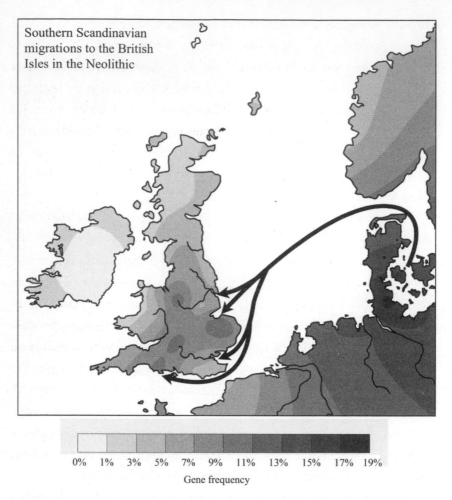

Southern Scandinavian
migrations to the British
Isles in the Neolithic

0%   1%   3%   5%   7%   9%   11%   13%   15%   17%   19%

Gene frequency

Figure 5.6b   Migrants from southern Scandinavia and Schleswig-Holstein targeted the
same regions of England during the Neolithic that the Angles did later in the Dark Ages,
but they visited Wessex as well (composite contour of: I1a-2'1b, I1a-7b, I1a-6b and
R1a1–1).

and, particularly, Frisia, with isolated instances in northern
Scandinavia, Switzerland and France. In frequency and diversity,[79]
I1a-7 is more British than Continental. It has a Bronze Age date in
Britain accounting for 9% of British Ian types.[80]

As we shall see in Chapter 11, there are a number of reasons for
rejecting the notion that the presence of Ian in England and eastern

Britain is simply a reflection of the much later Anglo-Saxon or Viking invasion. Not the least of these counter-arguments, apart from the genetic dates of the clusters, is the observation that, as with Ingert, where there are exact British Ian gene type matches with the Anglo-Saxon homeland these have overall *low* frequencies in England. Furthermore, unlike the rest of Ian's British distribution, the matches are limited in distribution to those areas of eastern England in which we might expect, from Dark Ages and Medieval chronicles, to find them (see Part 3).

This very limited vindication of the extravagant claims of monks such as Gildas in the Dark Ages in no way explains the even distribution of Ian throughout the rest of non-Welsh Britain. What is more, there is a clear differentiation between north and southern Britain in terms of their Neolithic Continental Ian source. For instance, as mentioned above, Scotland derives its two most characteristic Neolithic clusters, I1a-5 and R1a1-2, from northern Scandinavia (Figure 5.6a), appearing to match in frequency, date and origin the unique Neolithic J1b1 maternal entrant from the same region. A further argument against Scottish Ian being mainly a later entry, perhaps with the Vikings, is that the even distribution of Ian throughout Scotland is not matched by the uneven distribution of R1a1, which, like the Vikings, concentrated on the coast and offshore islands.

In summary, whether we look at eastern England – where Ian reaches his highest frequency of 32% in Fakenham – or the rest of England and Scotland, Ian seems to provide evidence of an extensive Neolithic gene flow from north-west Europe. This Neolithic influence shows a clear bias towards Scandinavian sources, both northern and southern, rather than north-west German or Frisian sources. Only cluster I1a-3 gives clear evidence of recent invasion. It constitutes 30% of British Ian types and between 8 and 11% of male types in parts of eastern England (see Chapter 12).[81]

These findings have significance for more than the Neolithic period. The Ivan group, of which Ian makes up a large part in the

West, is the main Y gene group signalling male gene flow through north-west Europe, but his presence in England is part of the non-specific evidence used for an interpretation of the Anglo-Saxon invasion as a massive ethnic cleansing event in England.[82] These new figures would suggest no more than a 4–12% Dark Age component for this particular male line in England (see Chapter 11).

The Neolithic trail from the Mediterranean

The other male gene groups associated with the Near East Neolithic are E3b and, confusingly, a male J to match the maternal J line. Like the latter, they both feature on the Mediterranean coast, but unlike maternal J they are absent from Germanic-speaking mainland north-west Europe.

E3b probably arose in East Africa and made its way during the Neolithic via the Near East between 14,000 and 7,000 years ago, and ultimately west along the Mediterranean coast of Southern Europe (Figure 5.7).[83] Farther up the Atlantic coast, E3b reappears again at appreciable rates of 5–10% in southern Britain (England, Wales and the Channel Islands, but excluding Ireland and Cornwall), with a high point of 33% in Abergele in Wales.

The male gene group J, in particular J2, has a similar history to E3b in that it came out of the Near East and moved west along the Mediterranean coast to Italy and southern Spain, probably during the Neolithic, and avoided the Basque Country.[84] Ireland again was missed out by this intrusive Mediterranean Neolithic marker, receiving nearly exclusively Ivan lines (and not many of them) instead of E3b and J. The effect of J2 on Britain was largely similar to E3b, in its size and wide southern British distribution, except that J2 did not miss out on Wales (Figure 5.7). Multiple J clusters show founding episodes dating to the Neolithic period in southern Britain.[85] Such a distribution, and the age of the founding British E and J clusters, suggests that there was a long trail of Neolithic gene flow into mainland Britain from the central and eastern Mediterranean via Spain, perhaps bypassing France.[86]

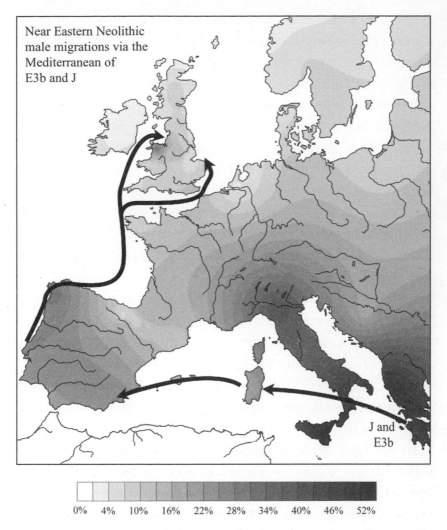

Near Eastern Neolithic
male migrations via the
Mediterranean of
E3b and J

J and
E3b

0% 4% 10% 16% 22% 28% 34% 40% 46% 52%

Figure 5.7   Near Eastern male intrusions. The distribution of Near Eastern male migrants of the E3b and J2 gene groups are well mapped in Europe, thus allowing us a fuller view of the Neolithic genetic trail along the Mediterranean and Atlantic coasts to Britain, in particular to the region around Ormes Head in north Wales (combined European E3b & J2).

Ancient trails and ancient maternal DNA

If we recognize the contribution of the Balkan Ivan male gene group to Neolithic expansions within Europe, the Amazonian imbalance between migrating male and female Neolithic lines into north-west Europe and the British Isles is partially redressed. This still doesn't quite explain why Anatolian females apparently joined hands with Balkan males and took a trip up the Danube. The higher rate of drift in the Y chromosome with loss of Near Eastern lines en route, in a similar fashion to the Saami epic migration example, could be one reason.

A fascinating study of ancient DNA by Wolfgang Haak of Gutenberg University (and colleagues) recently published in the journal *Science* does offers us a window into this gender paradox of the LBK-associated expansion, but it introduces even more questions than it answers.[87] The sort of reconstruction I discuss in this book mainly addresses DNA obtained from modern populations. The obvious reasons for this are that there are thousands of times more living DNA sources than viable ancient samples, and fresh DNA is much easier to analyse. But every now and then ancient DNA can provide the sort of information required to validate or falsify the picture obtained from modern DNA distributions.

The Haak study obtained mitochondrial DNA from skeletons excavated from sixteen Neolithic sites in Germany, Austria and Hungary. Eighteen of the twenty-four mtDNA sequences they managed to extract belonged to well-recognized European maternal types, but their relative frequencies indicated a bias towards lines that entered Europe during the Mesolithic (T2 and K) and Early Neolithic (U3, T1 and J). The others (seven of the eighteen) belonged to the older Helina or Vera groups.[88]

Admittedly the sample size is small, but the result is consistent with a real spread from Anatolia up to the north-west, but straddling these two early periods. The Mesolithic-dated lines outnumber the Neolithic-dated lines by 7 : 2, reinforcing the idea that the LBK may have been more of a cultural spread than a pioneering migration. These people could have been following an existing network trail up the Danube from the Balkans which was already established during the Mesolithic. Even the first LBK settlements spread up the Danube very suddenly and did not initially exhibit a full Neolithic cultural package. On this basis, one could suggest that the pre-existing populations, who might possibly be expected to have carried Mesolithic gene lines as well,[89] received culture and maybe some genes from their cousins farther down the Danube. If the Neolithic dating of the maternal lines is confirmed and they were actually joining pre-existing Mesolithic settlements, they might not be expected to dominate the picture.

Apart from raising such Mesolithic–Neolithic quibbles, the real surprise is that 25% of the lines Haak identified belong to an unexpected migrant newcomer, N1a. There is no clear N1a homeland, and she is rare where she is normally found, in scattered parts of West Eurasia, the Red Sea and South and Central Asia. Interestingly for us, N1a made it to Scotland, providing supporting evidence for a northern-route Neolithic input. The N1a gene types found in the Neolithic graves belong to two related sub-clusters, which are distributed evenly, one through Europe and one through west Central Asia, and are extremely uncommon even in these regions. The Central Asian type was found in a Hungarian Neolithic sample in the region of the Lower Danubian (AVK) Neolithic cultural region. N1a is virtually absent from Anatolia, the putative source of migrating Near Eastern farmers,[90] but has been found in a Scytho-Siberian burial in the Altai region.[91]

The ultimate geographical source of N1a in these north Central European Neolithic burials is a matter of complete speculation. They might even have come from eastern Scythia (Kazakhstan), the reputed home of the female Amazon warriors, who Herodotus tells us intermarried with Scythians to found the famous warlike pastoral-nomadic tribe, the Sarmatians.[92] The latter were recently resurrected by Hollywood to serve under King Arthur in a Hollywood movie.[93]

But before I give false cause for a tale supporting a Wagnerian view of the Amazonian origins of flaxen-haired Brünhilde and the Valkyries, and other Norse heroines, I should point out that N1a is conspicuous by her absence in modern German populations, thus posing another question: what happened to her? Haak and colleagues tested and rejected the possibility of loss by simple drift, offering the default hypothesis that 'the surrounding hunter-gatherers adopted the new culture and then outnumbered the original farmers, diluting their N1a frequency to the low modern value'. The dilution–acculturation view is consistent with the non-N1a types found in the graves, most of which are themselves now much less common in the Austria–Germany region, the modern picture being dominated by Helina. This study provides *direct genetic* evidence supporting the view that the ultimate numerical impact of the Near Eastern Neolithic invasion on Europe was generally underwhelming. This view, which is argued in this book and many genetics papers of the past decade, is that the Near Eastern intrusion during the Neolithic accounted for not more than a third of modern European gene lines, and usually rather less.[94]

Another unexpected result is that the dilution and loss of the maternal N1a could explain to some extent why the invading male Balkan Ivan line outnumbers maternal Neolithic Balkan immigrants to Northern Europe. If there were some immigrant women from the Balkan region among the first LBK settlements, they might have remained in those first settlements while all-male parties moved on to found new settlements with local wives.

Neolithic invasions of north-west Europe and Britain: overview

At this stage, it looks as though the British Isles had a complex new input of female and male gene lines during the Neolithic period which amounted to between 10% and 30% of extant populations today. Several familiar patterns can be discerned as well as a couple of unexpected ones.

The first point, which has been stressed repeatedly throughout this book, is that the British Isles have always received two broadly separate inputs from the European mainland: one from across the North Sea and the other coming up from the south along the Atlantic coast. The Neolithic was no exception to this rule (Figures 5.4a and 5.4b), although the balance and ultimate origins of the gene flow were rather different from the preceding Mesolithic. The southern influence, ultimately from the Near East, came along the Mediterranean coastal route via Spain and southern France. This can be seen to have contributed nearly half the Neolithic male lines in southern England and the Midlands and about a third of those in eastern and northern England, but very few in Scotland (Figures 5.4–5.7).

Consistent with this, the northern Neolithic influence, as represented by Ian on the male side, appears to have come up the Danube with the LBK culture, spreading throughout England as far as the Welsh borders and to Scotland. The difference here is that whereas England's incoming Neolithic gene flow looks very similar to that received by nearby Frisia, Scotland and its islands may have received their main Neolithic input from Norway, on both the male and the female side (Figures 5.6a and 5.4b). These impressions persist even after making allowance for the later Viking invasions.

The second impression is that the overall impact of Neolithic lines on the British Isles is patchy. Male post-Mesolithic intrusions from any source – most of them Neolithic – range from a low of 7.5% in Irishmen with Gaelic names[95] to a high of 39% in Abergele, north Wales (Figure 5.4c).[96] Irish females have a low frequency of Neolithic lines (13%)[97] compared with the overall frequency of 22% estimated for north-west Europe.[98] The conservative Irish picture fits with the genetic evidence for substantial indigenous re-expansion during the Neolithic (Figure 5.5).

The map (Figure 5.4c) of combined putative invading male Neolithic lines shows the highest local British rate of 'intrusion' in Abergele, north Wales at 39%. Regionally, England had the highest rates (range 10–28%), followed by Wales (range 10–33%) Scotland (range 6–19%), Cornwall (10%), and Ireland (range 6–9% overall). These rates are in general lower than in the northern Neolithic source region in north-west Europe, where percentages for the equivalent intrusive Neolithic lines range from a low of 25% in Frisia to a high of 35% in northern Germany. Scandinavia has between 27% and 35% Neolithic male lines. In the southern Neolithic source region of Iberia, the highest rates of Neolithic male gene lines are in the coastal regions of Galicia (38%) and Valencia (36%), with the lowest rate in the Basque Country at 10%.[99]

The differences between England, Scotland, Wales, Cornwall and Ireland in terms of Neolithic and previous intrusions are sufficiently marked to indicate that a progressive pattern of regional genetic identity was already well under way by the Neolithic. We should remember, however, that the demographic division of Britain started as far back as the Mesolithic or earlier. On the east of the Welsh border, Ingert was present from just before the Younger Dryas, and R1b-7, 8, 11 and 12 from immediately after. Along the Atlantic fringe to the west, the presence of the 'Irish' Rory (R1b-14) cluster in Ireland and up the western side of Scotland, had already begun to separate Ireland, Cornwall Wales and the Scottish Atlantic coast from the rest of the British Isles. These regional identities persist in genetic distance maps whichever groups of markers are used, specific or non-specific, Palaeolithic, Mesolithic or Neolithic.[100]

Celtic, a new Irish language without a major invasion of people?

Do the divisive genetics of the British Neolithic help with the 'Celtic question'? Not really. The low rate of intrusive Neolithic lines for Ireland (6–9%), which is found even among Irishmen with impeccably Gaelic surnames, is consistent with previous results for maternal lines, and demands an explanation. The simplest and most obvious explanation is one I have suggested already: if pre-existing Irish Mesolithic populations, observing new fashions from afar, undertook their own Neolithic Revolution, importing more culture than people, they might only have absorbed minimal 'Neolithic' gene lines from the Mediterranean to the south, rather than suffering a major elite invasion.

The real puzzle is finding a conservative Atlantic-coastal / Mesolithic genetic make-up (i.e. an undiluted founding hunter-gatherer gene pool) specifically as a feature among Irishmen carrying Gaelic names. 'Gaelic' names should imply celtic language intrusions. But surely Gaelic did not arrive as the glaciers first melted? Some (e.g. archaeologist Colin Renfrew[101]) have suggested that the branches of Indo-European languages spread from the Near East on the back of agriculture. But how could a new language arrive during the Neolithic without people? There are clear examples for language changing rapidly without major population replacement; the change from celtic Gaulish to French as a result of the Roman occupation is one. But there should be some traces of intrusion. Was 6% invasion enough to change culture and language? As we shall see in Chapter 11, 5.5% might have been enough to introduce Anglo-Saxon culture and language, although I keep an open mind on whether the Germanic language was taking its first steps in Britain.

If the Irish Neolithic represented an invasion which brought celtic languages ultimately from the Near East via Spain or France, why do so few Irish males carry intrusive Near Eastern Neolithic male gene lines? Most studies of male names and the Y chromosome show strong associations. From the evidence associating Rory with male Gaelic names,[102] it could be that these Gaelic names were adopted by non-Gaelic-speaking indigenous people. If this were the case, why is there a significantly *higher* frequency of specific Atlantic coastal Ruisko Y markers, such as Rory, among Irishmen with Gaelic names than among Irishmen with names of other linguistic derivations?[103] A brief examination of this paradox shows that if the males with non-Gaelic names were more recent immigrants to Ireland, then it would not matter whether the Gaelic names were inherited, adopted or given to individuals in the vernacular – they would still be associated with the older male lines.

As I have been suggesting, the uncommon Rory cluster, which accounts for one-third of Gaelic-named Irishmen, may well have moved north up the Atlantic Coast from Iberia into Ireland even before the Younger Dryas, and re-expanded afterwards. The rich Mesolithic lifestyle continued in Ireland until about 6,000–5,000 years ago, when much of the rest of Europe was already moving fully over to farming. This possibility was even allowed for by geneticist Emmeline Hill of Trinity College Dublin and her colleagues, in demonstrating the association of Ruisko generally with Gaelic names.[104] Whether the Gaelic language branch actually moved up to Ireland at the same time is of course a matter of irresolvable speculation, but it again brings us back to the language problem I began the chapter with. That problem is of course the frustration of comparing language reconstruction with the archaeological record when stones and pots do not talk and most linguists 'don't do dates'.

Another simplistic approach to fixing the arrival of celtic languages might be first to define the period of greatest immigration from a genetic perspective and then to try to hook celtic languages to that time. However, when we add up all the genetic lines coming into the British Isles,[105] we find that what is left over after the Neolithic is insufficient to call a real mass immigration, male or female, Celtic, Anglo-Saxon or Viking. The founder work by Martin Richards and colleagues is sufficiently precise to confirm that post-Neolithic Near Eastern maternal immigration accounts for no more than 5% of modern indigenous maternal DNA lines in north-west Europe.[106] For the British Isles at the Atlantic fringe, this figure is likely to be even less. My analysis of ages and numbers of post-Neolithic British Y-chromosome lines is similarly low (discussed in Part 3, and see Appendix C), but suggests an overlap in migration events to Britain between the Late Neolithic and Early Bronze Age.

The Neolithic period in Britain lasted nearly two millennia, from the first cultural arrivals over 6,000 years ago up to the elite Wessex Bronze Age cultures starting 4,000 years ago, well after the arrival of Bell Beakers (see Chapter 2 and Fig 5.11). I would love to have been a fly-on-a-megalith and really know which of the four cultural sweeps that occurred during the British Neolithic (which I shall shortly summarize), or those even later during the Bronze and Iron Ages, brought celtic or any other languages.

I suspect that if celtic languages did arrive during the Neolithic, they came rather early, and to Ireland and the British Atlantic fringe first. This would certainly be more consistent with the deeper esti-mated age of the celtic-language branch discussed in Chapter 2. As we have seen, the actual proportion of 'fresh Neolithic' lines in this western region was rather small, although they derived from the same general Iberian direction as did nearly all the older lines. As with the spread of new languages into the Pacific over the past few thousand years, an alternative view to their being part of a Neolithic juggernaut is that they were the new maritime trade-net language allowing people who already had ties of culture and genes to exchange prestige materials such as copper, gold and beaker pots more easily.

Other languages in Britain during the Neolithic?

A less obvious problem is the fixation in all the academic literature on celtic languages to the exclusion of any others. Compared with the Atlantic fringe, England and northern Britain received a much larger influx of genes from north-west Europe across the North Sea during the Neolithic. Were they accompanied by new language? Or, in line with the Renfrew language–farming hypothesis, did yet another branch of Indo-European arrive with the Norfolk Neolithic? Apart from English and Brythonic, there are no linguistic relics left, Indo-European or otherwise, except perhaps in some incomprehensible place- and river-names, such as London,

Kent, the Thames and the Severn, and the vexed Pictish question (see Chapter 2).

So it is difficult to speculate on what languages might have been brought across the North Sea during the Neolithic. The orthodox view is, of course, that there is no need to look for other languages since insular celtic was universal by the time the Romans arrived (but see the discussion in Chapter 7). However, given that since the Early Neolithic there were consistent cultural and genetic differences between eastern Britain and the Atlantic fringe, not to mention the two different sources of genetic entrants, the question of whether there was another Neolithic language in England does need asking.

To me, the obvious default candidate is an ancestral Germanic branch. As discussed in this and the previous chapter, if Renfrew's paradigm of Indo-European branches carried by different Neolithic spreads from Anatolia and the Balkans is correct, then the LBK culture may have carried the ancestral Germanic branch of Isidore Dyen's Meso-European cluster (see Chapter 2) up the Danube to Germany along with Y-chromosome group Ian. I shall discuss Forster's minority view, arguing for the antiquity of English, later in this book, but I quote here an excerpt from his conclusions, which allow the possibility of English diverging as a separate branch of Germanic during the Bronze Age or even earlier:

> The ... analysis reveals a Scandinavian influence on English and apparently a pre-Scandinavian archaic component in Old English. All Germanic lexica spoken today appear to converge in the network on an ancestral Common Germanic lexicon spoken at an unknown time, but constrained to before AD 350 and probably after 3600 BC.[107]

Forster's view is consistent with Dyen's lexico-statistical analysis, in which English separates early from Dutch, German and Scandinavian languages and forms a deep branch from core Germanic on a par with the Brythonic/Goidelic split.[108]

Before grasping the nettle of language again and trying to blame it on Neolithic farmers (see Chapter 6), I should like to put these speculations on the British Neolithic in a more cultural context, since I have so far only sketched the dates of its commencement in the British Isles. There are two more 'cultural ages' prior to the historic period which could have been associated with gene flow, and even language flow, into the British Isles and need to be considered. These are the Bronze Age and the Iron Age; in the orthodox view of course, the latter (discussed in Chapter 1) sees the arrival of celtic languages into the British Isles.

Neolithic megaliths: a tale of two coasts

The first Neolithic cultural development happened in the west on the *Atlantic* side of the British Isles, and took its initial form in a rapidly evolving monumental attention to the dead. From around 5,800 years ago, megalithic 'portal tombs', literally blind gateways to the other world, began to appear on the coastlines of Ireland and Wales. Today these tombs can be recognized on the landscape, shorn of their earth covering, as dolmens. Dolmens typically have at least four stones, two large uprights representing the gate, a smaller upright at the back, and these three propping up a large flat slab on top. Dolmens continued to be built for up to six hundred years.

Contemporary with these, but outlasting them in fashion, was a more elaborate industrial type of megalithic grave suited to repeated communal burials: the passage tomb, lined with massive, flat slabs. Northern Ireland saw a proliferatation of local variants named court tombs (or cairns) after the courtyard opening into the passage. Similar designs appeared across the short sea-gap from Antrim in western Scotland, where they are known as Clyde cairns (Figure 5.8). Although most of the numerous passage graves were relatively simple, several such as those at Knowth, Dowth and Newgrange in the Boyne Valley, Ireland, were built on a truly monumental scale: 85 metres in diameter and 11 metres high, with extraordinary attention

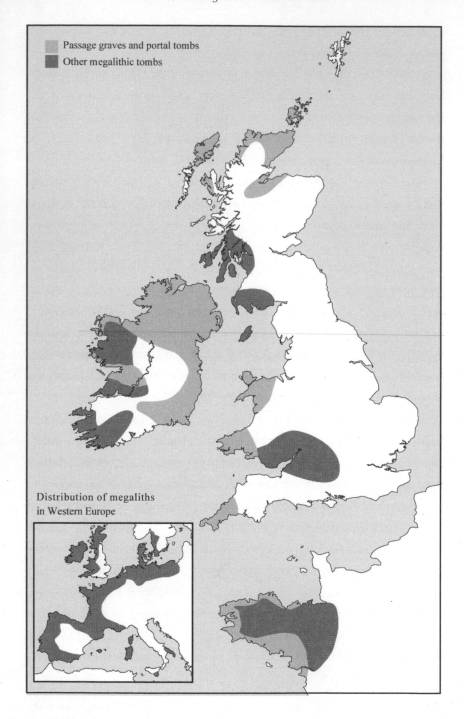

Passage graves and portal tombs

Other megalithic tombs

Distribution of megaliths
in Western Europe

to detail of celestial alignment and including decorations such as petroglyphs and quartz stone (Plates 6).[109]

Megalithic chamber tombs were fashionable along a vast coastal belt during the Early Neolithic, stretching from Corsica and Sardinia in the Mediterranean, along the south and west coasts of Iberia, through Brittany and Normandy to the south coast of England and the Atlantic fringe of the British Isles (Plate 7).

But chamber tombs also appeared to the *east* of Britain, as far north as Frisia, north-west Germany and Denmark on the Continental coast, where between 5,600 and 5,200 years ago they went through their own megalithic evolution, from simple then polygonal dolmens, through passage graves, to complex chambered and gallery graves. These were associated with so-called *funnel-necked-beaker* (in German *Trichterbecherkultur* or TRB pottery), derived ultimately from LBK. North-west European Dolmen burials may have evolved locally from earlier long barrows, a tradition which had, in turn, seeped into eastern England during its earliest Neolithic phase (Figure 5.8).[110]

Similarly, from 5,500 years ago the inhabitants of the Orkney Islands, north-east of Scotland, began to build a variety of small, chambered cairn graves. These, although numerous, were smaller than the glorious, later Orcadian chamber tombs. Given the Scandinavian genetic links, it is worth speculating whether these early north-eastern megalithic tombs derived from the Irish–Welsh–Cornish passage grave tradition or from the Scandinavian–German–Frisian chamber graves. Orkney, Caithness and the Outer Hebrides were later to build their own more elaborate slab-lined chambered tombs. The genetic evidence I have mentioned for

Figure 5.8 Big stones of the Neolithic. In addition to being the southern trail of Neolithic genetic spread, the Mediterranean- and Atlantic-facing coasts of Europe and the British Isles were home to a broad band of megalithic monuments around 5,000 years ago. Northern Ireland and Orkney were the main centres of innovation in the British Isles.

Neolithic male and female gene flow into this region, both from the nearby Continent and from Scandinavia, might point towards a northern Continental cultural influence.

Around 5,500 years ago a new type of megalithic grave, the gallery grave (with elongated slab-lined chambers), joined the Continental Atlantic scene, taking hold in Brittany and also farther east in Normandy and right up to Denmark. This date was, in Cunliffe's words, 'conventionally taken as the divide between the Middle Neolithic and Late Neolithic Periods – phases defined conveniently in terms of changing pottery'.[111] It also heralded the use of animals for traction and the first use of the plough in Western Europe.[112]

From 5,200 years ago, wheeled vehicles appeared in Hungary; much later they were to find their way through north-west Europe to Yorkshire. But other dramatic cultural innovations were happening among the megalithic structures of Western Europe, tending to emphasize the separate eastern and western cultural inputs to the British Isles. A new ritual style took hold in Brittany and the British Isles characterized by arrangements of large standing megaliths. Their function and patterning was, at least partly, associated with the heavens. The ritual complex included stone circles, stone alignments, and large stones (menhirs).

Stone circles appeared in large numbers between 5,200 and 3,500 years ago, particularly across the western half of the British Isles (Figure 5.9a). Some of these were built with a clear celestial orientation. For instance, as at Stonehenge, stones were placed to align with with midsummer or midwinter sunrise. Other stones were oriented in lines. These stone alignments are found particularly in Brittany, the most famous being at Carnac, but also in Cornwall, south Wales and southern Scotland (Figure 5.9a). Single massive standing stones known as menhirs featured as part of this ritual

5.9   Cultural influences on the British Isles from the Atlantic coast and from north-west Europe in the Neolithic.

Figure 5.9a The Atlantic influence. The distribution of stone circles, stone alignments and menhirs in the western British Isles and Brittany.

complex in Cornwall and Brittany. They are also famous in the latter region as a result of the Obelix and Asterix cartoons (Plate 3). The continuity in distribution of these megalithic stone arrangements between the Pyrenees, Brittany and the western half of the British Isles would be consistent with the southern Neolithic genetic input I have described.

Henges, a another widespread contemporary style of Neolithic monument in the British Isles (Figure 5.9b), had a rather more northerly representation on the Continent than the stone arrangements I have just described. Henges and timber circles may have evolved out of a more general enclosed earthwork, known as a causewayed enclosure. All these three types of enclosure are found throughout Germany and Alsace. Henges did not necessarily feature visible standing stone rings, obvious exceptions being Avebury and Stonehenge in Wessex and several in Orkney. Constructed over a long period, in general form they were large, circular, ditched enclosures. Several, such as Woodhenge and the recently exposed wooden circle on the sands of Norfolk (named 'Seahenge' or Holme-next-the-sea) (Plate 3) had circles made of massive tree trunks. The latter features an upturned tree in the centre of the ring with its roots exposed.

One of the least visited Neolithic monuments in Europe is Maes Howe in Orkney. Found in association with three huge stone circles (here classified as henges), various menhirs and chamber tombs, Maes Howe is 30–40 metres in diameter. The dominating feature is a central, vaulted, corbelled chamber tomb originally 5 metres high, making it as one of the finest prehistoric monuments in Europe (Plate 10).

The Orkney monuments were associated, from about 3,200 years ago, with a particular pottery, used only for ritual purposes, called grooved ware (see Plate 11 for designs very similar to those on grooved ware). Both henges and grooved ware spread rapidly as a complex ritual package, probably south, through the British Isles,[113]

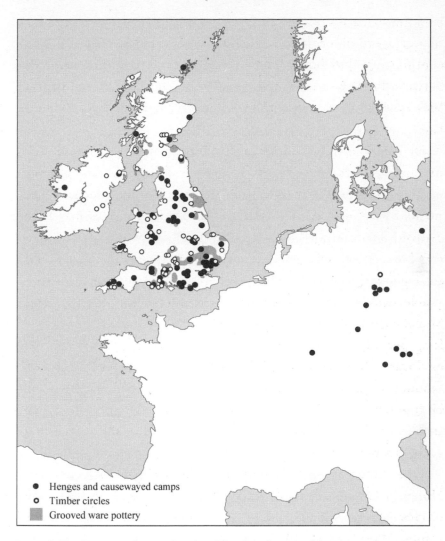

Figure 5.9b From north-west Europe? The distribution of henges, timber circles, causewayed enclosures and grooved ware pottery.

reaching an extraordinary concentration and development over the next six hundred years in Wessex (Figure 5.9b), with great henge complexes at Avebury, Marden, Durrington Walls, Knowlton, Mount Pleasant and Stonehenge. The earliest 'Stonehenge 1', however, was similar in design and age to the Orkney types, with a

110-metre diameter circular ditch-and-bank structure enclosing a large ring of fifty-six holes which probably held large wooden posts.

In contrast to passage graves, the spread of the cultural complex of henges, timber circles and grooved ware tended to favour the east coast of Britain as well as the south coast (Figure 5.9b). The distribution of grooved ware south of Scotland is exclusively in eastern and southern Britain and is not found on the Continent. As mentioned, the henge/timber circle/causewayed camp complex had a different distribution from the Atlantic passage grave zone, having its main cultural links across the North Sea in Germany, with long mounds for burial, strips of land defined by ditches or lines of pits described as cursus monuments and ditched enclosures called causewayed camps.[114] This east-side cultural picture could be suggested by the north-west European Neolithic genetic input I have described.

A cult of heroes from across the North Sea

Around 5,000 years ago came another cultural 'change' in Denmark and the North European Plain area, which later spread extensively across Europe and also to Britain. It did not have the pomp and majesty of megaliths, but had far-reaching effects, in particular for archaeologists' perceptions of the prehistory of the time. An important ritual change saw a move from communal to individual graves, and males began to be buried with stone axes perforated to take a haft, and distinctive one-litre drinking beakers often decorated with impressions from twisted cord. This is informatively called the *Corded Ware/Battle Axe Culture* (Figure 5.10).

As Cunliffe tells us, 'The origin of the Corded Ware/Battle Axe phenomenon is a matter of continuing debate.' And he explains why:

> The use of cord decoration was well known among eastern communities extending to the steppes, while the stone battle axes were evidently copied from metal forms already well established among

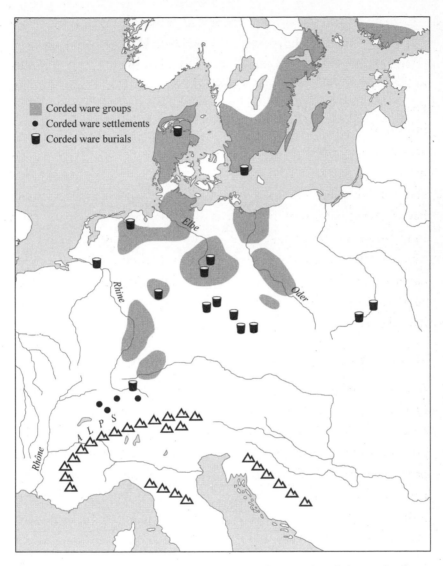

Figure 5.10 A cult of heroes from the East. Distribution of corded ware, battle-axe groups, settlements and burials in north-west and Central Europe.

the copper-using communities of south-eastern Europe. In the past it was conventional to explain large-scale culture change in terms of invasions. Thus some archaeologists argued that the Corded Ware/Battle Axe 'culture' reflected a mass migration of warriors moving into northern Europe from the Russian steppes.

Explanations of this kind are no longer in favour, and it is now generally accepted that the development is likely to have been largely indigenous, growing out of contacts between the local farming groups of the TRB (Funnel-necked Beaker) culture, the metal-using communities of the south, and pastoralists on the Pontic steppes where the domestication of the horse had taken place.[115]

I give this quote in full to give a perspective of the prevailing archaeological attitude. While we need a good dose of caution if we are to avoid seeing invasions in prehistory where there were only spreads of fashion, Cunliffe's phrase 'no longer in favour' records a change of academic fashion in itself. One reason why the 'Kurgan' pastoral migration from the steppes to Denmark and the Netherlands went out of archaeological fashion was that another migration was being promoted in competition during the late 1980s, and from a considerable height of academic authority. This was of course the hypothesis of a rather earlier movement of agriculturalists with their Indo-European languages from Anatolia.[116] Significantly, at the time, neither view was supported by any physical evidence for actual movements of people.

Now that a number of genetic studies have examined evidence for the Anatolian Neolithic invasion and have found it to be present – but not overwhelmingly so – at less than 25%,[117] it would seem fair at least to see whether the Kurgan hypothesis can be falsified on the genetic evidence. Usefully in this case, the genetic evidence extends from humans to the first domesticated ponies in north-west Europe, although the genetic dates for the latter are still uncertain.[118] In my view a Late Neolithic invasion of humans and horses from Eastern Europe (Kurgan or otherwise) cannot be falsified genetically – at least not yet.

In this chapter I have presented evidence for large-scale Neolithic gene flow from the Balkans/Ukraine area into north-west Europe marked particularly by the Ian male group. Ian is also represented in the Ukraine and farther east in the so-called Kurgan homeland, north of the Caspian Sea.[119] From the Late Neolithic in north-west Europe they are in fact coincident, so we cannot exclude the regions and dates (6,000–5,000 years ago) proposed for the source of the Corded Ware / Battle Axe cultural traits.

In addition there is the question of male gene group R1a1 (Rostov), which more convincingly characterizes the Balkans, Eastern Europe (in particular Poland)[120] and the steppe region. Zoë Rosser and colleagues at the University of Leicester proposed Rostov as the Kurgan marker.[121] In my analysis, Rostov has at least four clusters with different dates for expansion in north-west Europe, so they cannot all be 'Kurgan'. One of these, R1a1-2, moved down into northern Scandinavia in the Middle Neolithic around 5,700 years ago, appearing in small numbers in north-east Britain, particularly in Shetland and Orkney, around the same time, in other words contemporary with the earliest Neolithic in Shetland (Figure 5.6a).[122] This would have been rather earlier than Maria Gimbutas suggested for 'Kurgan' culture arriving in north-west Europe or eastern England,[123] and in any case Norway was not involved in the Corded Ware / Battle Axe phenomenon, and the ultimate source of this particular Rostov gene flow would have been Mesolithic folk from Lapland.

R1a1-3, another cluster with a marginally better location and date claim as a 'Kurgan' founder originated as a founding event in southern Norway, but is also found in northern Germany and Denmark and eastern Britain and Orkney at modest rates. This later cluster has two sub-clusters dating in Britain to between the Late Neolithic and the Early Bronze Age, around 4,000 years ago (Figures 5.14a and 5.14b).[124] Southern Scandinavia did share in the distribution of the corded-ware culture and traded copper with the Balkans. A little later during the Bronze Age, Baltic amber was traded along the same routes (used to trade tin from Britain and copper from the Baltic).[125]

What has all this to do with Britain, which did not have its own Corded Ware / Battle Axe culture? Britain may not have consistently had the full cultural package, including the original battleaxes and corded ware beakers with stepped feet, but there are lots of individual beaker burials, particularly along the east and south coasts, characterized by a type of beaker known as All-over-Corded (AOC) Beaker, which developed in the Netherlands around 4,700 years ago.[126] These beakers, decorated all over as their name suggests with cord impressions, are found commonly in the Rhine Valley, the Netherlands and Britain (Figure 5.11a). The predominantly east and south coastal distribution seems yet again to emphasize the geographical relationships of these regions of Scotland and England to the nearby Continent,[127] as suggested by the distribution of both the Rostov and Ian Neolithic intrusive male gene lines there.

The King of Stonehenge – from north-west Europe?

If there should be any doubt about the genetic evidence for Continental influence and movement of people into England between 5,000 and 4,000 years ago, we have some extraordinary direct non-genetic evidence in the form of a recently excavated set of grand beaker burials about 3 km from Stonehenge and a similar distance from the great temples of Woodhenge and Durrington Walls. In the general media excitement, the highest-status burial was dubbed variously the 'Amesbury Archer' and 'King of Stonehenge' (Plate 8). The latter appellation partly derived from the location and the fact that Stonehenge went through one of its main stone facelifts around that time.

Six other bodies subsequently turned up during excavation, including one which, bone analysis suggests, could have been a male relative of the Archer. Grave goods included eight beakers, seven of which were decorated all over (AOC), six with cord, one with plaited cord. Other typical beaker grave goods included a black sandstone wrist-guard (on the archer's forearm), a bone pin (which may have held a cloak), several copper knives, boars' tusks, a cache of flints, a shale belt ring, two gold 'earrings', and tools including an

5.11　The Beaker phenomenon and the British Isles.

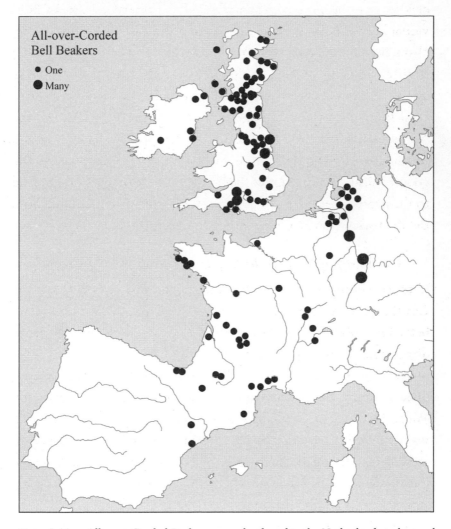

Figure 5.11a　All-over-Corded Beakers were developed in the Netherlands and spread throughout France and distributed more to the eastern side of the British Isles.

antler spatula for working flints. The radiocarbon dates show that the Archer lived between 4,400 and 4,200 years ago.[128]

Stripped of all the media hype and the rich artefacts, the most interesting aspects of this high-status interment were that it is a

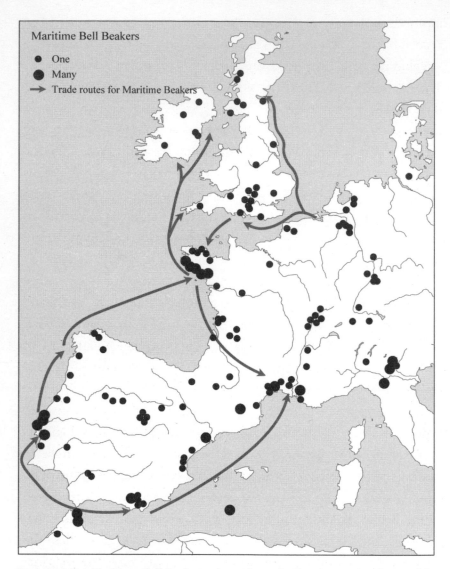

Figure 5.11b　Maritime Bell Beakers: the style evolved in Portugal and spread by maritime trade along the Mediterranean and Atlantic coasts (arrows) and also to the western rather than eastern British Isles (arrows include exchange networks across the English Channel for All-over-Corded Beakers as well).

typical Continental beaker burial, close to and contemporary with the greatest henge monument of all. And, most surprisingly, there is evidence that the Archer was not local at all, but from elsewhere in north-west Europe. The enamel on our teeth stores a chemical record of the environment where we have grown up. Scientists use a technique called oxygen isotope analysis to measure this record. The Archer's teeth show that as a child he lived in a colder climate than that of Britain today, in Central Europe, and possibly close to the Alps.[129] It may be that there was too much desire among archaeologists to place him in Switzerland and the Alps, but the isotopic study shown as a contour map (similar to the genetic maps in this book) gives a band of possible locations with similar isotopic records, stretching from the Alps, north-east through Poland, to Finland. Chemical analysis of his knife shows that the copper came from Spain.

Clearly, one dead rich archer and six relatives do not an elite migration make, but they do anything but rule it out. The status and geographical location of this individual, at origin and death, hints at an elite network of power stretching well beyond the North Sea rim and is consistent with the male genetic evidence for Neolithic migration that I have discussed.

Cunliffe traces a different distribution and history for another type of beaker, called the Maritime Bell Beaker (Figure 5.11b). As mentioned above and in the previous chapter, he sees these as markers for an Atlantic façade trade and culture network through which metal-mining and possibly celtic languages were introduced to the British Atlantic fringe from 4,400 years ago. He argues that the Maritime Bell Beaker form evolved in Portugal, rather than the Netherlands, from about 2,800 BC.[130] So although the original genesis of Bell Beakers, as a general pottery type, was in north-west Europe, they evolved in the different parts of the west and eventually entered Britain as two broad types, reinforcing the now long-standing west/east division. Whether celtic languages arrived at this time or were already established as a trade network language in the earliest megalithic phase is a matter of speculation. I prefer the latter view.

Both these routes of entry of beakers to the British Isles had some connection with the use of copper. For the Atlantic coast there was the attraction of the copper mines in Ireland and Wales, while for the east coast there was no local copper and prestige items had to be traded probably from the Balkans, where the copper age had started long before, during the Early Neolithic.

The Bronze Age

A Spanish colony in north Wales

There was considerable cultural overlap between the Neolithic and the Metal Ages. At the point of leaving the Neolithic and moving on to mix newer metals with Neolithic copper, I would like to mention one of the sample locations for the Y-chromosome analyses, the one that showed the closest genetic resemblance to a Late Neolithic Spanish colony, although the archaeology points to the British Early Bronze Age. This is the small town of Abergele, already mentioned as the British location with the highest Neolithic genetic input, on the north coast of Wales, situated near Llandudno. Until the nineteenth century, the nearby rocky promontory of Great Ormes Head had working copper mines. Recent archaeological excavations at Ormes Head reveal evidence of copper-working going back more or less continuously to the Early Bronze Age, 3,700 years ago. Pottery is notably absent from the mine workings, but is present in 'Kendrick's Cave', situated on the east-facing cliffs of the Great Orme overlooking Llandudno town, where shards of Beaker pottery and Peterborough Ware have been found.[131]

Abergele stands out in all the genetic distance maps in my analysis since it always stands away from the rest of the British samples, being better matched to Iberian sets. The first reason for this is that it holds a high proportion and diversity of Near Eastern/Iberian Neolithic lines, such as E3b (Figure 5.7), which makes up 33% of the total. Furthermore, the selection of Ruisko lines (56% of the total) is more typical of two Spanish metal-rich regions of Valencia and Galicia (and to a lesser extent Catalonia) than of the Basque Country types. Just to demonstrate its diversity and lack of drift, Abergele also has 11% of the North European Neolithic Ian gene group.[132] The point of mentioning this unique genetic connection between the Neolithic and Bronze Age here is that this evidence of skilled metalworkers arriving from Spain supports the ample archeological evidence of metal trade and bears out Barry Cunliffe's prediction of the *Longue Durée* – long-term continuity of the Atlantic coastal trade network.

Apart from this tantalizing gem of Welsh mining prehistory, there is no other specific or convincing genetic evidence for Bronze Age or Iron Age gene flow into the British Isles from Iberia. This of course does not mean there was none, just that the genetic evidence indicates that the majority of male and female lineages for which dates can be ascertained arrived in the British Isles before the Bronze Age.[133]

I have looked for evidence of such a grouping on Principal Components Analysis (for a description of this method of measuring genetic distance, see Chapter 11) in other metal-rich areas of the western British Isles, but although Llangefni in north Wales shows some similarity, there are no sample locations from southern Ireland, and Cornwall shows only a small trend, rather less in fact than in south-coast genetic sample locations such as Dorchester in Dorset, near the Bronze Age ritual centres of Wessex (Figure 5.4a). In fact, the south-coast centres still show a slightly greater male Late Neolithic genetic input from north-west Europe than from Iberia (Figure 5.4), but none from either source specifically during the Bronze Age. The 3% Bronze Age input to the British Isles from Northern Europe distributed further north and east (Figure 5.14).[134]

Blurring the division in southern England

The Bronze Age saw a dramatic increase in trade between the British Isles and Europe, at least to the extent that the archaeological visibility can tell us. In Chapter 2, I discussed the tin trade, which covered all the stations on the Atlantic coastal trade network and gave the western British Isles a unique importance to the Mediterranean. There was, however, a progressive overlap between the Atlantic trading zone and the north-west European one, with centres along the major rivers such as the Rhine, Seine, Loire and Thames joining in the network. As with the Neolithic, this overlap was seen particularly along the south coast of England. Here it centred on the Wessex Bronze Age elite cultures, in the same region as Stonehenge, which may have acted as a common market between eastern and western influences. Zones manufacturing prestige goods developed in small regions along the Atlantic fringe; they later specialized and became part of a trade network linking north-west with south-west Europe.[135]

Wessex, already the ritual centre of England during the Neolithic, continued its dominance during the Bronze Age, between 4,000 and 3,400 years ago, with lavish, rich individual burials replacing communal graves (Figure 5.12). The fact that these were in the same locations as the earlier Neolithic monuments is cultural evidence for the continuity of an indigenous population taking on new fashions rather than being replaced by invasion and this is consistent with the quiet genetic picture.[136]

Between 4,000 and 3,500 years ago, precious metals joined the Neolithic trade routes, with items such as gold lunulae (ornaments in the shape of a crescent moon) moving from Ireland to Cornwall and Brittany. Bronze Age gold hair rings made in Ireland later in the Bronze Age, around 3,000 years ago, appear to have been trans-shipped in Wessex on their way to destinations on the Rhine (Figure 5.13a).[137]

Around the same time, but in the reverse direction, so-called carp's tongue swords (Figure 5.13d), a long bronze sword with a slotted hilt, originating in Brittany, started to appear along the Atlantic coastal fringe, penetrating up major rivers from western Germany to western Iberia. They were also common in England, but

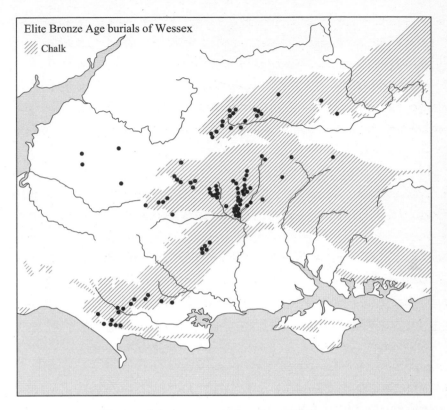

Figure 5.12   Distribution of elite Bronze Age burials in Wessex.

here, breaking with the western tradition associated with the Atlantic tin trade, they are found concentrated in the east and south-east of the country and quite far up the Thames.[138] Another bronze sword, the Ballintober type, traded out from Ireland to south-east Britain and to the Seine and Loire Valleys 3,100 years ago (Figure 5.13b).

Bronze Age barbecues with elaborate articulated roasting spits and cauldrons became quite the thing along the Atlantic coast from 3,000 years ago (Plates 12 and 13). The distribution of these beautiful objects includes not only western Iberia, north-west France and Ireland, but also south-east England and the Thames (Figure 5.13e). Slightly later, 2,700 years ago, Armorican (i.e. Brittany and Normandy) socketed bronze axes did similar rounds, spreading

5.13 During the Bronze Age, prestige items were traded from the British Isles across the English Channel and North Sea, both ways, in a variety of networks, usually including Wessex.

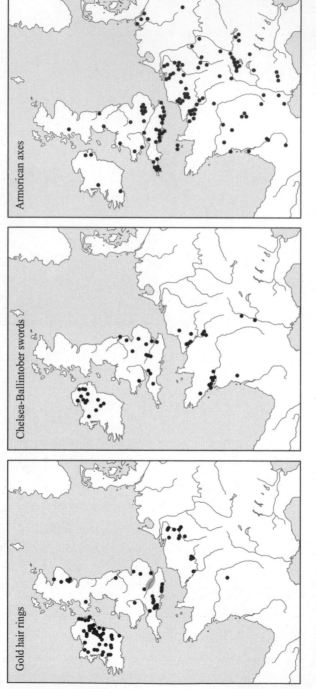

Gold hair rings

Chelsea-Ballintober swords

Armorican axes

Figure 5.13a Thick, slotted gold hair rings were made in Ireland in the Late Bronze Age. They were traded via Wessex to nearby northern Europe.

Figure 5.13b Chelsea-Ballintober swords. Swords of the Ballintober type were made in Ireland around 3,200 years ago and found their way to south-east England and the Seine and Loire valleys in France.

Figure 5.13c Armorican socketed axes. Tens of thousands of these were made in Brittany and Lower Normandy. Most were buried in hoards locally, but some were exported throughout France and to southern Britain, particularly the south coast.

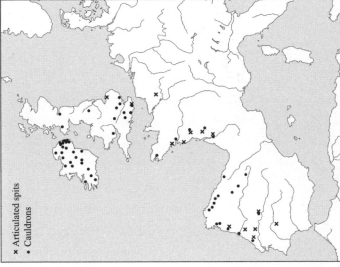

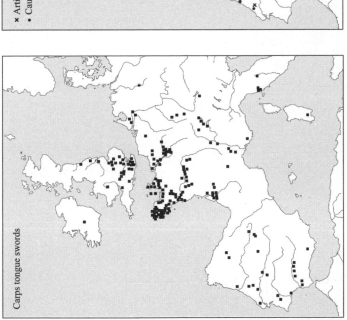

Figure 5.13e Elaborate barbecue spits and small bronze bowls, used for elite feasts, were introduced into Iberia around 3,000 years ago and spread along the Atlantic coast to eastern Britain and Ireland.

Figure 5.13d Carp's tongue swords were long with a narrow point and slotted hilt; made in western France, they are found in eastern England and along the Atlantic seaways.

from the south of France to the River Elbe in northern Germany. Although they were widespread in the British Isles, hoards of these axes, almost like deposit banks, concentrated along the south coast (Figure 5.13c).

The Eastern axis and its genetic counterpart

Between 2,700 and 2,500 years ago, the balance of the two networks connecting the British Isles to the Continent tipped towards the east. Britain, and to a lesser extent Ireland, was now linked predominantly with powerful chiefdoms in Northern Europe. Chieftains in eastern England may have been receiving gifts of horses and four-wheeled vehicles, not to mention swords, and even learning the dwarvish craft of iron-working.[139]

The Rhine connection with Britain developed in the Late Bronze Age. In return for the Ballintober sword, west Central Europe created the Hemigkofen sword type, which found its way down the Rhine and across to East Anglia and the Thames Valley. Later, from 2,700 years, ago the bronze Grundlingen sword of the Hallstatt C warrior aristocracy became popular in eastern Britain, especially the Thames Valley. Copies may even have been made locally. This weapon also appeared in Ireland, although without the horses with which it was associated in Britain. It is possible that Irish gold flowed in the opposite direction, but there is no evidence of east–west people flow, either to or from Ireland.[140]

Are there any genetic parallels for the intensive Bronze Age interchange between Europe and the British Isles? Yes there are, but the gene flow intensifies the North Sea links and is nearly exclusively from north-west Europe, including Scandinavia, to the east coast of Britain up to Orkney and Shetland, and even down to the Channel Islands. Wessex in spite of its central position in trade, does not feature at all in this gene flow. Again, the two main north-west European gene groups, Ian and Rostov, contribute the clearest evidence in the form of two dated clusters each. This was by no means a dramatic migration, contributing in all only 3% to the British gene pool.[141] The two relevant Ian clusters date to around 4,000 years ago (Figures 5.14a and b), and characterize, respectively East Anglia and north-east Britain, including Orkney and Shetland.[142] On the Continent, the former derives from northern Germany, the latter from Norway. The two related Rostov clusters date to 3,700 and 4,100 years ago, and derive respectively from northern Germany and Norway. The former is found all along the east coast of England, while the latter targets the islands off the north of Scotland.[143]

It is worth speculating on a possible Bronze Age equine genetic link between Shetland, Britain and southern Scandinavia. German-based geneticist Thomas Jansen and colleagues from Cambridge University analysed mitochondrial DNA in domesticated horses both ancient and modern. Their work revealed two particular gene groups strongly associated with North European ponies. One of these groups, C1, identifies North European pony breeds on both sides of the North Sea (Exmoor, Fjord, Icelandic, Connemara and Scottish Highland) and is restricted to the Balkans, the British Isles and Scandinavia, including Iceland. The other group, E, consists entirely of Icelandic, Shetland and Fjord ponies.[144] The devil is again in the dates, because the mtDNA mutation rate has not been fully established for horses, but the link across the North Sea is definitely older than the historical Viking raids.[145] However, the ponies were unlikely to have been imported by the Vikings. Ponies were used in Pictish times (about AD 550–800) in eastern and northern Scotland, and are shown on carved stones. They presumably passed on some of their genes, including the C1 maternal type, to today's Scottish Highland pony. Shetland ponies (related maternally to Scottish Highlands though their modern-day appearance is very different) have been around for over 2,000 years. If one adds to this the fact that ponies were present in Scotland by at least 2,700 years ago, it is certainly difficult to credit genetically related pony imports to the Vikings, and more likely that they arrived earlier with other Scandinavian visitors.

5.14   In spite of the evidence for massive trade links, little Bronze Age gene flow can be detected from Europe into the British Isles – all of it from Scandinavia and north Germany.

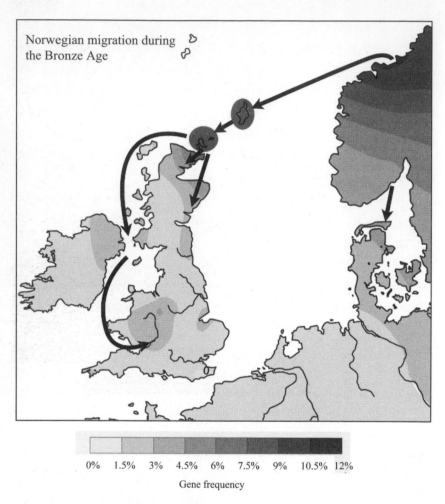

Figure 5.14a   Contour map of Norwegian gene flow into Shetland, Orkney, and northern and western Britain during the Bronze Age (composite of I1a-6a and R1a1–3a).

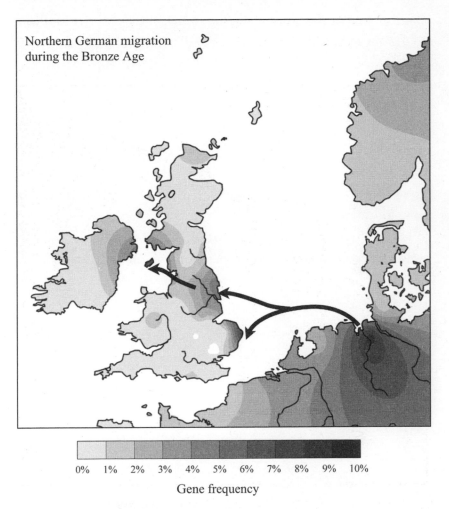

Figure 5.14b Contour map of gene flow into eastern England from northern Germany during the Bronze Age (composite of I1a-7a and R1a1–3c; arrows indicate direction of gene flow based on gene tree and geography).

The Iron Age

Before leaving the prehistoric cultural landscape, I should like briefly to mention the Early Iron Age, without repeating all the La Tène/Hallstatt discussion of Chapter 1. Study of the British parallels to the La Tène cultural package of the Central European Iron Age has not revealed any archaeological support for large-scale

Iron Age migrations into the British Isles. If we take Ireland as a 'Celtic' example, La Tène artefacts are relatively rare, and were almost always locally made.[146] It is becoming difficult to find archaeologists and historians who still accept the idea of a Celtic migration to Ireland.[147] There is, in any case, no genetic evidence for an Iron Age migration from my analysis. Even if my dates for very small-scale Bronze Age human gene flow between north-west Europe/Scandinavia and Britain were overestimates, the specific lines I have identified do not reach the Atlantic fringes of Britain, let alone Ireland.

The big question remains: what language group was used by traders on each side of the North Sea and both north of the Channel and along the south coast of England during the Neolithic and Bronze Ages?

6

WHO SPREAD INDO-EUROPEAN LANGUAGES?

The documentary evidence on prehistoric British languages is fragmentary, relying as it does on a few indirect remarks by Caesar and Tacitus, and can be of little help in telling us which came from where, and when. Our Germanic, celtic and other languages are unique cultural repositories and have some potential of their own to reveal past cultural movements. Indeed, moves to preserve celtic languages as a cultural identifier on the Atlantic coast should not be dismissed lightly as nationalism. Language is more important for cultural identity than most genes, except perhaps for those mutant genes found in Northern Eurasia, which interfere with the normally dark skin pigmentation of humans.[1] Nearly all languages spoken in the British Isles are Indo-European in origin, and it is of great interest to know when they each arrived. It is not possible to fulfil the dreams of migrationist archaeologists and match skin tones with genes, but it is of interest to everyone to be able to compare culture and descent. In other words, Indo-European languages have their own unique relevance for the peopling of the British Isles.

Indo-European is almost the only language family in Europe

The secret fascination of the Indo-European language family for prehistorians is that there are very few extant languages in Europe that belong to other families. The exceptions are famous in that they break the rule. Apart from some European members of the Uralic family (Hungarian, Finnish, Estonian and Saami), Basque is the most widely touted exception since it has no known relatives at all and has a special pride of place for geneticists. The Basque Country is not only one of the central locations of the West European Ice Age refuge, but there are clear genetic and cultural differences between Basques and the surrounding populations. As I have mentioned, these differences have been overstated – the Basques are a genetically representative population for south-west Europe who were conserved and isolated and largely bypassed during the Neolithic. However, in Roman times they were not the only linguistic exception: Iberian was another, totally different, non-Indo-European language.

Archaeologists have long sought to link the spread of Indo-European languages with one or other of the east–west cultural sweeps across the Continent detected in the archaeological record of the last 10,000 or so years. As with all such cross-disciplinary matching games, the problem is in the dates. Historical linguists tend towards a generally sceptical view, given the current lack of agreement on language dates and wide confidence margins on existing ones. Until there is robust dating of the Indo-European language family, their initial expansion could as well have been earlier in the Mesolithic or even Late Upper Palaeolithic as, in the more popular theories, in the Neolithic or the Bronze Age.[2]

In my previous books, I have promoted the view that prehistoric population expansions were, in general, more likely to have resulted from climate change than any killer cultural advantages possessed by invaders, for instance farming, with mass-migration and population replacement. Europe is no exception to this generalization,[3] as I hope I have shown so far in this book. The expansion of a language family signals mainly cultural spread rather than population migration. Although both necessarily have to be involved to some degree, it is vital to separate the two processes. Conventional archaeology is essentially a record of culture. Genetics addresses migration, however imperfectly. Language, being in between, needs to be treated on its own, ideally with its own internal dates.

Although it is interesting to make comparisons, language history should not lightly, or forcibly, be 'fitted up' to any other proxy for human prehistory such as the archaeological record, physical anthropology or genetics. The stories that each of these proxies tell may be similar, but more often than not they are different. I take a more cautious view in this than American biological writer Jared Diamond, who argues that, with exceptions, 'The simplest form of the basic hypothesis – that prehistoric agriculture dispersed hand-in-hand with human genes and languages – is that farmers and their culture replace neighbouring hunter-gatherers and the latter's culture.'[4]

In the past, archaeologists have tended to stress the migrational implications of language reconstruction. For part of the last century, warlike, battleaxe-wielding, horse-riding Kurgan nomads riding out of southern Russia in the Late Copper/Early Bronze Age were the favoured 'Aryan' harbingers of Indo-European languages.[5] Later, it became fashionable to invoke farmers arriving from the Near East, bringing their own languages, to compete with Mesolithic folk.[6] Enthusiasts have recruited several genetic patterns to underwrite both migrationist views.

Some of the best-known examples of these genetic patterns arose from the work of the grand-daddy of genetic studies, Luigi Luca Cavalli-Sforza. His group used a well-known technique, *Principal Components Analysis*, to reduce the analysis of frequency of numerous common 'classical genetic markers', such as blood groups in populations round the world, into a few simple measures. In spite of its grand title, this method combines complex data into an easy-to-view visual format. For instance, it splits up the variance in frequencies of numerous genetic markers between populations into a series of parcels –the components – of decreasing statistical importance, labelled 'First Principal Component', 'Second Principal Component', and so on. In effect, each component gives a composite measure of gene frequency, which can be plotted on a map (Figures 6.1b–6.1d) like the temperature in a weather forecast or like the gene-frequency maps in this book. The first and second components are generally the most important.

Each of these components assigns numerical values to each population and thus gives its own independent measure of the genetic distance between populations being studied. So, for instance, one might expect Turks to be more distant genetically from the English than they are from Greeks, which is indeed what we find when we look at the value of their Principal Components. It would not help to go further into the details and quirks of the method here; suffice it to say that the first three components may be regarded as giving, *by analogy*, the genetic equivalent of a latitude, a longitude and an altitude for genetic distance between populations.

The *First Principal Component* of variation in European populations in Cavalli-Sforza's analysis gives a gradient of increasing genetic distance starting around Iraq, moving through Turkey and north-west towards Scandinavia (Figure 6.1b). This just happens to fit the postulated north-westward spread of farmers into Europe from the Near East (Figure 6.1a). The *Third Principal Component* alternatively just happens to fit a similar pattern radiating out from southern Russia, which could fit the Kurgan invasion hypothesis (Figure 6.1c).[7] These patterns all sound very neat, but there are problems with their interpretation and the message carries no date-mark, as we shall see.

In his influential book *Archaeology and Language*, first published in 1987, Colin Renfrew eloquently and lucidly trashed the Kurgan hypothesis along with other theories based on horses, skull shape,

Beaker people and cord-impressed pottery, using a number of arguments, not the least of which was an authoritative injunction against using unfashionable concepts of migrationism. Yet a form of migrationism called 'wave of advance' was and still is at the heart of Renfrew's own preferred model for the spread of Indo-European languages. Although he does not insist on population replacement, he champions the view that Indo-European languages were carried by farming people ultimately from Anatolia in Turkey and spread into Europe carried by agriculture and the inexorable advance of the Neolithic.[8]

In addition to the dating dilemma, Renfrew saw an important issue in the question of where the huge Indo-European language tree was rooted. Where was its homeland? After all, its branches reached from Ireland in the west, north to the Arctic Circle, east to Central Asia and south to Sri Lanka. Large water bodies such as the Mediterranean, the Black Sea and the Caspian, not to mention mountain ranges such as the Caucasus, Taurus and Zagros, limit the possible corridors of travel in the Near East. So the Kurgan nomadic hypothesis would predict a very different language tree from the Anatolian farmer hypothesis. Renfrew acknowledged that there was a degree of controversy over the linguistic evidence, especially the limited information available on dead languages, such as Hittite, in his preferred Anatolian homeland. Support for the Anatolian root has increased since his book appeared.[9]

As genetic support for this view, Renfrew was very impressed with Cavalli-Sforza's previous mathematical analysis, his 'wave of advance', which suggested that although migratory movements of individual farmers of Near Eastern origin might be relatively small, their large families would change the balance of the local population against those of the hunter-gatherers as an advancing wave. In particular this avoided the need to talk about massive migrations, which are now 'unfashionable' in the archaeological world. Before going any further, I should point out that Renfrew now takes a rather sceptical view of the meaning of Cavalli-Sforza's Principal Components Analysis, in particular their lack of dates.

6.1   Fitting genes and archaeology using Principal Components Analysis.

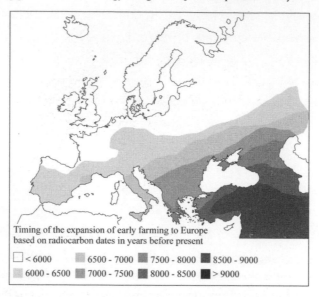

Figure 6.1a   The early spread of farming from the Near East through Europe. In the past it was thought that Neolithic farmers largely replaced pre-existing Mesolithic hunter-gatherers in Europe. Genetic studies ought to help determine the degree of that replacement.

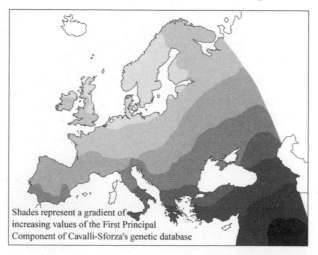

Figure 6.1b   Cavalli-Sforza's genetic First Principal Component (PC). The similarity between this PC pattern and the spread of farming (Figure 6.1a) was previously thought to indicate massive Neolithic gene flow. But there is no date-mark on such PCs, and this image could represent any, or all, movements into the European peninsula from the Near East during the past 45,000 years (each PC gives a composite measure of regional genetic association, in decreasing order of importance from the first PC).

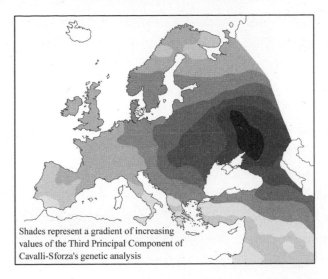

Shades represent a gradient of increasing
values of the Third Principal Component of
Cavalli-Sforza's genetic analysis

Figure 6.1c Cavalli-Sforza's genetic Third Principal Component. This PC appears to show a pattern radiating out from southern Russia, and has been thought to represent the 'Kurgan invasion'. Again, there is no date-mark, and this pattern may simply signal the post-glacial re-expansion from the Ukrainian refuge – or all gene flow between Eastern and Western Europe.

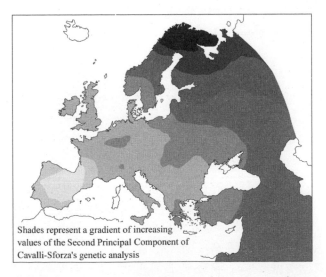

Shades represent a gradient of increasing
values of the Second Principal Component of
Cavalli-Sforza's genetic analysis

Figure 6.1d Cavalli-Sforza's genetic Second Principal Component. This PC appears to show a pattern radiating out from Lapland, for which there is no archaeological counterpart. Recently geneticists have pointed out that PCs do not specify a 'direction' of gene flow, and this PC may well represent the post-glacial re-expansion from the Iberian refuge in the other direction.

The trouble with Principal Components Analysis (PCA) is that although it is a good method of displaying genetic distance graphically and separating different genetic gradients and vectors, it is impossible by this method to date these components as genetic events. So while the Principal Components demonstrate dramatic gradients and obviously represent some migratory phenomena, they could result from multiple migrations at greatly different times along similar trails. Furthermore, although each PCA map shows gradients, the direction of gene flow cannot be automatically inferred from them. Europe is effectively a west-pointing peninsula, so the net vector direction of the First and Third Components is most likely to have been east–westward rather than west–eastward. Not so for Cavalli-Sforza's Second Principal Component (Figure 6.1d), for which the vector stretches from the Basque Country to Lapland. One could not predict from the PCA analysis whether this vector represented the southward Norse invasions of Europe from Roman times onwards or, as is much more likely, that it reflected the earlier and larger northward post-LGM re-expansion from the south-west European refuge (see Chapter 3).

Dating language again

Of course, unless one is a modern archaeologist shocked by the past crimes of anthropologists and politicians, there is no particular reason to be coy about using the word 'migration' when that is what you mean. After all, migration is what happened after the last Ice Age as people gradually moved to the north and west to fill up the empty spaces. The problem is with inferring migrations largely on the basis of language reconstruction, when language has such a mercurial relationship with the people who speak it. And more important than this is the lack of consensus on the timing of language splits, which gives at best dodgy dates and at worst no safe dates at all.

No one is more aware of the weakness of linguistic dating than Renfrew himself. It is, after all, crucial to his thesis. During his long

leadership of the McDonald Institute of Archaeology in Cambridge
he organized a number of workshops on issues connected with the
Kurgan warrior and farming/language dispersal hypotheses. These
have covered a variety of topics, including an examination of when
people changed from only eating horses to riding them as well
(quite recently, as it turns out). Several workshops have looked at
the prehistory of language and attempts to date language splits.
Being aware of the potential circularity of dating language from the
archaeology, Renfrew would dearly like to see linguists estimate
some of their own dates. Linguists who refuse to countenance direct
dating are equally convinced that Indo-European cannot be as old as
the Neolithic. Curiously, their arguments are based on memories
from old theories on the rate of language decay, based on methodology they have long since discarded.

After decades of wilderness years, during which historical
linguists have shied away from date estimation and its associated
uncertainty, some sort of activity has restarted, with research on
language dates published in journals such as *Nature*. However, with
the exception of some articles written in Russian, these papers
emanate not from linguists themselves, but from a motley
assortment of geneticists, psychologists and mathematicians. One
might wonder how non-linguists could manage in such an arcane
field on their own. Well they do not, quite. The interesting accom-
modation is that the same old warring linguists provide the
linguistic data – the all-important sets of cognates – and, apparently
in return, avoid having their names as authors on the papers.

I have already mentioned some of this language-dating research
in Chapter 1 in the context of celtic splits, but the estimates
attempted by these researchers also go right back to the roots of
Indo-European. As before, I shall discuss the same two groups:
Forster and Toth, whose dates and methods are bold and yield old
dates, and Gray and Atkinson, whose methods are more conventionally linguistically and produce younger splits.

Russell Gray and Quentin Atkinson have not only used the largest body of data, but also compared two independent datasets and followed a conventional linguistic approach, of sticking to well-attested cognates. Their mathematical approach attempted to reduce a variety of sources of error and bias.[10] They first analysed a set of cognates constructed originally by Isidore Dyen and colleagues.[11] Dyen was both famous and notorious amongst linguists for his previous use of a dating technique called glottochronology (literally, 'dating sounds'). Although they do not use glottochronology, much of the re-analysis performed by Gray and Atkinson does not actually alter the broad relationships Dyen noted among Indo-European languages, although the latter expressed them graphically in terms of proportions of shared cognates.

In constructing their language tree (Figure 6.2a) from Dyen's dataset, Gray and Atkinson adopted Renfrew's favoured Anatolian root, although they acknowledge that this could introduce bias:

> rooting the tree with Hittite could be claimed to bias the analysis in favour of the Anatolian hypothesis. We thus re-ran the analysis ... rooted with Balto-Slavic, Greek and Indo-Iranian ... This increased the estimated divergence time from 8,700 years BP to 9,600, 9,400 and 10,100 years BP, respectively.[12]

Using a different Indo-European dataset, provided by linguist Don Ringe of the University of Pennsylvania, Atkinson, Gray and others reconstructed a very similar tree with very slightly younger dates (Figure 6.2b).[13]

Do major Indo-European language branches fit male genetic groups?

Before discussing those dates, I should like to demonstrate how the European branches *could* be structurally interpreted as fitting the big, archaeo-genetic Neolithic arrows explored in previous

Figure 6.2 Dating Indo-European language splits.

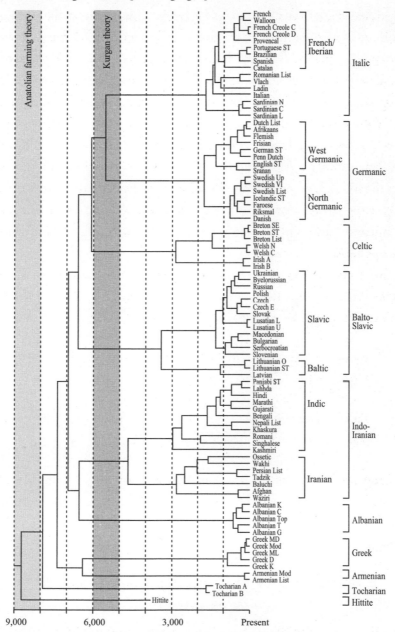

Figure 6.2a Indo-European language tree constructed mainly from living languages, with dated branch splits and lengths (dates from 9,000 years ago to present; grey timebars represent the relevant periods for the two alternative origin theories of Indo-European languages).

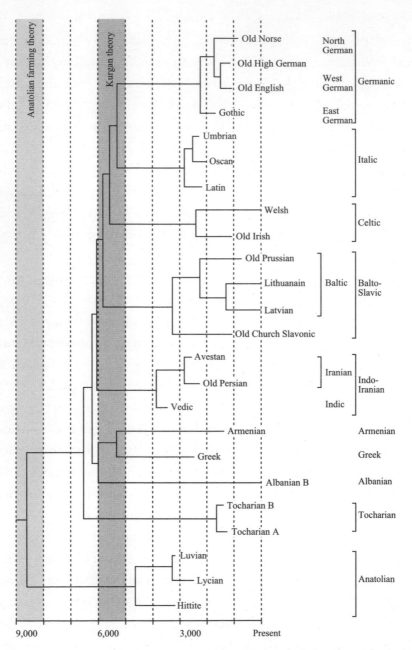

Figure 6.2b   Indo-European language tree constructed mainly from dead languages, with dated branch splits and lengths (dates from 9,000 years ago to present; grey time-bars represent the relevant periods for the two alternative origin theories of Indo-European languages).

chapters that point to farmers moving west and north out of Anatolia and the Balkans. I keep an open mind, however, as to the real significance of these interpretations. The first thing to note from both trees is that the eastern Mediterranean Indo-European languages such as Greek (and probably Albanian) and Armenian split off early, apparently before the Indo-Iranian branch headed off down south-east to South Asia. Earliest of all are the Anatolian languages spoken by Hittites and Lycians, whose sophisticated civilizations anticipated Greek and Cretan cultures. We can also note that Greece and Turkey are noticeably lower in their frequency of the European Y gene group Ian (I) than the Balkan countries immediately to their north.[14]

In Dyen's original analysis, three of the remaining four European-based Indo-European language branches (Romance, Germanic, celtic and Balto-Slavic) belonged to a super-branch which, with the interesting exception of celtic, he called *Meso-Europeic*. Populations speaking Meso-Europeic languages all have high rates of European Y gene group Ian, thus setting them apart from the insular-celtic speakers and Greeks. In Gray and Atkinson's rc-analysis of Dyen's data, however, the deepest split in the Meso-Europeic branch yields the 'Balto-Slavic' group rather than celtic. Balto-Slavic covers the whole of Eastern Europe with the exception of Rumania's and Hungary's recent linguistic intrusions. This branch includes the eastern Baltic languages Lithuanian and Latvian, and the huge Slavic group stretching from Bulgaria and the Balkans in the south to Poland and the Czech Republic in the west and Russia in the east. The most obvious Y-chromosome branch that describes this Balto-Slavic distribution best is Rostov (R1a1). The other is I1b*,[15] which I described in the previous chapter as probably moving north and east out of the Balkans, and fits the Balto-Slavic spread as well, except that it does not feature strongly in Poland or in the eastern Baltic states.

So far, so good for Eastern Europe – what about the west? The rest of Europe divides neatly into the three remaining Indo-European branches, all splitting apart in roughly the same millennium, 6,100–5,500 years ago, the deepest being celtic, with the Germanic and Italic (Romance) branches showing a marginally closer relationship. (At least that is what these trees show; there is an alternative view that there was an Italo-celtic branch, which would have been Mediterranean.[16])

Surprisingly, this packaging seems to fit the genetic story well. It would be difficult not to suggest LBK pots carrying Y-group Ian up the Danube for Proto-Germanic languages. We could then match the spread of Cardial Ware along the Mediterranean with I1b-2, E3b and J2 for Proto-Italic, and ultimately Maritime Bell Beakers and the same Y markers for Proto-celtic spreading along the Atlantic coast into the British Isles.

But neat associations are not proof, and the Neolithic is not the only period in which such a pattern could have occurred. Certainly the two Neolithic genetic trails to the British Isles that I have described, one north via Germany and one round the south via Spain and France, could fit the spread of both languages and farming. But, as I have shown, the first such post-Younger Dryas trails could have started during the Early Mesolithic, carrying Ingert to Germany and Rory (R1b-14) to Ireland.

Or is it a palimpsest problem?

The problem is that the reasons for two main routes existing from the Balkans to Britain are not specific to the Neolithic. They are the result of certain unchanging geographical features providing corridors such as the Danube, the Mediterranean and the Atlantic, and barriers such as the Alps. After all, the big opportunity for Holocene expansion north was not only the Neolithic, it may also have been the climatic amelioration at the end of the Younger Dryas, 11,500 years ago. Archaeologists and geneticists have reused an old classical word, *palimpsest* (meaning a manuscript or papyrus that has been written on and erased repeatedly), to describe this phenomenon of pattern overlaying. It applies equally to the trails derived from Cavalli-Sforza's famous Principal Components maps (Figure 6.1).[17]

If, as climatologist/archaeologist pair Jonathan Adams and Marcel Otte have suggested,[18] Indo-European could have begun to spread in the Mesolithic, then where would that leave the Indo-European farming-linguistic hypothesis? Has celtic been in Ireland for 10,000 years rather than the conventional Iron Age, 2,300 years ago, or the Bronze Age in the second millennium BC, or the Neolithic, 6,000 years ago? While I find the Neolithic language hypothesis most attractive for Europe and the Near East in general, this sort of confusion and the overwhelming temptation to fit language branches to a favourite migration hypothesis make it imperative that splits in language trees have their own dating rather than that imposed by archaeologists to support their own particular theory.

Which takes me back to the dates obtained by Gray, Atkinson and their colleagues. These are systematically younger than would be expected from the actual dates for the Neolithic cultural spread as indicated by the archaeological record. They claim that their estimate of 8,700 years for the break-up of Anatolian, derived from both the Ringe and Dyen datasets, suits the origins of agriculture neatly. This is not the really case, since the Neolithic as measured by the earliest traces of grain domestication in Anatolia is rather older than that – more like 13,000 years ago.[19]

However, the revolution did not happen overnight, and hunting and gathering continued among the Anatolian villagers.[20] So, it could be said that they did not become fully committed farmers until about 8,500 years ago, but this is uncomfortably close to the first Neolithic evidence in the Balkans. This creates real problems for their estimates for the break-up of Dyen's Meso-Europeic group, with dates of 6,100 years ago for the separation of the ancestor of celtic and 5,500 years for the split between Proto-Germanic and Proto-Italic (6,000–5,000 years ago in the Ringe set) (Figures 6.2a and 6.2b).

These are about two thousand years too late for the archaeologically dated movements of Cardial Ware and LBK.

Gray and Atkinson comment on the younger age of these West European splits, suggesting that the western branches could be evidence for a Kurgan nomadic expansion from the Caspian Sea succeeding the Neolithic expansion. There doesn't seem to be any geographical sense or archaeological evidence for this rationalization. When they come to calibrating and dating the break-up of these three major Meso-European branches into their extant modern European languages, it all seems to be flat-packed into the millennium and a half since the Romans left Britain. This foreshortening of virtually the entire history of diversification of Meso-Europeic languages into less than two millennia is, to a great extent, a consequence of linguists concentrating on the limited window of written sources.

Glasgow husband-and-wife linguist-and-geneticist team April and Robert McMahon have reviewed these attempts at dating linguistic splits. They discuss a foreshortening effect, resulting from undetected word borrowing between related languages (e.g. 'beef' in Modern English was borrowed from Norman French rather than inherited from Old English), which could have systematically underestimated the dates of splits in this kind of analysis, even with a carefully winnowed dataset:

> These effects are also minimized by Gray and Atkinson's use of a carefully prepared dataset, based on the best available linguistic knowledge, in the shape of Dyen, Kruskal and Black (1992) – though even here, as we have shown and as Gray and Atkinson (2003) suspected, there are individual miscodings which are mainly internal to subgroups, but which could again have a foreshortening effect on individual branch lengths.[21]

Peter Forster and Alfred Toth take the opposite approach to Gray, Atkinson and colleagues. Instead of trying to approximate the ideal language tree or network by studying only well-attested cognates, they initially include in their network any and all lexical changes within their chosen word list, including both those resulting from borrowing and those 'genetic' changes resulting from regular intrinsic sound changes. They clean up afterwards by removing terms with too many synonyms. To analyse this soup of linguistic 'mutations', they then incorporate methods used for dating mtDNA branches with imperfect or ambiguous information. In their words:

> a network may contain reticulations when convergence (i.e., through historical loan events and chance parallel changes, or even through data misassignments incurred by the researcher) has obscured the evolutionary tree. The linguistic network approach is therefore expressly intended to search for treelike structure in potentially 'messy' data.[22]

Two outcomes may be predicted from such an approach. One is that, by explicitly including borrowing as a 'language mutation', their dates would avoid that source of foreshortening and perhaps turn out dates older than those obtained by the more conventional, exclusively cognate-based methods. The other is that, given their smaller dataset – including poorly known, dead celtic languages with many more unknowns, linguists would scream 'Poor data!' In the event, their dates for the fragmentation of Indo-European in Europe are very old, possibly 10,000 years ago, suggesting an Indo-European expansion in the Early Mesolithic, just after the end of the Younger Dryas freeze-up, or maybe just consistent with the leisurely onset of agriculture.[23] As for the screaming, an extensive critique of their methods and results may be found in the same recent publication 'Why linguists don't do dates' by April and Robert McMahon.[24]

Summary

What relevance do all these contentious considerations of language dating have for the peopling and cultures of the British Isles? Potentially, they offer a different perspective on the Germanic/celtic divide in Britain. As I have stressed, the possible relationships of language to archaeology and genetics are all speculative and can be misleading. But the genetics tells us the Welsh/English genetic border is perhaps of Neolithic or earlier antiquity, and the archaeology tells us that there were always two separate sources of cultural flow into the British Isles, with Ireland and the British Atlantic coast relating southwards, and the east coast of England relating across the North Sea. If the prehistoric language split followed the same Neolithic geographical divide, the cultural and linguistic divisions between the English and the rest could be just be as old too, or at least older than the Roman invasion. In Colin Renfrew's picture, which has Indo-European languages spreading on the back of the Neolithic, the best candidate Indo-European branch to accompany the English Neolithic and Bronze Age would be Germanic.

Which brings us to the last of the main questions asked in this book: who are the English, and when did they arrive?

Part 3

Men from the north: Angles, Saxons, Vikings and Normans

Introduction

THE SAXON ADVENT

[A] council was called to settle what was best and most expedient to be done, in order to repel such frequent and fatal irruptions and plunderings of the above-named nations [Picts and Scots] ... Then all the councillors, together with that proud tyrant Gurthrigern [Vortigern], the British king, were so blinded, that, as a protection to their country, they sealed its doom by inviting in among them (like wolves into the sheep-fold), the fierce and impious Saxons ... to repel the invasions of the northern nations ... [N]othing was ever so unlucky ... A multitude of whelps came forth from the lair of this barbaric lioness, in three cyuls, as they call them, that is, in three ships of war, with their sails wafted by the wind ... Their mother-land, finding her first brood thus successful, sends forth a larger company of her wolfish offspring, which sailing over, join them-selves to their bastard-born comrades ... The barbarians being thus introduced as soldiers into the island, to encounter, as they falsely said, any dangers in defence of their hospitable entertainers, obtain an allowance of provisions, which, for some time being plentifully bestowed, stopped their doggish mouths. Yet they complain that their monthly supplies are not furnished in sufficient abundance, and they industriously aggravate each occasion of quarrel, saying that unless more liberality is shown them, they will break the treaty and plunder the whole island. In a short time, they follow up their threats with deeds.[1]

This is our earliest British text, written around AD 540–560, describing an unwise invitation made to Saxon warriors in the previous century, and their subsequent turning on their hosts. The story, as told by Gildas, continues with further incursions, some battles won and some lost, and culminates in the domination of England by people we now call Anglo-Saxons and who gave England its name and language. Gildas, or St Gildas as the Welsh later knew him, describes an inferno of rapine, bloodshed and genocide which has formed the basis for a persisting view of the Dark Ages ethnic cleansing of the 'Celts' from England. Significantly, some modern authors infer a population replacement by a combination of peoples from the coastal Germanic-speaking mainland of north-west Europe, including the Saxons, the Angles, the Frisians, the Jutes and even the Franks. Despite Gildas' nationalist agenda and endless religious ranting, this extreme view can still be regarded as an orthodox position, held as it has been by a number of historians, not to mention linguists, archaeologists and some geneticists.

Of course, there is a sceptical camp, particularly of archaeologists, who view any extravagant claims for migrations as unwise and prefer to point to the more fashionable option of elite takeover and dominance by small groups of nobles. The problem with this 'soft option' is the *apparently* overwhelming body of evidence for cultural, linguistic and genetic change after the Dark Ages, with little in the way of cultural carry-over from the conquered peoples.

In this third part of the book I hope to show that, rather than supporting a sudden replacement, this striking evidence is also consistent with a more prolonged cultural and genetic interchange between England and its neighbours across the North Sea, one that began even before the Roman invasion. Elite takeover by small groups of nobles is made easier if cultural links already exist. There are other, older explanations than sudden replacement for the clear genetic differences between England and the Atlantic fringe of Britain, as I have laid out in Part 2. The elite linguistic, cultural and genetic incoming influence may also have included southern Scandinavians as much as people from the more traditional Anglo-Saxon homelands of Schleswig-Holstein and north-west Germany.

I shall be discussing linguistics, historical texts and archaeological evidence, followed by genetic analysis; but I should like first to recap briefly on the background to the British east/west divide. To avoid repetition in this introduction, I shall not differentiate male from female sources of gene flow, although most of the geographical detail I shall later present refers to the Y chromosome.

Much of the genetic input of north-west Europe derives from re-expansion from Iberian refuges after the Last Glacial Maximum (LGM), and before the start of the European Neolithic 7,500 years ago. The persisting Iberian influence is more evident in the British Isles than in the neighbouring Germanic-speaking regions of Europe. While the older Iberian post-LGM influence reaches its highest rates in Ireland and the west coast of Britain, it is still present in 60–75% of males, even in England. This English genetic conservatism tends to undermine the idea of complete recent replacement by Anglo-Saxons, since the effect is seen whether one looks at broader gene groups or exact matching gene types between the two potential sources of gene flow – Iberia or the 'Germanic-speaking' regions.

In the Neolithic or before, an east/west division, or differenti-ation, began and progressed. The dividing line stretched from the Scottish Grampians in the north to Wessex in the south, between the west and east coasts of Britain. The differences are apparent in male genetic lines and cultural influence coming in from two distinct sources. The Atlantic regions of Ireland, Cornwall and Wales all received cultural input from Brittany and Iberia. Ireland and Wales received rather little Neolithic genetic input, but what there was all came via the southern source. Cornwall was a little different as far as its people are concerned, in that it shared in a general south-coast genetic melting pot of eastern and western Neolithic influences. The south coast and Channel Islands as a whole received a modest (15–25%) genetic input from *both* the north and south Continental sources during the Neolithic, although rather more from north-west Europe.

On the east coast of Britain, the Neolithic cultural influence was more clearly derived from north-west Europe, which is consistent with the introduction of new male gene lines to England from that part of the Continent. Norfolk and York and the islands of Orkney and Shetland received slightly more Neolithic input than did other parts of the east coast, but together with the earlier post-LGM period this leaves even less space for any post-Roman invasion. What is striking about these regions, especially the north-east of Britain, is the relatively high proportion of the Neolithic input, both cultural and genetic, from Scandinavia.

During the Bronze Age, cultural interchange with the Continent intensified, particularly with the north-west part of Europe. The genetic picture of ancient British pony breeds gives us another tantalizing glimpse of connections between northern Britain and southern Scandinavia. In the south, the Wessex elite Bronze Age cultures acted as a cultural and trade centre between the east and west of the British Isles in their contacts throughout the Continent. The main evidence for genetic inflow, however, is in eastern Britain from across the North Sea, although even that is relatively small. My overall estimate for further male gene flow into the British Isles during the immediate pre-Roman period (Bronze and Iron Ages) is about 3%.

7

WHAT LANGUAGES WERE SPOKEN IN ENGLAND BEFORE THE 'ANGLO-SAXON INVASIONS'?

Along with my queries about English identity and origins, I left a language question hanging at the end of the last chapter. It may seem academic, but this question goes to the heart of Anglo-Saxon cultural identity. Before plunging into the quagmire of unreliable historical texts from the Anglo-Saxon period, I shall restate it: given the increasingly close cultural and genetic relationships between north-west Europe and the British North Sea coast before the Roman invasion, what languages were spoken in eastern Britain at that time?

Linguists interested in this period, who by default are mainly celticists, assume that a form of celtic, most likely 'British' (Brythonic), was universal at the time of the first Roman accounts (see p. 280). Language is as much a cultural as a genetic marker, so it might be reasonable to ask whether any languages of the Germanic branch of Indo-European were spoken in some parts of England even before the Romans came.

The classical writers are not much help here, mainly because Britain and Ireland were islands off the far end of the Continent with their own names, and most authors were not very interested in languages. As I pointed out in Chapter 1, no classical author referred to any Britons as being Celtic, let alone celtic-speaking. Strabo in fact *contrasted* the Britons racially with the Celts – in his usual disparaging style, when writing at the limits of his knowledge. It is obvious that his direct experience was confined to seeing a few slaves from England being paraded half-naked in the markets of Rome:

> The men of Britain are taller than the Celti, and not so yellow-haired, although their bodies are of looser build. The following is an indication of their size: I myself, in Rome, saw mere lads towering as much as half a foot above the tallest people in the city, although they were bandy-legged and presented no fair lines anywhere else in their figure. Their habits are in part like those of the Celti, but in part more simple and barbaric.[1]

So, if in these lines Strabo is referring to lads from England as different from his idea of Celts, what about the other tribes in Britain during the Roman occupation? What did they look like, and what sort of cultures and languages did they share? We do have some first-hand accounts from other Roman commentators, in particular Tacitus, who wrote about the British tribes in a laudatory biography (AD 98) of his famous father-in-law, Agricola. In the following passage, the Silurian part of which I have already cited (in Chapter 1), he asks the questions, but acknowledges how few of the answers he has. Even the gaps, however, are revealing:

> Who were the original inhabitants of Britain, whether they were indigenous or foreign, is, as usual among barbarians, little known. Their physical characteristics are various and from these conclusions may be drawn. The red hair and large limbs of the inhabitants

of Caledonia point clearly to a German origin. The dark complexion of the Silures, their usually curly hair, and the fact that Spain is the opposite shore to them, are an evidence that Iberians of a former date crossed over and occupied these parts. Those who are nearest to the Gauls are also like them, either from the permanent influence of original descent, or, because in countries which run out so far to meet each other, climate has produced similar physical qualities. But a general survey inclines me to believe that the Gauls established themselves in an island so near to them. Their religious belief may be traced in the strongly marked British superstition. The language differs but little ...[2]

Taken together, Tacitus' three examples have a curiously anachronistic ring (Plate 22). Even then, it seems, Scotland (Caledonia) was notable for red hair; and the suggested Germanic connection anticipates modern research on the specific genetic types associated with redheads in Scandinavia and Europe and my discussion on Neolithic Scandinavian genetic influence during the Neolithic in Chapter 5.[3] If heard today, such remarks about redheaded Scots, Mediterranean-complexioned Welsh and the Dutch sounding a bit like the English might well be dismissed as ethnic stereotyping, but I would take it seriously. Hearing the same remarks repeated from a time capsule two thousand years ago might suggest that whatever the intervening political upheavals and ethnic label-changing, some things like genes may not really have changed much at all.

Perhaps the most tantalizing opinion Tacitus records here is the last comment quoted in the extract above – that between Britain and Gaul 'the language differs but little'. This statement is somewhat opaque when taken out of context, since he did not specify which of the peoples and languages of Gaul he was referring to. But in the context of the second half of the extract, we can see that he is referring specifically to those tribes living in what we now call the south of England, and by the 'nearest Gauls' it is more than likely that he was referring to the Belgae, Continental Gauls who lived north of the Seine.

Caesar seems to confirm this interpretation in his *Gallic Wars* when he describes the Britons of south-east England whom he met on his second expeditionary invasion in 54 BC, as migrants. So for Wessex and West Sussex we read:

> The interior portion of Britain is inhabited by those of whom they say that it is handed down by tradition that they were born in the island itself: the maritime portion by those who had passed over from the country of the Belgae for the purpose of plunder and making war; almost all of whom are called by the names of those states from which being sprung they went thither, and having waged war, continued there and began to cultivate the lands.[4]

And for Kent (the Cantiaci tribe), the area later supposedly occupied by the Jutes:

> The most civilized of all these nations are they who inhabit Kent, which is entirely a maritime district, nor do they differ much from the Gallic customs. Most of the inland inhabitants do not sow corn, but live on milk and flesh, and are clad with skins.[5]

What languages did the Belgae speak?

So, both Tacitus and Caesar seem to be saying that the people of coastal south-east England spoke like and were culturally like the Belgae, and were both agriculturalists and pastoralists, while those of the interior were aboriginal in some way and practised pastoralism. From the distribution of languages elsewhere in the British Isles, it seems likely that the languages spoken by those 'aboriginal Britons' of south-east England to which Caesar refers were celtic. There is certainly evidence for celtic-derived place-names and personal names in some parts of south-east Roman Britain, especially north of London and the Thames, although they are thinner on the ground along the south coast of England than inland when compared with Caesar's Celtic part of Gaul (see p. 42, and Figures 2.1b and 7.2).[6] However, for the coastal Belgic colonies in England, the language type is not obvious. We have already noted that Caesar reserved the designation '*celtic* in their own language' for Gauls living *south* of the French rivers Seine and Marne, so it is not immediately clear what language (or group of languages) was being spoken by the Continental Belgae, who lived north of this boundary.

Personal names

As we shall see, one possibility is that the Belgae spoke Germanic languages, perhaps ancestral to Dutch or Frisian, which they carried to England even before the Roman invasion. Unfortunately there is little in the way of direct evidence or clear indications from Caesar or Tacitus to determine what languages were spoken in North Gaul or in the parts of south-east England occupied by the invading Belgae. While several personal and tribal names in Belgica described by Caesar have a clearly Gaulish derivation, a larger proportion do not, and some may have belonged to the Germanic branch of Indo-European.

One of the most important linguistic surveys of West European names is *Gaulish Personal Names* by David Ellis Evans, former Oxford University Professor of Celtic. Evans acknowledges that only one of the three main regional dialects of Gaul, 'which Caesar specifies … ("Belgic, Celtic, Aquitanian")' may correspond to the modern celtic-language division at all, and there may have been other ancient dialects in addition.[7] He prefaces this sober qualification with the gloomy comment that he has 'been particularly worried by one problem, namely that of identification of Celtic in Ancient Gaul, of deciding whether a particular proper name is Celtic or not'.[8] Taken out of context, these bleak statements might make one wonder if his study is of any value, so I should stress that his academic, self-effacing qualifications hide extraordinary scholarship and numerous rigorous proofs. His source material includes Caesar and other authors, and many ancient inscriptions. The evidence, particularly from ancient inscriptions, gets progressively more solid as he moves south towards the southern Roman province of Narbonensis. So, for the Belgic part of Gaul, if there was a single or main language, the message from personal names is that it cannot be assumed to have been celtic.

Caesar and Tacitus on the Belgae

Caesar tells us some interesting things about the 'Belgic' inhabitants of northern Gaul, although not much directly about the languages they spoke. For instance, at the beginning of his *Gallic Wars* he notes:

> Of these [the Gauls] the Belgae are the bravest ... and they are the nearest to the Germans, who dwell beyond the Rhine, with whom they are continually waging war ... The Belgae rise from the extreme frontier of Gaul, extend to the lower part of the river Rhine; and look toward the north and the rising sun.[9]

Caesar goes on to say that he has heard that the Belgae are forming a confederacy against the Romans, so he goes up to northern Gaul to investigate:

> As he [Caesar] arrived there unexpectedly and sooner than any one anticipated, the Remi, who are the nearest of the Belgae to [Celtic] Gaul, sent to him ... ambassadors: to tell him that they surrendered themselves ... to the protection and disposal of the Roman people: and that they had neither combined with the rest of the Belgae, nor entered into any confederacy ... and were prepared ... to obey his commands, to receive him into their towns, and to aid him with corn and other things; that all the rest of the Belgae were in arms; and that the Germans, who dwell on this [southern] side of the Rhine, had joined themselves to them; and that so great was the

infatuation of them all, that they could not restrain even the Suessiones, their own brethren and kinsmen ...[10]

The Remi then gave Caesar the low-down on the Belgic opposition:

> When Caesar inquired of them [the Remi] what states were in arms, how powerful they were, and what they could do, in war, he received the following information: *that the greater part of the Belgae were sprung, from the Germans* [my italics], and that having crossed the Rhine at an early period, they had settled there, on account of the fertility of the country, and had driven out the Gauls who inhabited those regions ... The Remi said, that they had known accurately every thing respecting their number, because being united to them by neighbourhood and by alliances, they had learned what number each state had in the general council of the Belgae promised for that war.[11]

The Remi told Caesar that the tribes of the Belgic confederacy included their brethren the Suessiones, and

> the Bellovaci [who] were the most powerful [and] promised 60,000 picked men ... the Nervii, ... as many; the Atrebates 15,000; the Ambiani, 10,000; the Morini, 25,000; the Menapii, 9,000; the Caleti, 10,000; the Velocasses and the Veromandui as many; the Aduatuci 19,000; that the Condrusi, the Eburones, the Caeraesi, the Paemani, who are called by the common name of Germans [had promised], they thought, to the number of 40,000.[12]

Caesar's description here of the Belgae as largely descended from Germans seems to have been forgotten; yet he confirms it again later in his book. The last four tribes he mentions in this passage he clearly identifies as German and as occupying Belgic territory to the west of the Rhine, an area rather larger than is now included in modern Germany west of the Rhine (Westphalia and the Rhineland)

(Figures 2.1a and 7.1a). Caesar also later identifies those tribes occupying the part of western Germany to the east of the Rhine (the Sugambri and Ubii), as solidly German. In other words, if we follow Caesar's use of the term 'Germans', their west-Rhineland territory roughly followed the modern north-west German borders down south as far as Koblenz, where the Moselle joins the Rhine.

This whole area west of the Lower Rhine, later to be called Germania Inferior by the Romans, included modern Luxembourg, the Ardennes and eastern Belgium, where a dialect of Low German is still spoken. These present-day Germanic-language areas previously constituted the north-eastern quarter of Caesar's Belgica. In contrast, to the south of this region, part of the lower Moselle was the former territory of the Treveri, who *do* seem to have had a couple of celtic-derived personal names and were centred around modern Triers (or Treves, after the Treveri) on the Moselle.[13]

South of Koblenz, the tribes on the west of the Rhine were also 'Belgic Germans'. Tacitus identifies a further three Belgic tribes as Germanic: 'Such as dwell upon the bank of the Rhine, the Vangiones, the Tribocians, and the Nemnetes, are without doubt all Germans' (see Figure 7.1a).[14] This block of the west upper Rhine, divided from Germania Inferior by the Treveri region, was later reclassified as part of the province of Germania Superior.

The direct evidence for German tribes in Belgica at the time Caesar was writing does not stop in the eastern part of Belgica. The Nervii, whose numerical contribution to the Belgic confederacy matched that of the Bellovaci, Caesar also later identifies as descended from the Cimbri and Teutones,[15] who invaded Belgica at a much earlier date, probably from the Danish (Cimbrian) Peninsula. These are both regarded as more likely Germanic- rather than celtic-speaking tribes, in spite of the superficial similarity between 'Cimbric' and 'Cumbric'.[16] Tacitus seems to confirm this (albeit in an ambivalent way) and even extends the label to the

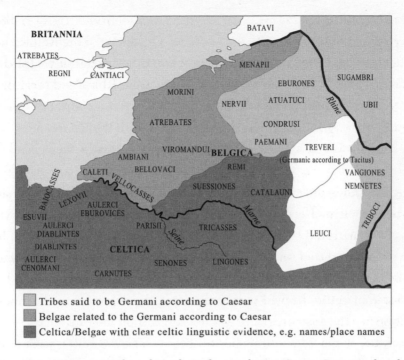

Figure 7.1a Germanic tribes of North Gaul, according to Caesar. Caesar implies that the areas in north-east Gaul shown white on this map were Celtic.

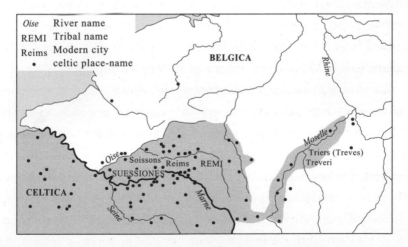

Figure 7.1b Evidence of celticity in North Gaul from place-names. This is consistent with Caesar, making most of Belgica non-celtic, with the exceptions of the central-eastern areas occupied by the Remi, Suessiones, Treveri and Catalauni. (Place-name terms: *briga* (mountain); *kondate* (joining of rivers); *nant* (valley); *Noviantum*; *vern* (alder).)

Treveri: 'The Treverians and Nervians aspire passionately to the reputation of being descended from the Germans'[17] (Figure 7.1a).

As for the other confederate tribes, further west in Belgica, there are suggestions for some celtic-derived personal names. For instance, the Belgic tribe the Atrebates had a twinned tribe in southern England with the same name, which means 'settlers' in celtic. However, the Atrebatean that Caesar made their leader, Commius, had a name which cannot be securely derived from celtic roots.[18] Commius would later become a thorn in Caesar's side when he turned-coat and led Gauls against Caesar in the famous battle of Alesia. It seems that he was the same 'Commios' who then moved to Britain, as yet unconquered, to become leader of the Atrebates there (Figure 2.3) and started a dynasty, even stamping his own name on minted coinage (see p. 284).

As for the rest of the Belgic tribes, although they have a few personal names with a celtic derivation, such names belong mostly to only three tribes: the Treveri, the Remi and their 'brethren' the Suessiones. Given Ellis Evans' ambivalence over the celtic nature of languages spoken by the Belgae, it is worth looking at other evidence.

Place-names in Belgic Gaul

A common last resort to get a handle on what languages people spoke in such unknown situations is to analyse ancient and surviving place-names linguistically. Place-names can preserve the identity of older languages as bits or even whole words even after those languages have otherwise been completely replaced. River-names are even more resistant to change and can tell of older linguistic affiliations. This persistence, however, can make it impossible to tell when such names were *given*. Furthermore, there are often doubts about the strength of the language attribution, or whether the derived language represents that of the bearer or informant. For instance, the eminent Celticist Patrick Sims-Williams of the University of Aberystwyth points out that 'Many Seans and Kevins do not speak Irish, while many Ryans and Kellys speak Welsh.'[19] He adds that 'Some care is needed in interpreting the presence or absence of such names, since place- and ethnic names can be taken over from foreigners, while foreign personal names can be adopted for reasons of prestige.'[20] Even tribal labels such as 'Belgae' and 'Atrebates', found in Latin texts referring both to northern Gaul and to south-east Britain, can have celtic derivation without necessarily defining the language they spoke (see above).

Given that these cautionary words can cut either way, it is worth looking carefully at a study of place-names in the former Belgica by German linguist Hans Kuhn, entitled (in translation) 'The People between the Celts and the Germans', in which he finds only a very limited and specific distribution of shared celtic names to the north-east of Paris (Figure 7.1b).[21] He uses eight different Celtic name-roots to map the former area of celtic-linguistic influence. Although Kuhn has other linguistic agenda, the presence of these names in Celtic Gaul, their geographical congruence on his map, and their absence from most of Belgica tends to provide support for their relevance in this discussion. These celtic place-name roots are found only in two small regions of southern Belgica, where they are associated with particular river systems. Nearest to Paris, we see a cluster of celtic names bounded by four tributaries of the Seine, going from west to east: the rivers Oise, Aisne, Ourcq and Marne. This area of celtic name-clusters coincides with the territories of the Celtic Remi and of their recalcitrant Celtic brethren the Suessiones, as can be seen from Caesar's description and, of course, confirmed by the modern city names Reims and Soissons. The other, much more limited cluster is found further north-east, on the River Moselle, and is associated with the territory of the Celtic Treveri and Treves, the modern city of Triers (Figures 7.1a and 7.1b).[22]

This localized 'celtic' place-name evidence in the Reims, Soissons and Triers regions is much more abundant than any found in south-east

England (see below). Kuhn's analysis is consistent with the impression given by Caesar's description of Belgica: that only a small minority of the peoples there were actually celtic-speaking, and that these comprised mainly the inland Treviri, Remi and Suessiones, with the territories of the Suessiones spreading directly into Celtica.

From Kuhn's map we can also see that the rest of Belgica, in particular the maritime regions, were devoid of celtic place-names. While the cluster of preserved celtic place-names spreads south of the Marne and the Seine into Celtica, it notably does not spread west of the Oise towards the Belgic coast (i.e. towards those parts nearest England, including Calais on the French side), where the emigrants to England were supposed to have come from. These 'non-celtic-place-named' areas were the same coastal territories occupied by Belgic hostile confederates such as the Caleti, the Bellovaci (near Beauvais), the Atrebates, the Ambiani (near Amiens), the Morini (near St-Omer), and the Menapii and Nervii (both in Belgium) (Figure 7.1a).

So what was the language spoken in the part of the rest of Belgica that wasn't actually occupied, in Caesar's words, by 'Germans'? Early Germanic place-name roots are found throughout a large part of northern Belgica south of the Rhine.[23] Kuhn is more concerned with a narrow 150 km wide strip north of the Seine, for the place-names of which neither celtic nor Germanic derivations are possible, only 'pre-Germanic'. This area is what he calls 'The People between the Celts and the Germans'. The language imported by the Belgae and mentioned by Caesar and Tacitus could have belonged to either Kuhn's 'early Germanic' or 'pre-Germanic' group, without necessarily being identical to that spoken by the tribes the Romans called the Germani.

Evidence for non-celtic names in Roman England

Any language locally imported into Britain by the Belgae before Caesar's first landing in 55 BC had an odds-on chance of being non-celtic, and about an evens chance of having a Germanic origin.

So, we need to look more closely at the evidence on the spot in south-east England. Unfortunately, use of Latin as the military and elite language tends to obliterate the linguistic evidence from inscriptions in Roman England, and neither confirms a universally celtic-speaking Ancient England, nor reveals any clear trace of a Germanic import in personal or place-names, except perhaps from Germanic-speaking legionaries imported by the Romans.

The study of place-names in Roman Britain started with George Buchanan in the sixteenth century, and over the last two centuries has grown into a minor industry supporting several professorial chairs in the British Isles. Three standard works bring together much of the best scholarship invested in this subject in the last fifty years: *Language and History in Early Britain* by Kenneth Jackson of Edinburgh University, *Gaulish Personal Names* by David Ellis Evans and *The Place-names of Roman Britain* by Leo Rivet and Colin Smith. A magnificent new work of erudition by Patrick Sims-Williams published just after I delivered my manuscript draws all this celtic scholarship together in a comparative analysis and composite map covering all Europe and Anatolia (summarized in Figure 2.1b).[24] It is clear from these works that there were celtic-derived place-names and personal names in England during Roman times.

While there is no doubting the quality of place-name evidence for the presence of the celtic language family in Roman England, there are questions of how much was celtic and specifically where, which require systematic comparison with other regions of the British Isles subjected to the same scholarly analysis. The main questions one should ask of such place-name surveys are how much celtic was *spoken* in Roman England, and which dialects and what other languages were spoken at the same time, or had been in the past. This might seem like setting impossible goals, but opinions vary so much among linguists, some saying that nearly everyone in Roman England spoke Latin, and others that nearly every indigenous inhabitant primarily spoke celtic.

It is because of this controversy that these questions need to be asked, if only to gauge the strength of evidence as to whether English replaced celtic or Latin, or whether some Germanic language was already present in eastern Britain and later evolved into Old English. Old English inherited less than a couple of dozen celtic words, and those were mainly from celtic-speaking areas such as Cornwall and Cumbria, whereas the Latin bequest was more like two hundred.[25] In comparison with the huge French component brought into English by the Normans, even the early Latin borrowings are few and the celtic borrowings notable for their near absence. To me, the lack of Latin or celtic borrowing in Old English points to one of two extremes. One is the traditional wipeout theory; the other is the possibility that there was a Germanic language already being spoken in the areas invaded by Anglo-Saxons, which more naturally hybridized with 'Anglo-Saxon'.

Mining Ptolemy's Geography *for celtic place-name frequency*

Is it possible to make valid quantitative comparisons between the classical place-names of Roman England and contemporary ones in other European regions with celtic-speaking populations? It is, and there is one way of doing it – by using a comprehensive place-name survey performed throughout Europe by the same survey team during ancient times. The Greek geographer-astronomer Claudius Ptolemaeus (Ptolemy), who lived in the Egyptian city of Alexandria in the second century AD, carried out just such a survey during the later part of the Roman occupation of Britain. His *Geographia*, in seven volumes, became the standard text on the subject until the fifteenth century. Two chapters in volume 2 deal with the British Isles. Chapter 1, 'Hibernia island of Britannia', is concerned with Ireland and the smaller isles; Chapter 2, 'Albion island of Britannia', covers mainland Britain (England, Wales and Scotland). Ptolemy used a comprehensible grid reference system analogous to longitude and latitude, and gave lists of the prominent coastal landmarks, rivers and estuaries, as well as the names of the British tribes and main towns. While the ancient names given by Ptolemy in his map are far fewer than place-names gleaned by celticists from scouring modern maps or the classical texts, they have the two great advantages that the process of survey he used was consistent throughout Europe and it was contemporary with the Roman occupation of Britain.

A publication based on an international workshop organized and edited by Patrick Sims-Williams of the University of Aberystwyth and David Parsons of the University of Nottingham has laid out Ptolemy's place-names region by region with standardized linguistic derivations.[26] Parsons wrote the chapter on England, and moves through the data systematically, explaining his exclusions and inclusions. Ptolemy mentioned 88 names relating to England in his chapter on Britain. Of these 37, were place-names, 18 names of rivers, 16 of tribes and a further 17 of coastal features. After removing 7 duplications and the 16 tribal names, Parsons is left with 65. Of those, 13 (20%) are Greek or Latin and 22 (34%) are clearly British celtic; a further 14 (22%) are 'very difficult', 8 (12%) 'uncertainly' British celtic and another 8 (12%) '?Pre-Celtic Indo-European'.[27]

So, only a third of Ptolemy's English place-names are British, and the concept that there could have been other Indo-European languages in Roman England is supported. Now, while Parsons is relieved that a third of Ptolemy's place-names from England are celtic-derived, he acknowledges the large gap that needs explaining:

> It is gratifying that the names explicable as Celtic make up the largest single class, since there is a generally held assumption that the language of all the population of present-day England is likely to have been British [i.e. Brythonic-celtic speaking] at the time of the Roman conquest. It is striking, however, that nearly half of the names are nonetheless more or less opaque, at least to the modern

outlive the language that created them. This fact can be used both ways in dating arguments. Archaeologists are now good at dating material culture, but artefacts (pots, tools, jewels, weapons, etc.), their most abundant form of evidence, are usually silent on language. One kind of artefact can, however, give us useful direct clues about language in the past. That is the written word, inscribed on stones, ornaments and coinage, and transcribed from original texts.

As we have seen, the distribution and diversity of Continental celtic languages towards the south of Europe can be clarified archaeologically from inscriptions in Etruscan, Latin and other scripts, some going back to the sixth century BC. Arguments against this inference – that the southern distribution of inscriptions merely reflected these alphabets' spread from the south – cannot explain why Continental celtic inscriptions continued to remain restricted to the south of Europe even after the spread of the Roman Empire and writing.

Can we use this approach to determine and date the presence, absence and spread of insular celtic in the British Isles before and after the Roman occupation? Well we can, but only up to a point. That point is determined by the spread of writing, but also by who actually commissioned the inscriptions. Roman script seems to have been the first to enter the British Isles, but it did so long before the Roman occupation, and even at the time of Caesar's sortie in 56 BC it was present in the form of coins struck with the names of local chieftains which are most prevalent in southern England, along the Thames Valley. The earliest of these coins were struck by rulers of the first century BC, such as Tincomarus of the Atrebates region and Commios, who may have been the same 'Commius' who rebelled against Caesar in the Gallic War. The problem with this sort of information is that the text on the coinage was limited and mainly in the form of names, which does not add much to personal name study from ordinary written texts.

As mentioned in Chapter 2, perhaps the most interesting aspect of early British coins is not where they were made but where they were *not* made. The latter areas are those for which there is the most evidence for celtic-language use and in the form of inscriptions on stone, and include southern Britain, Wales, Cornwall and the West Country. So the distribution of early coins in Roman and pre-Roman Britain is almost an exact negative image of the distribution of post-Roman celtic inscriptions (Figure 2.2).

Apart from the general point that the main influences on early coinage in Britain came from Belgic rather than Caesar's 'celtic-speaking Gaul', there is a more specific clue to how close, culturally and politically, the people of southern England were to the Belgae. The clue lies in the British distribution of Gallo-Belgic coins dating to the period of Caesar's all-out war against the Belgae.

Gallo-Belgic coins are found in Britain in more or less the same areas as were affected by the early 'Saxon invasions' (before c. AD 500). Notable concentrations occur in Kent, along the Sussex Coast and either side of the Thames. The density of Gallo-Belgic coin finds falls off sharply in areas occupied by Angles in the fifth century AD, in Norfolk, Suffolk and north of the Ouse. Gaps in the South-east include a band north of the Sussex coast stretching westwards from the Weald Forest (Figure 7.3a).

Gallo-Belgic coins found in Britain are divided into six classes, denoted by the letters A–F. Archaeologists generally agree that, with the exception of class B and probably A, which arrived earlier, there was a significant influx of Gallo-Belgic coinage around the time of Caesar's all-out war on the Belgae. Specifically, Gallo-Belgic E (and probably C) coins of the Ambiani may have flooded in during the years 56–55 BC prior to Caesar's attack on North Gaul in order to help pay for the Belgic insurgency.[30] Cunliffe suggested that if all Gallo-Belgic coins except B arrived during the Gallic War, 'the gross distribution of the Gallo-Belgic coinage should reflect in general terms the extent of the territory over which the war had its effects'.[31]

The geographical distribution of the time-focused Gallo-Belgic E coins (and, to a lesser extent, of C coins) is nonetheless broad and

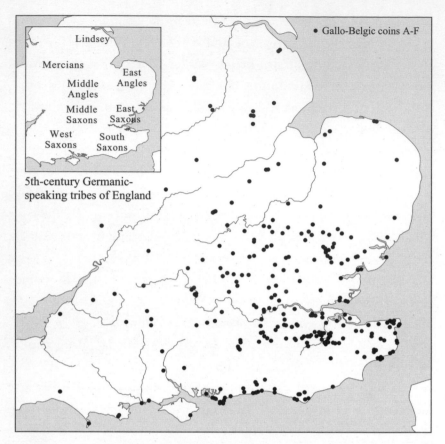

Figure 7.3a   Gallo-Belgic coin distribution in Britain (all types, A–F). This distribution
thus excludes Wales, the West Country, Cornwall and the North, and is coincident with the
territories occupied in the fifth century by 'Saxons' and, to a lesser extent, Angles (inset).

matches the overall British distribution of Gallo-Belgic coinage
during the Roman period (Figure 7.3b). This suggests to me that
the relationship between Belgic Gaul and what was later mainly
Saxon England was more than just between treaty allies, and more
as Caesar suggested – one of cultural continuity and common
concern. If both the Belgae and the British south-east were celtic-
speaking, that closeness might be expected, but if the Belgae were
not celtic-speaking or, as Caesar suggests, were largely descended

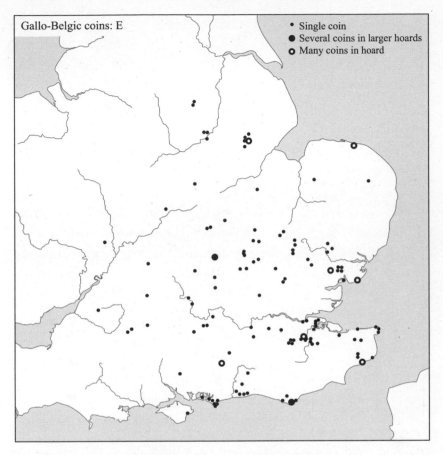

Figure 7.3b Map of Gallo-Belgic type E coins, which appeared in southern England in 55 BC in payment for English support against Caesar's Belgic campaign.

from Germans, then it would make less sense – unless some tribes of south-eastern England also spoke Germanic languages.

Another conclusion may be drawn from the wealth of indigenous name-stamped coins throughout the Roman occupation of England, namely that powerful Britons in England were able to make strong statements about their local status and identity in Roman script, but without feeling the necessity to Romanize their names or any other written content. This implies they were not so Romanized that they had lost their own sense of cultural identity. One should therefore not interpret the near-total absence of celtic inscriptions of any other kind in England during and after the Roman occupation, as evidence that Britons had lost their indigenous languages.

Curiously, the strongest evidence for the presence of celtic language among the middle classes in England is almost anecdotal and comes from a comic metallic source in the West Country, at the edge of the area where coins are found:

> In the second to fourth centuries the metal curse-tablets from Roman Bath tell a … story. Typically, they were deposited by angry people who had lost their clothes while bathing because they were too poor to pay a slave to guard them. On them, we find a tremendous number of British Celtic names [and text]: about 80, compared to 70 Roman names.[32]

… *And in stone*

As far as the text content is concerned, as opposed to just celtic names, all stone inscriptions in south-east Britain during the Roman period are in Latin; significantly, *none* are written in the celtic vernacular. The names are mostly Latin as well, in stark contrast to the names on coins. Patrick Sims-Williams has some interesting observations on this phenomenon and points out that the language of names varies with the medium on which they are recorded. On stone inscriptions in Roman Britain, he says:

> we tend to see monotonous Latin names … Celtic-looking names are few and far between, especially those distinctively Celtic compound names of the sort which Julius Caesar encountered in Britain (*Cassi-vellaunus*, *Cingeto-rix*, and so on). The number of such compound names in Romano-British stone inscriptions is almost the lowest in the Latin-speaking Empire …[33]

The last sentence in this statement seems to raise doubts about the strength of the celtic influence in England. Alternatively, Sims-Williams explains that the dominance of Latin in stone inscriptions may reflect a bias towards the military, since we do find celtic names in signatures of artisans such as potters, tile-makers and metal-smiths.[34]

When the Romans withdrew from Britain, something very strange happened in south-east England: virtually no inscribed stones were set up at all, either in the immediate post-Roman period or afterwards. One stone with an Ogham (an Irish script imported into Britain during this period) celtic inscription at Silchester is likely from the late Roman period; one in Wareham is in Roman script, probably celtic, from the sixth or seventh century. The relative difference in numbers of inscriptions could be the relative paucity of good stone in the south-east, unlike the West Country, but this would hardly produce the near-complete void in England (Figure 7.4). By the time inscriptions started up again, much later, they were in Anglo-Saxon or Medieval Latin, not celtic.

This sudden down-tools might have been a consequence of soldiers leaving, since the inscribed stones are argued by some to have been set up by the Roman military. However, many of the inscriptions are civil. Even if the military explanation applies, it is not as simple as that, because inscribed stones continued to be set up in previously Roman-occupied parts of Wales, Cornwall, Cumbria and Scotland,[35] for at least the next seven hundred years, and *their* Roman soldiers had presumably left too. Furthermore, taking Wales as representative of these regions of celtic post-Roman Britain, these inscriptions initially included Latin names, but over the next couple of centuries a diversity of celtic names came overwhelmingly to predominate, with Latin names either disappearing or becoming celticized.[36]

But there is clearly a problem of logic here. If there are plenty of Latin *and* celtic inscriptions from after the Roman collapse in all *other* Romanized regions of the British Isles, why are there no post-Roman

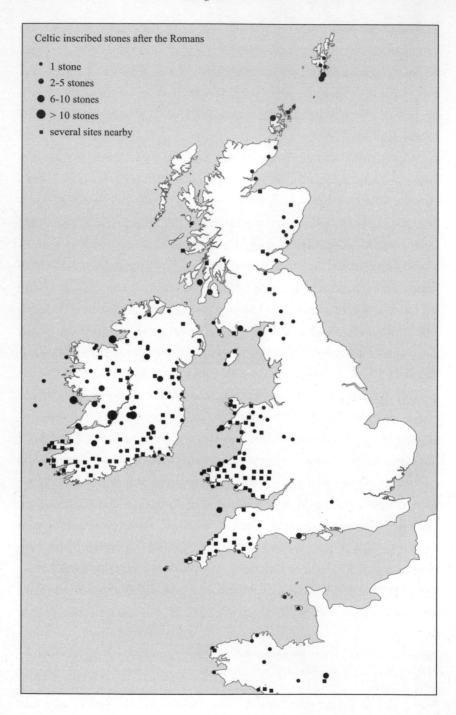

Celtic inscribed stones after the Romans

- 1 stone
- 2-5 stones
- 6-10 stones
- > 10 stones
- several sites nearby

inscriptions, Latin or celtic, to speak of in England until the first runic (i.e. non-celtic) stones, which appear later? Surely the Anglo-Saxon invasions were not a complete and instantaneous blitzkrieg. Even the most avid supporters of the Anglo-Saxon wipeout theory accept that it could have been possible only over several hundred years. If celtic languages were spoken universally in England throughout the Roman occupation, surely they would have initially persisted, as in Cornwall and Wales, for several hundred years before succumbing to Anglo-Saxon?

This seemed such an extraordinary 'English' anomaly to me that I searched the literature for explanations. Not surprisingly, ancient stone inscriptions are an intensively researched area, and British scholarship has much to say on it.[37] There is a very accessible website set up by the Department of History and the Institute of Archaeology of University College London, which has recently completed a major project to catalogue celtic inscriptions on stone in the British Isles and Brittany, known as the Celtic Inscribed Stones Project, or CISP.[38]

This project catalogues over 1,200 celtic inscriptions made between the years AD 400 and 1100 in the celtic-speaking regions of the Early Middle Ages: Scotland, Ireland, Wales, Brittany, the Isle of Man and parts of western England. Included are all stone monuments inscribed with text, whether in the celtic vernacular or in Latin, in the Roman alphabet or Ogham. Excluded are runic inscriptions, which are Germanic and appear in England only after a brief brief hiatus in the early fifth century (see Chapter 9). As can be seen from the map drawn up by this project (Figure 7.4), with the exception of the West Country, the lowland and most densely populated areas of England are dramatically notable by the absence of celtic inscriptions.

Figure 7.4   Distribution of over 1,200 celtic inscriptions on stone in the British Isles and Brittany (AD 400–1100). During the Roman period, all inscriptions in south-east Britain were in Latin, none in celtic. By the time they started up again in England after the Roman exit, they were in Medieval Latin or Anglo-Saxon. This leaves a near-total absence of celtic inscriptions in England and also large parts of Scotland, *at any time*.

Not only do we see a void of celtic inscriptions in England, but
there is also an almost total absence of any relict evidence for celtic
languages in Old or Modern English. The same applies to the bulk of
regional dialects from most parts of England, except in Cornwall
and Cumbria. One would normally expect to find some incorpo-
rated remnants of a language that has been replaced, a so-called
linguistic substratum. There are, however, only four attested celtic
words to be found in English (or perhaps a couple of dozen if local
dialect terms are included, such as *tor* and *pen*, both meaning 'hill',
and *crag*).[39] These numbers do not add up to a significant substratum.
In other words, this lack of evidence is not consistent with celtic
language having any original role in the evolution of English.

One explanation of the dramatic difference in numbers of celtic
inscriptions between England and the rest of the British Isles is that
by the time Romans left, the inhabitants of England no longer spoke
celtic, having changed over to some version of Latin, perhaps like
the French. This rationalization is becoming increasingly popular,
although it is still a controversial minority view. The evidence for it
includes British Latin loanwords in Anglo-Saxon and a small
number of Latin place-names adopted by the Anglo-Saxons. The
main problem with this explanation is that one would expect to see,
in England as in Wales, Latin inscriptions continuing after the
Roman withdrawal – which is not the case. Also, the low number of
borrowings needs some other explanation. Although there are
considerably more Latin than celtic loanwords in early English, the
figure of two-hundred-odd is still very small when compared with
the massive effect of the Norman invasion on English.

Given that neither Latin nor celtic words intruded much into
Old English, there is the possibility that a third, pre-existing and
more closely related (i.e. Germanic) language survived in England
during Roman times, one which could hybridize more easily with
the incoming Germanic influences from the Continent. I shall look
at the evidence for this in the next chapter.

Was the First English nearer Norse or Low Saxon?

English as West Germanic

Imagine that English was spoken during Roman times. What would it sound like, and which languages on the Continent would it affiliate to? Nearly all linguists would view this is a silly question. The orthodoxy sees 'Old English' not as indigenous to England, but as the direct descendant of a Continental Germanic language, 'Anglo-Saxon,' introduced for the first time into England by Germanic-speaking Anglo-Saxon invaders in the fifth century AD. As such, English is classified as a West Germanic language. This group includes modern language groups such as German, Frisian and Dutch, as distinct from North Germanic (Scandinavian) languages, Swedish, Danish, Norwegian, Icelandic and Faroese, and East Germanic languages, represented best by the now extinct Gothic, known from Biblical translations, but including also Vandalic, Burgundian and Lombardic (Figure 8.1a).

According to this classification, two thousand years ago, or possibly earlier, the West Germanic group underwent a distinct split, producing a new branch called High German, a group now

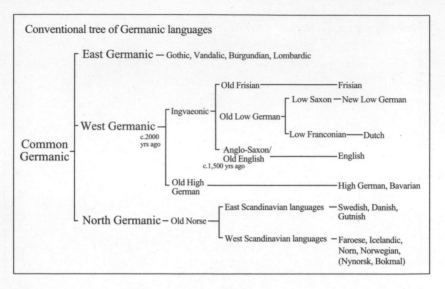

Figure 8.1a   Conventional Germanic language tree. In the conventional reconstruction, based on sound changes, Old English arose around 1,500 years ago from an 'Old German' continental version of 'Anglo-Saxon'. The latter groups with the main West Germanic branch of Germanic languages and sub-groups with the 'Low Germanic' variants now found on the North Sea coast. The closest of these to modern English, on this basis, is Frisian.

spoken farther east by the majority of Germans, Austrians and Swiss. The change happened in the ancestor of Old High German, which underwent a systematic sound-shift differentiating it from the rest of the group, namely Low German (or Low Saxon), Dutch-Flemish (and its ancestor Low Franconian) and Frisian, which preserved the relevant ancestral West Germanic sound values.

A second overlapping set of geographically based differences occur in West Germanic languages, the so-called coastal features, which can be shown to exist between Old English, Old Frisian and Old Saxon as a group on the one hand, and Old High German on the other. These coastal features are also sometimes used to invoke a language group variously described as North Sea Germanic or Ingvaeonic. The latter term, in spite of the misspelling, refers back to a classification by the first-century AD authors Pliny and Tacitus,

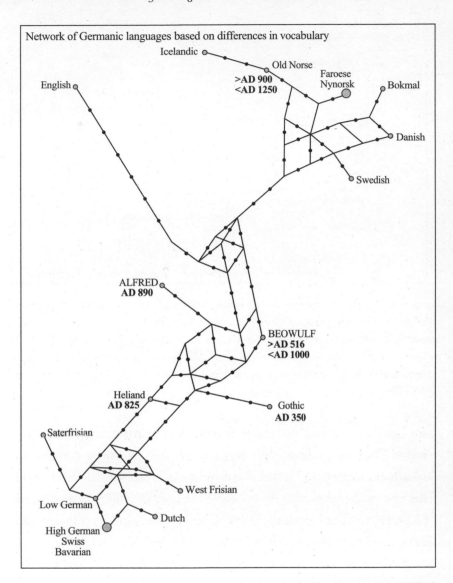

Figure 8.1b Germanic vocabulary network suggests English as a fourth Germanic branch. Based on vocabulary similarity, Forster's network groups all Continental West Germanic languages close to each other and relatively near to Old German (Heliand poem). However, Old English (*Beowulf*/Æfred) was already diverse, and as far from Old German as the latter is from Frisian — even beyond the Gothic branch. English appears to form a fourth branch splitting off closer to Scandinavian languages than the others.

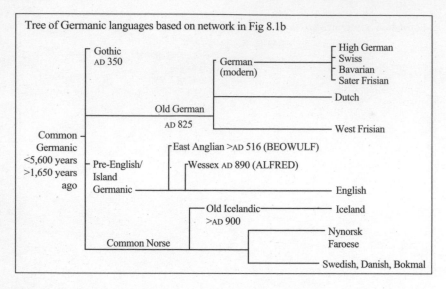

Figure 8.1c   Germanic tree four-branch reconstruction based on Figure 8.1b. When the Germanic network is viewed as a tree, there are four branches rather than the conventional three, with a new pre-English 'Island Germanic'. Forster argues that the break-up of these four can be no younger than the Gothic Bible (AD 350), and thus older than the 'Anglo-Saxon invasion', and possibly as much as 4,000 years old.

who divided the West Germans into three groups: 'the Ingaevones, dwelling next the ocean; the Herminones, in the middle country; and all the rest, Instaevones'.[1] 'Ingaevones' referred to the tribes who lived along the coast from North Gaul to Denmark.

The North Sea Germanic group contains several branches. Low German (known as Low Saxon in the Netherlands) is spoken on the North German Plain in Germany and the Netherlands. Frankish is now extinct, but was the language formerly spoken in northern or Belgic Gaul and the Low Countries, and is thought to have spread within the past two thousand years. Low Franconian, an approximate

Figure 8.1d   Early historical locations of Germanic languages with the 'Anglo-Saxon homeland' enlarged. Ptolemy located the Saxones at the base of the Cimbrian Peninsula in AD 150 and confirmed older classical sources of the location of 'Old Saxony' far to the north-east of Lower Saxony.

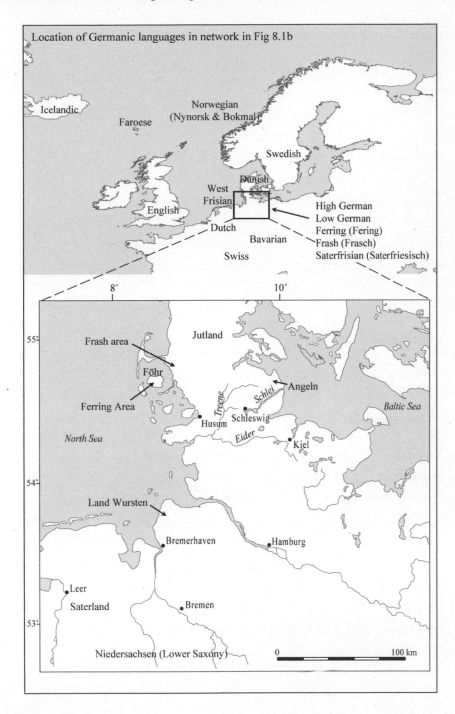

Location of Germanic languages in network in Fig 8.1b

ancestor of Dutch-Flemish, was closely related to Frankish. Depending on exactly when Germanic languages first arrived in the Netherlands, there may even have been another Germanic language spoken by those Belgae with whom Caesar had so much trouble, and whom Hans Kuhn calls 'The People between the Celts and the Germans' (see p. 276). Frisian is a distinct contemporary group of languages spoken in the northern Netherlands and Germany. When the method of comparing systematic language changes is applied, Frisian, of all the Germanic languages – rather than Low Saxon – is found to be most closely related to English. This is surprising both from the point of view of the 'Anglo-Saxon invasion' and, as we shall see, the genetic evidence. The same structural proximity applies to Old Frisian and Old English (Figure 8.1a).[2]

There are problems with these seemingly neat divisions into High and Low German languages and the simplistic view that Low German spawned English quite recently. Apart from the point that Old English should, on the evidence of Gildas and Bede, be more closely related to Low Saxon and/or Anglian than to Frisian, English is actually surprisingly dissimilar from all its neighbours, whatever its elevation above the sea (high or low) or its structural position in the West Germanic language tree. This is the case whether one compares ancient with ancient or modern with modern.

Old English and Old Frisian both changed their treatment of vowels compared with other Low German languages such as Old Saxon. But by the time the first texts appeared in Old English, only a couple of hundred years after the 'Anglo-Saxon' invasion, there had apparently been further vowel shifts (changes), resulting in a dramatic reduction in the number of distinct endings (inflections) that could be added to words when compared with Germanic languages on the other side of the North Sea. Not only that, but the first texts in 'Old English' also differed markedly from one another, according to region, for there were three distinct dialects: Anglian, West Saxon and Kentish.

The reason for this apparently rapid divergence into dialects might be the different languages spoken by the founding groups. But the languages of the three putative Low German invaders, the Angles, Saxons and Jutes, were supposed to be very similar, if not the same, at the time of invasion. In any case, this still does not explain the rapid divergence of King Ælfred's West Saxon or of Beowulf's Anglian from contemporary Continental Low Saxon texts such as the ninth-century Old Low German epic poem *Heliand*.[3] An alternative explanation might be local hybridization with celtic or Latin, but as has already been pointed out, there is scant evidence for that.

The greatest differences between Old English and other North Sea Germanic languages, however, are in their vocabularies rather than just their structural ancestry. Even Old Frisian, the 'closest relative' of Old English, shares five times more uniquely derived forms of words with German than with Old English.[4]

If we now compare modern descendants of Germanic languages for shared cognate words, as the Hawaiian linguist Isidore Dyen did, English again turns out to be the odd one out. English branches off deeply compared with both the surviving West and North German (i.e. including Scandinavian) language branches.[5] This ancestral position of English with respect to its putative parent cannot be explained simply by the intrusion of a substantial minority of French words following the Norman invasion.

North Germanic words in Old English

New Zealanders Russell Gray and Quentin Atkinson analysed the same data that Dyen had created, and confirmed the deep rooting of English in Germanic languages (Figure 6.2a). English linguistic-genetic researchers April and Robert McMahon have also addressed this question by re-analysing Dyen's data using a number of different tree-building methods which again consistently put English as an early split.[6] They came up with a fascinating possible explanation –

that there was significant under-detected borrowing from North Germanic (i.e. Scandinavian languages) at an early stage. Obvious borrowing of words such as the French loans can be adjusted for in analysis of relationships; but less obvious loanwords, such as those occurring between the main Germanic groups, can cause problems in tree-building. In their words,

> for English, although the trees are less informative, we know that there is considerable early borrowing from North Germanic ... (Embleton ... finds sixteen word loans in the Swadesh 200-word list from North Germanic to English). English does not quite move into North Germanic altogether ...[7]

What they seem to be saying is that, while English is structurally most similar to its neighbour Frisian, its vocabulary was more strongly influenced at an 'early' stage by Scandinavian languages. This effect was already apparent in Old English before the Vikings arrived on the scene to give the Anglo-Saxons some of their own invasion medicine. There are some pieces of textual evidence which are consistent with this idea and which I shall describe shortly, but the early Scandinavian loanwords are attested and stand by themselves.

The main problem this evidence raises for the Anglo-Saxon orthodoxy is that there is no specific record of Scandinavian invasions between the first arrival,[8] in the fifth century, of the Saxons, who were supposed to have introduced Old English as Anglo-Saxon, and the much later Viking raids, before which time the Old English texts were already showing strong Scandinavian influence on vocabulary. So, the Scandinavian influence on Old English should have been earlier than the Saxon invasions. But in that case there would have been no Anglo-Saxon to influence (and Latin and celtic had insufficient effect on Old English to matter). So, what other language was around during the late Roman occupation to be influenced by Norse? Could it be that other Germanic languages, or even

Norse, were already in residence in England before the fifth century AD? I shall come back to this possibility later.

Old English as Norse-influenced?

An alternative explanation is that the language of Gildas' invading 'Saxons' may not have been quite as 'Saxon' as he claimed. Were they, for instance, actually speaking the same language known from surviving Old Low German Continental texts such as the *Heliand* poem?[9] Could the 'Saxon-invaders' have already been influenced by Norse, or even be speaking it – or something else? Perhaps the oldest *possible* example of Norse words entering England (rather than Old English) comes from a single Germanic word Gildas himself interpolated when writing in Latin in the mid-sixth century AD. This is also the first extant written clue to the invaders' language. The context is highly relevant since he applied it to the 'Saxon' invaders and their language in his famous line:

> A pack of cubs burst forth from the lair of the barbarian lioness, coming in three 'keels' (*cyulas*), as they call war-ships in their language ...,[10]

That these were high-prowed, sailing longboats, like those of the later Vikings, is explicit in the Latin of this and the next sentence. Forster argues from the *Oxford English Dictionary* (*OED*) etymology for 'keel' (*cyulas*) that 'it is not clear that a Scandinavian origin for this word can be ruled out'. The relevant entry, among several in *OED* for 'keel', cites Old English *céol* (found in *Beowulf*) and Old Norse *kjl*.[11] From my reading of the *OED* entry it could be either North Germanic or Old Low Saxon.[12] The Norse–English ship connection, however, pops up again in a rare Old English synonym for ship, *fær* (derived from the Norse word *faer*, meaning 'journey' – as in 'seafarer'), which is found in the sense of 'ship' only in two very early Old English poetic texts (*Beowulf* and Genesis A) and in

Old Icelandic/Old Norse. Dennis Cronan, an expert on Old English poetry, points out that 'Since Old Icelandic fær exhibits a similar semantic range, the meaning "ship, vessel" was probably inherited from Common Germanic' (i.e. from the common ancestor of all Germanic languages).[13]

The geneticist Peter Forster and colleagues have taken this speculation on Norse vocabulary's influence on Old English rather further than many linguists would like. To do this, he has adapted the established genetic network methods I have used throughout this book and applied them to data on vocabulary (Figure 8.1b). He explains:

> Firstly, the network diagram which our method generates (Bandelt et al. 1999) visualizes the changes of individual word use along the language branches, rather than providing only an abstract branch length or difference measure. This incidentally means that at its best, the network approach can reconstruct the lexicon, i.e. the specified word list, of a language which no longer exists. Moreover, our network method makes allowance for borrowing and convergent evolution by not forcing the data into a tree.[14]

In other words, by comparing the origins of synonyms for common words, Forster is charting the ancestry of the use of individual words in different Germanic languages and how they got to their present locations. This helps to separate the history of Old English and later English vocabulary from the more conventional comparison of systematic structural and sound changes. By using a network without specifying a tree-root, Forster avoids forcing English, ancient or modern, to associate with any particular ancestor, such as Frisian. This is important, because English is more profoundly affected by word borrowings than any other Germanic language, thus violating the assumptions of the simple tree-model of language evolution used to make conventional Germanic language trees (as in Figure 8.1a). Forster goes on:

Secondly, we have added four extinct lexical [vocabulary] sources from the first millennium AD (*Beowulf*, Ælfred [three Latin texts personally translated by King Ælfred], *Heliand*, and the Gothic Bible) into the analysis along with modern Germanic languages to investigate whether their inclusion would change the deep branching of English reported by Dyen *et al*. (1992), Gray and Atkinson (2003) and McMahon and McMahon (2003).[15]

The results are revealing, but to see why, first we have to orient ourselves. To help comparison with the conventional Germanic tree, I have simulated an equivalent tree from Forster's network with his help (Figure 8.1c). The network (Figure 8.1b) groups itself largely into two major poles, corresponding exclusively to North Germanic (i.e. Scandinavian) languages in the top right-hand corner, and exclusively to Continental West Germanic languages in the bottom left-hand corner, while some of the older extinct languages are scattered in between. Gothic, in an intermediate position, represents the extinct third or East Germanic branch.

To an extent this broad pattern of lexical divergence, or changes in actual word values, between the two surviving major branches of Germanic languages on the periphery of the network, with the more archaic ones rooting in the middle, is consistent with the conventional evolutionary linguistic tree shown in Figure 8.1a. But, as may be apparent already, there the similarities between the two depictions stop. For instance, English forms a fourth deep branch of its own. This is consistent with, and would be predicted from, the studies on Dyen's cognate set cited above, which all show English rooting deeply. In contrast, the other modern West Germanic languages on the Continent form a tightly knit group at the bottom of the network (Figure 8.1b) with a very similar choice of words in their vocabularies, while the High–Low German split and other structural differences apparently count for little. This separation of the English branch on the other side of the network from its theoretical colleagues would not be expected at all from the Anglo-Saxon orthodoxy and cannot simply be explained by later borrowings from Vikings and Normans and new words introduced by Latin scholars.

It may help to add a little perspective to the non-English part of the network. Based on the conventional tree of Germanic languages shown in Figure 8.1a, the centre or root of the network should be at the three-way division between North, West and East Germanic. On the network this is logically somewhere between Gothic and *Heliand* and the Old English roots. From the perspective of the languages in the modern West German pole, this root makes sense because the language of the *Heliand* poem (AD 825) was Old Low German, and should be very close to their theoretical common ancestor. From the perspective of this inferred root, however, the vocabulary of one contemporary Old English text, *Beowulf*, ends up leaning a little towards the Scandinavian pole.

If Old English is so diverse, maybe it is even older

While this might explain why the tale of *Beowulf* is set entirely in a Scandinavian context (see later), it does not explain why two 'dialects' of Old English, *Beowulf* (Anglian) and Ælfred (Wessex) were already diverging so much from each other in England so soon after the invasion. Not only that but, within several centuries of the Anglo-Saxon invasion they are already each further from their supposedly close relative in *Heliand* than the latter is to Gothic (Figure 8.1b). As Forster says, 'The lexical diversity which formerly existed in Britain [more than a thousand years ago] was comparable to the present diversity within mainland Scandinavia, or within the present Dutch/German language area.' This complex picture is hardly consistent with the aftermath of a sudden Dark Age founding event by languages all closely related to Old Low Saxon. It is not as if they had picked up much in the way of celtic or Latin words on arrival – a known reason for rapid change. An alternative, deeper timescale and relationship needs to be sought in Old English dialects.

Or a fourth Germanic branch?

The position of modern English on the network is several word-pegs closer to Scandinavian than to *Beowulf* but it also forms a deep branch of its own, taking off from a different part of the network slightly nearer to the Scandinavian than to the Old Saxon part (Figure 8.1b). In this context, Forster suggests that 'the Scandinavian relationship may date back at least as far as *Beowulf* if the word for "sleep" is taken as an indicator.' As far as the marked divergence of the English branch is concerned, much of this happened surprisingly early. Again, Forster claims:

> Thus, in a time span of around 1,200 years, English seems to have changed about ten words within the shortened 56-word list used for the network ... Moreover, according to the *Oxford English Dictionary* (1989), which documents the earliest appearances for each of these words, all 10 changes originated before AD 1300. This compares with only five changes in the 1,200 years between Heliand (approximately AD 825) and modern Low German, and two changes in around 1,000 years between Icelandic and 'Old Norse'.

Given that the roots of English vocabulary appear, like the two 'Old English' texts in the network, to have diverged so early, it is not clear whether either of the latter are directly ancestral to modern English. This might mean that the connection between Old English and the ancestral Common Germanic root predated the arrival of the Angles and Saxons. English would then be neither directly descended from Old Saxon nor from Scandinavian (i.e. Old Norse). In Forster's words:

> The network analysis reveals a Scandinavian influence on English and apparently a pre-Scandinavian archaic component in Old English. All Germanic lexica spoken today appear to converge in the network on an ancestral Common Germanic lexicon spoken at an unknown time, but constrained to before AD 350 and probably after

3600 BC. The results presented in this study are based on only fifty-six words and must be considered preliminary pending more detailed networks using longer word lists.[16]

In other words, Old English, rather than just borrowing a few Scandinavian words at the start, may have been around for some time before the Anglo-Saxon invasion *and* these effects are reflected in modern English. I have included Forster's note of caution in the quote, because this is a very short word list from which to draw such profound historical conclusions. I understand that he and his colleagues are in the process of extending the word list to 200, which might strengthen their conclusions. Their date range is also hedged by extreme academic qualification, but starts to look as if it could include some of the period of the Later Neolithic and Bronze Age, when, as we have seen, there do appear to be genetic and cultural influences coming into eastern England from southern Scandinavia and north-west Europe.

... and the Picts?

Before leaving this web of tenuous linguistic clues to Scandinavian influence on Old English, I should, at least speculatively and anecdotally, mention an internationally recognized modern English dialect known as Lowland Scots or Lallans, which seems to have retained more Norse and Common Teutonic vocabulary than has English farther south. For examples, a few words I jotted down on an envelope before checking in the *OED* include *bairn* ('child'), *burn* ('stream'), *ken* ('know' or 'suspect') and *kirk* ('church').[17] Lallans is Germanic, is distinct from Scottish Gaelic, and is attested back to Late Medieval times. Northumbrian Middle English is thought to have strongly influenced Lallans. That may be so, but that would not explain the Norse vocabulary, given the conventional view that Northumbrian Anglian was supposed to be West Germanic. Viking influence has also been suggested.

I also wonder about the linguistic affiliations of the mysterious 'other' Pictish language referred to by Kenneth Jackson as non-celtic (see Chapter 2). As mentioned earlier, Adamnam said at the end of the seventh century that St Columba, a *Gaelic*-speaker, had used an interpreter to converse with the Picts in the previous century.[18] This may have meant that they spoke a non-celtic tongue. Picts were still apparently a vigorous nation during the eighth century, in Northumbrian Bede's time. Apparently, the Picts did not speak a language Bede recognized, celtic or Anglian. Was this the 'non-celtic Pictish' Jackson refers to? Bede, being Anglian, ought to have been able to identify a Germanic tongue. Pictish seems to have disappeared without trace in the past thousand years. Could it have a substratum in modern Scottish dialects? Whatever its origins, the written evidence for Lallans goes back no further than the Late Medieval period, so cannot form the basis for more than speculation.

All talk: what of other evidence?

But surely, if there had been previous Norse invasions and/or there were pre-existing Germanic-speaking communities living in England before Gildas' three longboats arrived, there should be some record of this, in ancient texts or inscriptions or place-names? As it happens, there are several records of foreign and even Saxon presence in Britain during the Roman period. These can be found in contemporary Roman texts, and later confirmed by Bede. And there is other evidence too. Some of it is archaeological and, depending on historic preconceptions, may be less obvious and more contentious. We now move into this literary and archaeological minefield.

9

WERE THERE SAXONS IN ENGLAND BEFORE THE ROMANS LEFT?

By the third century AD, the regular Roman army was partly Germanized. Any archaeological evidence for a Saxon presence in England in Roman times has usually been put down to mercenaries or legionaries from across the North Sea, the so-called *foederati*, but there are other interpretations.[1] The Venerable Bede, for instance, tells us about much earlier 'infestations' by presumably Germanic-speaking Franks and Saxons in his *Ecclesiastical History*:

> In the year of our Lord's incarnation 286, Diocletian, the thirty-third from Augustus, and chosen emperor by the army, reigned twenty years, and created Maximian, surnamed Herculius, his colleague in the empire. In their time, one Carausius, of very mean birth, but an expert and able soldier, being appointed to guard the sea-coasts, then infested by the Franks and Saxons, acted more to the prejudice than to the advantage of the commonwealth; and from his not restoring to its owners the booty taken from the robbers, but keeping all to himself, it was suspected that by intentional neglect he suffered the enemy to infest the frontiers. Hearing, therefore, that an order was sent by Maximian that he should be put to death,

took upon him the imperial robes, and possessed himself of Britain, and having most valiantly retained it for the space of seven years, he was at length put to death by the treachery of his associate, Allectus. The usurper, having thus got the island from Carausius, held it three years, and was then vanquished by Asclepiodotus, the captain of the Praetorian bands, who thus at the end of ten years restored Britain to the Roman empire.[2]

For this part of his text, Bede, by nature a careful historian, was not obliged to repeat wild harangues about fifth-century Britain garnered from the prophet St Gildas, but comfortably used more reliable texts. But it is not clear here whether the infestation of Saxons and Franks meant unwelcome mobile parasitic residence or overseas invaders, nor which coasts they were infesting and attacking.

To do him justice, Carausius may have been forced into rebelling by the accusation of corruption. Carausius belonged to the Menapii, one of the Germanic tribes in the far north of Belgic Gaul in Holland, just south of the Rhine, who gave Caesar so much trouble (Figure 2.1a). He operated out of Boulogne. What seems extraordinary is that, as a 'non-Celtic' usurper, he found it so easy to move across the Channel and then dominate Britain for seven years – unless he already had very good contacts on the other side.

Bede's text suggests that for a short time in Roman Britain of the third century AD there may well have been a Germanic ascendancy as well as a chronic Germanic 'infestation' in southern England, and that only the return of the Romans prevented the retreating Franks from sacking London.

Forts of the Saxon shore

During his short rule of Britain and northern Gaul, Carausius is credited variously with building, starting, or at least continuing a network of huge forts or walled enclosures along the so-called Saxon shore. Parts of these buildings can still be seen today (Figure 9.1).

There are eleven or twelve of them strung along a stretch of coast from Brancaster (Branodunum) in Norfolk, round East Anglia, Essex, the Thames Estuary and Kent, to Portchester (Portus Adurni) in Wessex, opposite the Isle of Wight. Although extensively reconstructed in Medieval times, the Portchester fort forms a huge enclosure and is regarded as one of the finest Roman monuments in Europe.[3]

The popular English TV archaeologist Francis Pryor has questioned the orthodox assumption that these enclosures were necessarily all forts, let alone part of a system of defence against attacking Saxon pirates, and suggested that they were probably fortified trading or distribution stations. In his recent book *Britain AD*, he spends a whole chapter arguing that the semantic embellishment of

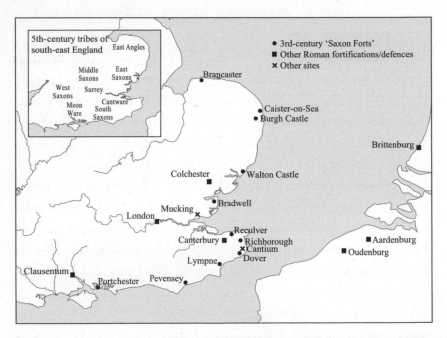

Figure 9.1   Forts of the Saxon shore. Listed in a late Roman military inventory as under the command of the 'Count of the Saxon shore along Britain', these were probably fortified trading stations rather than defence against Saxon pirates. As such, they described the distribution of fifth-century Saxons (inset) rather well, thus suggesting the Saxons were already in residence in Roman Britain.

the *anti*-Saxon fort system is part of millennial, circular, self-fulfilling mindset of 'proofs' used to support the original Gildas Saxon blitzkrieg story.

The name 'Saxon shore' first appeared in a late fourth-century Roman military inventory, the *Notitia dignitatum* ('Register of Offices'), as *Litus Saxonicum*. In this document, various military buildings on both sides of the English Channel were listed as 'under the command' of the 'Count of the Saxon Shore'. Pryor points out that 'just because the forts were listed under the same heading in the *Notitia*, it does not mean that they were part of an integrated system of coastal defences across the Channel, as has been suggested'.[4] He continues:

> The strangest aspect of the 'Saxon Shore' forts is the paucity of archaeological remains across their interiors ... the Romano-British [occupation] in the third century ... was not at all what was expected: there were no barrack blocks, granaries or headquarters buildings ... within the massive walls ... [O]ne would doubt whether Portchester was even a military site. It seemed to have most of the hallmarks of a civilian settlement, and not a very organized one at that.[5]

Pryor also repeats another archaeologist's views (Andrew Pearson,[6] with whom he agrees), that the location of the three earliest 'forts' 'makes no military sense, because they would have left Suffolk, Essex, southern Kent, Sussex and Hampshire undefended'. And 'at least half of the "Saxon Shore" forts were abandoned before the end of Roman rule, at precisely the time when Gildas and Bede tell us that Saxon raids were becoming a problem.'[7]

Instead of trying to make military sense of forts built at the wrong time in the wrong place for defence, Pryor turns the problem round and asks what purpose these bases actually served. He provides the answer by building up a picture of these 'forts' as 'part of long-distance trading networks ... in effect secure stores'[8] – a view which much better fits their locations. Pryor summarizes his perspective on the defence question:

> Gildas greatly exaggerated the severity of Anglo-Saxon piracy, which mainly seems to have been confined to Gaul and the southern side of the English Channel. Far from being cut off from the Continent, Britain would appear to be actively trading overseas throughout the fourth century, and perhaps later.[9]

So, apparently there was not much need for defence during the fourth century against marauding 'Saxons' on our side of the Channel. Elsewhere Pryor points out that 'The *Notitia* ... does not provide us with grounds to suppose that British "Saxon shore" forts had counterparts in Gaul, and formed part of a cross-Channel system of defence ...'[10]

The coast of New Saxony?

To my mind, Pryor just misses tackling the key semantic question head on, although it is implicit in his whole discussion. The question is this: why, if the 'forts' were nothing more than a chain of bonded warehouses, tax-collection points, or whatever, were they called 'Saxon', and what did 'Saxon shore' actually mean? In other words, who and where were the Saxons of the Saxon shore? The only thing we can be sure of is that the fourth-century Roman clerks who compiled the *Notitia dignitatum* referred to a group of enclosed, fortified buildings along the coasts of Essex, Sussex and Wessex as under the command of *comes litoris Saxonici* – 'Count of the Saxon Shore'. The meaning of 'Saxon' remains ambiguous in this title, mainly because of Bede's text (above), written hundreds of years later, which was clearly influenced by Gildas. Three interpretations have been suggested: it could mean 'the coast defended *against* the Saxons', or 'the coast defended *by* the Saxons', or 'the coast *settled* by the Saxons'. Only one of these interpretations casts such coastal Saxons as invaders. The simplest translation of the full Latin military title, *comes litoris Saxonici per Britannias* – without the Gildas/Bede spin – is Count/Companion of the Saxon Coast along Britain'.[11]

If a few Saxon pirates were not perceived as sufficient threat to build proper defence forts in the fourth century, why should the Romans have named the busiest trading coast of Britain, fronting a quarter of the island, after a few pirates, rather than after the current local inhabitants of that coast? Given the lack of other reliable context, the view of Saxons as already settled on the 'Saxon coast' of south-east England during Roman times seems another reasonable inference.

Several hundred years later, the inhabitants of the 'Saxon shore' were still known as East Saxons, South Saxons and West Saxons. Given our complete reliance on Gildas for the intervening period, and his lack of dates or useful context, the coincidence seems remarkable. The only named non-Saxons in this region were the inhabitants of Kent. Kentish people, supposedly derived from Jutes and Britons, spoke a dialect of Old English heavily influenced by Norse. The post-Roman self-identified name of the men of Kent, 'Ceint', is argued to be a Jutish version of the original Romano-British Cantiaci of Cantium ('Cantia' in Bede). Authorities differ on the derivation of the original Roman place-name 'Cantium', but one point on which they all agree is that some version of the root *Cant-* was already old, even in Roman times, since Pytheas and Diodorus used it.[12] Ptolemy's *Kantion* promontory in Kent is regarded as one of the 'Harder ?British' place names to identify etymologically in the recent analysis of Ptolemy's map (Figure 7.2) of Britain.[13] Jute is cognate with Jutland, part of modern Denmark. Why the new Jutish overlords should hang on to an old, possibly immigrant name as their tribal designation is not clear. Caesar clearly regarded the Cantiaci as non-aboriginal, although whether they were Belgic/Germanic-speaking or Gaulish-speaking he did not say (see above).

Pryor sees Anglo-Saxon communities as already present in the southern English archaeological record of Roman times, and points, for instance, to the Saxon site of Mucking in Essex, where Late Roman bronzes are found in both Saxon graves and Saxon huts dated to around AD 400. He argues strongly against the standard rationalization that these were just random homes of immigrant Germanic mercenaries, the *foederati*. I shall come back to the archaeology of the transition later; but returning to the question of identity and timelines, it is worth examining several consistent differences between Gildas and Bede's 'rendering of Gildas', which are revealing as to how recently the Saxons had immigrated and who Bede's 'Britons' and 'English' really were.

Who invaded England in the fifth century – *Angles, Jutes, Saxons or Frisians?*

Gildas identifies the occupants of the three invited warships as Saxons, who first landed on the eastern side of the island to help fight the Picts and the Scots.

Bede, although clearly trying to follow what he has read in Gildas, introduces a note of doubt in his transcription in chapter 15, book 1 of the *Ecclesiastical History of the English Nation*. He writes 'Angles or Saxons' (*Anglorum sive Saxonum*), as if to question Gildas' identification of the Saxons as the invitees-turned-invaders. In fact, Bede's title for this chapter refers only to *Angle* invitees, which would seem more appropriate, since the eastern side of England was indeed where we find them in 'Anglo-Saxon' times. This is one of several places where Bede seems implicitly to doubt Gildas' accuracy, without saying so explicitly. What grounds he had in this instance we may never know, but there are clues elsewhere in Bede's text. Again, at the end of the previous chapter, Bede repeats Gildas' tale of King Vortigern's invitation to the Saxons as '*Saxonum gentem de transmarinis partibus in auxilium vocarent*' (i.e. to those Saxons who lived overseas), thus not specifying whether the general designation

'Saxon' meant Saxons living in Europe or included others *in Britain* at the time of Vortigern.

Later, in chapter 16 Bede repeats Gildas' description of the famous fifth-century battle of Badon Hill, where Britons led by Roman citizen Ambrosius Aurelius – of royal descent, and thought by some to be King Arthur – defeated their enemies. However, where Gildas identifies the battle as taking place forty-four years and one month after the Saxon landing, and thus presumably with Saxons as the enemies, Bede identifies Angles as those invaders of forty-four years before and the enemy in the siege of Badon Hill. Curiously, Bede calls the home team 'Bretons'. He uses the term 'Brettones' in his (Latin) text rather than his usual 'Brittaniae', probably after Pytheas' archaic term 'Pretani' (from P-celtic Pritani[14]) in order to identify them more clearly as Britons (and presumably celtic-speaking) rather than just 'inhabitants of Britain'. This variation in an otherwise consistent text implies that Bede believed that even at the time of the battle of Badon Hill, at the end of the Roman occupation, there were other tribes in Britain, including *in his view* the 'Scythian' Picts, who would not and could not claim 'Pretannic nationhood'.

Gildas was, in Bede's words, the 'Britons' own historian'. Bede automatically excluded himself from such potential *British* bias by his statement. Saxons only occupied southern Britain. Bede, who spent his whole life in Northumbria, knew this and acknowledged northern England as an Angle-dominated region. Although he probably knew more about books than about the southern peoples, Bede describes the distribution of the Saxons in southern Britain accurately and in detail. So it is difficult to see why he should be so concerned about Gildas' accusations that the Saxons were the fifth-century invaders, attributing the intrusion instead to Angles, unless he knew otherwise and was honestly convinced that Gildas was plain wrong in this instance (as Bede knew he was elsewhere).

Elsewhere in chapter 15, Bede does give a breakdown of the ultimate Germanic origins of their various English equivalents, the

Saxons, the Angles and the Jutes. The former he specifies were from
'Old Saxony'. However, he gives no dates for migration, and it is
clear from the title and beginning of the chapter that he prefers to
link the invitation episode specifically to the Angles.

Being a well-read, polyglot Angle in a senior ecclesiastical
position, Bede should have known the truth about the Angles – gory
or otherwise. We have to assume that Bede was native to
Northumbria, having been taken into the monastery at Wearmouth
at the tender age of seven, according to his own words.[15] Bede was
proficient in Early Anglian. His disciple Cuthbert wrote that Bede
was 'learned in our song',[16] in a letter written in Latin, but quoting
Bede's moving death-bed song in the Anglian vernacular as part of
the same text.

Gildas' origins are shadowy, and we do not even know for sure
where he lived, although somewhere in the south-west seems
likely.[17] Gildas is generally regarded as more of a religious than a
political zealot. However, as a Briton described in Welsh biographies
as a saint,[18] it would be consistent for him to have had political as
well as religious agenda associated with the celtic-speaking regions
of his Britain. He is supremely and inventively offensive in his use of
adjectives for Saxons, describing them as 'a multitude of whelps
coming forth from the lair of a barbaric lioness',[19] and worse elsewhere, so there is no doubting the partiality of his feeling.

It must have been, to excuse the pun, galling for the celtic-
speaking nation to have seen the gradual cultural encroachment of
'their south Britain' over hundreds of years. So annoying, perhaps,
that it might explain why Gildas was creative with Saxon history.
Exaggerating Saxon people as recent interlopers in the south, and
thus illegitimate, may have had something to do with his own
agenda. The Saxon kingdoms of the south could have simply represented linguistically related elite takeovers which happened either
during or before the Roman occupation. These takeovers could have
been of long-standing settlements originally settled from the

Continent (e.g. those referred to long before by Caesar and Tacitus). If so, Bede may have had good reason to correct the error.

Rune stories: how different were Angles and Jutes from the Saxons?

I shall return shortly to the archaeological evidence for cultural continuity in the Saxon kingdoms of the south, but there is another set of clues to be found in the Old English landscape which tend to enlarge the gap between the linguistic-cultural histories of Angles and Saxons. These are runes and rune-stones.

Before discussing them, I should recap on the history of the main sources of stone inscriptions and other more portable, non-perishable pieces of text, including coins. Writing first appears in Britain in the form of Roman characters on engraved coins in the century before the Roman invasion. Non-Romanized names of local rulers were used. Coin use was largely restricted to south-east England, and before local minting started, coins were imported, originally from Belgica. As far as the text language is concerned, practically all other inscriptions in England throughout the Roman period were in Latin. Significantly, *none* were in celtic or any other vernacular. The names in stone were mostly Latin as well, in contrast to the naming of coins.

Then there is a hiatus in the immediate post-Roman period. South-east England stopped inscribing stones, and by the time inscriptions started up again they were in Anglo-Saxon or Medieval Latin, with still practically no evidence of celtic use in any part of England except the West Country (Figure 7.4). In contrast to the near-complete absence of celtic inscriptions from any period in England, the non-English parts of the British Isles, in particular Ireland, Wales, Cornwall, Cumbria and Argyll, had many inscriptions in celtic. These used mainly Roman script, but also Irish Ogham characters, starting from the end of the Roman occupation and covering the Medieval period.

It is against this geographical background that a new script appeared on stones and ornaments in certain parts of England, replacing Roman script in the post-Roman period for up to two hundred years and then lasting for a further four hundred after Roman script reappeared. This script is *runic*, and consists of an alphabet with characters made up mostly of short lines set at angles, such as could be cut into wood with a knife. Unlike the celtic inscriptions of the Atlantic coast, inscriptions that used Roman script and Latin were slower to return to 'Anglo-Saxon' England. The Cambridge expert on runes Raymond Ian Page brackets the gaps in time and space: he 'suspects the earliest example of Roman script used in an Anglo-Saxon environment to be the gold medalet issued by Bishop Leudhart, who accompanied the Frankish princess Bertha to England (c.580) on her betrothal to King Æthelberht of Kent'. Even this was Frankish and imported. 'In the fifth and most of the sixth centuries the only recorded script for Anglo-Saxons seems to have been Runic.'[20] Runic inscriptions, in their sparse distribution (Figure 9.2), are thus the earliest and only direct written record of the new people from over the water, and they give clues to how those people spoke in the first two centuries after the 'invasion'.

Runes are most likely of Scandinavian origin

The runic script is often associated with magic, the occult and archaic ritual, and was exploited in a whimsical manner to great popular effect by Tolkein in his wonderful *The Lord of the Rings*. Runes also have a more prosaic meaning in the *Oxford English Dictionary*: 'A letter or character of the earliest Teutonic alphabet, which was most extensively used by the Scandinavians and Anglo-Saxons'.

Figure 9.2a   Runes dated to before AD 650. These follow the early Scandinavian runic orthography (H) and are found exclusively in the regions of initial Anglian and Jutish settlement, although most of England had been 'conquered' by AD 660. This is consistent with Bede's Anglian invasion – and only a limited elite one at that. (The only exception is the 'Watchfield fitting' (inset), in west Oxfordshire near Badbury Hill.)

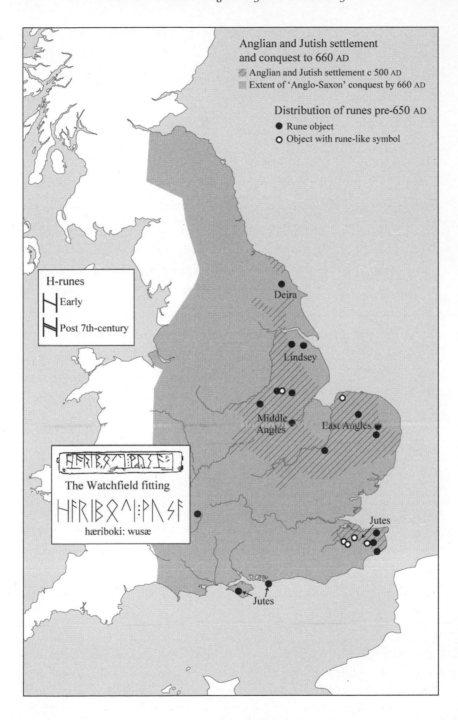

Anglian and Jutish settlement
and conquest to 660 AD

Anglian and Jutish settlement c 500 AD

Extent of 'Anglo-Saxon' conquest by 660 AD

Distribution of runes pre-650 AD

● Rune object

○ Object with rune-like symbol

H-runes

Early

Post 7th-century

Deira

Lindsey

Middle
Angles

East Angles

The Watchfield fitting

hæriboki: wusæ

Jutes

Jutes

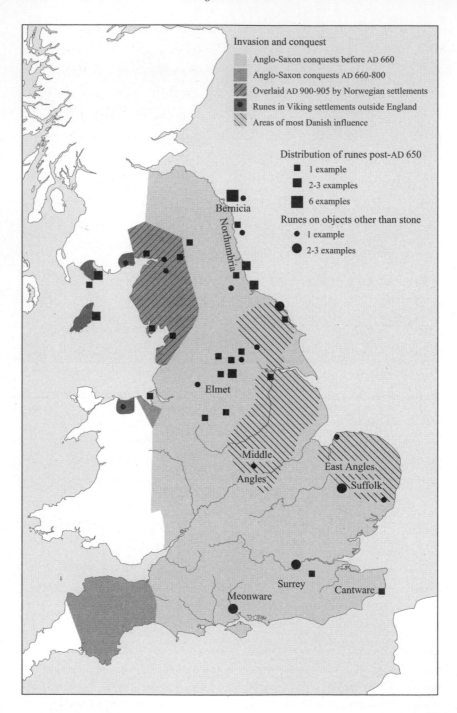

Invasion and conquest

Anglo-Saxon conquests before AD 660

Anglo-Saxon conquests AD 660–800

Overlaid AD 900–905 by Norwegian settlements

Runes in Viking settlements outside England

Areas of most Danish influence

Distribution of runes post-AD 650

■ 1 example

■ 2–3 examples

■ 6 examples

Runes on objects other than stone

● 1 example

● 2–3 examples

Bernicia

Northumbria

Elmet

Middle
Angles

East Angles
Suffolk

Surrey

Cantware

Meonware

Exactly when and where runes were invented is unknown, but the most likely location, based on frequency and age, is southern Scandinavia, probably Denmark, and certainly before AD 400.[21] The relationship between Scandinavian and English runic inscriptions is a fascinating historical puzzle, but before going any further I should point out that, in their distribution in England, they should be more strictly described as Anglo-Jutish, rather than Anglo-Saxon, since in Page's words, Wessex and the west Midlands are virtually empty of runes. This exclusion also extends to Essex, Middlesex and Sussex, but not to the Jutish kingdoms of Kent and the Isle of Wight. I shall come back to the significance of the 'Saxon voids' shortly.

British runes limited to eastern territories of Jutes and Angles

Page places a dividing line of AD 650 between the earlier and later runic monuments. His choice is based on what he sees as a number of relevant factors, including 'There are no Old English Manuscripts [for linguistic comparison] before 650 AD' and 'Surviving Old English manuscripts only cover part of the country, omitting large areas such as Lindsey, western Northumbria and East Anglia.'[22] Furthermore, datable runic-inscribed coins appear only after this point. The earliest runic monuments before 650 appear in all the locations normally associated with the early Anglo-Jutish intrusions, that is to say *limited* regions either side of the River Humber (Deira and Lindsey), the Wash (Middle Angles), Norfolk (East Angles), and Kent and the Isle of Wight (both Jutish) (Figure 9.2a).

There is only one curious exception to this tight eastern Anglo-Jutish runic distribution before 650. This is a single copper-alloy fitting belonging to a leather bag for a hanging scale-set, dating to

Figure 9.2b   Rune finds more recent than AD 650 have a more complex type and distribution, being found away from the initial Anglian settlements in other Anglian sites in Mercia and Northumbria and in later Norwegian Viking settlements in the west. Anglo-Frisian style runes are now found, but Saxon areas are still nearly empty of datable runes in this later period.

the sixth century and found at Watchfield, west Oxfordshire. Watchfield is roughly halfway between the nearby Uffington White Horse and Badbury Hill, near Faringdon. The latter is one of several Badburys in western England which are candidates for the location of the famous sixth-century battle of Badon Hill. Watchfield would have been near the border between Wessex and the early independent Kingdom of the Hwicca between the Severn and the Avon. The Hwicca have been thought by some to represent continuity with the coin-making Dobunni of Roman times. The names of their royalty were Saxon, according to Bede, who states that a West Saxon queen, 'whose name was Ebba, had been christened in her own island, the province of the Wiccii'[23]. However, from a few 'pagan' burials and place-names, we know that there was a small early Anglian presence in Hwicca during the fifth and sixth centuries.[24] As we shall see, this western scrap from Watchfield has a tantalizing bearing on the web of confusion about Saxons spun by Gildas.

Early English runes more Scandinavian

One of Page's other reasons for drawing a line between early and late English runes is found in details of their forms (orthography), which distinguish runes derived from the the North Germanic and West Germanic geographical and linguistic regions. Although runes were invented most probably in the North Germanic (i.e. Scandinavian) region, which, as far as this type of evidence is concerned, includes Jutland (in Denmark) and Angeln (in Schleswig-Holstein), they appear considerably later in the West Germanic Frankish, Allemanic and Frisian territories. In between these two runic regions are the parts of northern Germany/Lower Saxony that are thought by some to be the original homeland of the Saxons. These Continental Saxon areas are practically devoid of runes (Figure 9.3).[25]

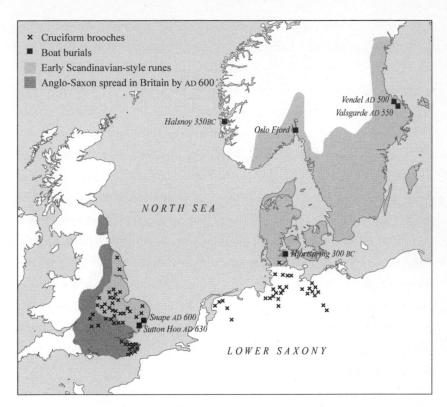

Figure 9.3 Multiple links of material culture, such as cruciform brooches, exist between the 'Anglo-Saxon homelands' at the base of the Danish Peninsula and Anglo-Jutish territories in England, but these do not extend to Lower Saxony. Early runes and boat burials, however, suggest more of a connection to southern Scandinavia and Angeln in Schleswig-Holstein.

The earlier runes differ in their orthography from the later ones. For instance, the character for 'H' in Norse/Scandinavian runes shows a single-barred sloping crosspiece between the uprights (ᚺ), while the later equivalent in the West-Germanic Frisian region has a double bar (ᚻ). The early runes in England before 650 are single-barred (i.e. Norse), while the later forms are double-barred, as in Frisia (Figures 9.2a and 9.2b). The double-barred variant could have had an English origin and then spread to Frisia, or a Frisian origin and then spread to England. Page prefers the first model, mainly on the clear basis of comparative dating. But surprisingly, this suggests an English influence on Frisia rather than the other way round.

Anglian trophy from Arthur's battle at Badon Hill?

Significantly, the single sixth-century western runic example from Watchfield mentioned above is of the Early Norse type. This copper-alloy fitting bears a runic text beginning *hariboki* … (Figure 9.2a). The H is of the Norse single-barred form (ᚺ), and the first element *hari-* means 'army'.[26] Given its date, location, lack of geographical association with Angles and proximity to several putative south-western Badon Hill battle-sites, we might pause to speculate on the identity of the combatants in that symbolic last battle, which the Britons actually won.

As we have seen, where Gildas has the enemy implicitly as Saxons, Bede again demurs and prefers to call them 'Anglians' in his chapter title for the same event. There were probably plenty of Saxons in southern England in the sixth century, but the Roman record of the Saxon shore in the *Notitia dignitatum* could imply that they had resided there long before the battle. Resident Saxons in Wessex would hardly have constituted a pushover for Ambrosius Aurelius (or King Arthur, if that is who he was). The Anglians, on the other hand, had only a brief residence in the south-west in the fifth and sixth centuries evidenced by some graves and place-names. Their disappearance until later centuries is consistent with them being on the losing side in that particular conflict. It would, of course, be taking this particular rune-story too far to suggest that this copper bag fitting was a trophy or detritus of King Arthur's great stand.

Are early runes evidence that Anglian and Jutish were Norse?

To oversimplify and summarize Page's favoured explanation: runes originated in Denmark and spread from the North Germanic or

Scandinavian region to eastern England in the fifth century (i.e. contemporary with Gildas' and Bede's invasion chronology), and then after 650 mutated to a modified form which appears to have spread across the North Sea to Frisia. This story of the runes gives no obvious support for Gildas' story of Saxons being invited to England and landing on the east coast. But it is clearly consistent with Bede's preferred version of Anglian invitations and invasions in the east of England in the title of chapter 15 of his *Historia*. It fits with a Jutish cultural intrusion too, although Gildas does not mention the Jutes, and Bede, although he mentions all three groups, is less specific about the Jutes' early role.

Almost more questions have been raised than answered so far by the map of earliest runes (Figure 9.2a). Looking at their source, the distribution of the earliest runes is exclusively southern Scandinavian, including Jutland and Angeln (for these locations see the map in Figure 8.1d). This in itself might suggest that the separation of language and culture between North and West Germanic places not only Jutish but also Anglian in the North (Scandinavian/Norse) group rather than the West Germanic group; and that the split happened long before the 'invasion'. Consistent with this is the fact that, although Anglian/Jutish runes were carried early over to England, they are absent from Saxon regions.

This Norse perspective has enormous implications for those parts of more northerly Old English we know least about from manuscript sources (namely Lindsey, western Northumbrian and East Anglian). In his summing-up, Page stresses the potential linguistic importance of runes: 'In manuscript there is nothing so early by two or three centuries as the Caistor-by-Norwich [East Anglia] or Loveden Hill [Lindsey] texts [runes].' He points out that we cannot be sure that the former was not Norse; in his opinion,

> modern linguists appear ignorant of or nonchalant about this primary evidence. The Scandinavian affinities of these and of the

Spong Hill [East Anglia] and Welbeck Hill [Lindsey] inscriptions remind us that a rigid division into North and West Germanic language groups is outdated and unrealistic, and that we must reckon on the likelihood of some northern influence on the forms of Old English.[27]

Procopius and his Angle informants: why Frisian invaders?

Much of the argument for all Old English dialects being part of the West Germanic group, rather than Norse, is based on much later texts, particularly the more abundant West Saxon ones in Roman script. We can ask whether the same linguistic division existed in Britain during the sixth century and earlier. Since there are no dated Wessex runes, no comparisons can be made with early Anglian runes. There is one tortuous clue to a linguistic division existing in the sixth century from the historian and commentator Procopius of Caesarea.

Writing at around the same time as Gildas, but from the other side of Europe in Asia Minor, Procopius was reliant on hearsay for information about the former western arm of the Roman Empire, so several of his garbled pieces of gossip about British geography and history are demonstrably inaccurate, and even internally incon-sistent. One confusing text has always puzzled English historians, raising as it does the possibility of additional Frisian invasions of Britain, perhaps in the fifth century. Significantly, this snippet was said to have been obtained from some Angles who were part of a Frankish delegation to Constantinople (c.553):

> Three very populous nations inhabit the Island of Brittia, and one king is set over each of them. And the names of these nations are Angles, Frisians, and Britons who have the same name as the island. So great apparently is the multitude of these peoples that every year in large groups they migrate from there with their women and children and go to the Franks. And they[28] are settling them [sic] in

what seems to be the more desolate part of their land, and as a result of this they say they are gaining possession of the island. So that not long ago the king of the Franks actually sent some of his friends to the Emperor Justinian in Byzantium, and despatched with them the men of the Angles, claiming that this island [Britain], too, is ruled by him. Such then are the matters concerning the island called Brittia.[29]

First, this is a unique use of the term 'Brittia' for Britain, which Procopius elsewhere refers to as 'Britannia'. Procopius seems to identify sixth-century Britons as an aboriginal nation sharing their island from some unspecified time. In that detail the text echoes Bede's distinction between 'Britannia' and the sixth-century 'Brettones'. The Frankish attempts to claim control over the 'Frisian' part of Britain can probably be dismissed as a political agendum, although it does imply intimate family contacts across the Channel – presumably non-belligerent! For me, however, the real clue here is unconscious of politics and possibly more revealing. The Angles tell Procopius that three populous nations inhabit Britain, two of them being Angles and Britons. So far, the Anglian version agrees with most historians, including Bede (except that Bede includes the Picts as a fourth nation). However, the third nation in the sixth century ought to be Saxons, not Frisians. Saxons are, surprisingly, not mentioned at all. The Angles are not *adding* Frisians in their story to Procopius, they are *substituting* them for Saxons.

The simplest explanation for this anomaly, short of ignorance or further political agenda, is that the Anglian informants conflated English Saxons and Frisians as culturally, and possibly linguistically, similar. This again seems to echo Caesar's comments about the links between the southern coastal tribes of Britain and Belgic Gaul six hundred years before. More credibly, it implies that, during the sixth century, Angles were sufficiently different linguistically from 'English Saxons' not to know the difference between Saxons and Frisians. Given that the Continental 'Old Saxony' homeland, in north-west Germany, adjoined Angeln, this seems odd. But it would at least be consistent with a significant divergence of Old English Anglian from Old English Saxon already existing at a time not long after the 'invasion'.

Rune-spread: why did the Saxons not use runes?

Getting back from language to the English geography of early runes, if the distribution of North Germanic runic script *before* 650 was a proxy marker for fifth- to seventh-century Germanic invasions, it was very limited geographically. This would be more consistent with small elite Anglian and Jutish invasions than the wipeout theory.

As far as the regions of the English Saxon kingdoms are concerned, with the exception of a couple of finds in London itself there are effectively no securely dated runes before or after 650 (Figures 9.2a and 9.2b). It could be argued that since there were no runes in the Saxon homeland in north-west Germany, then there wouldn't be any in English Saxony – but this does not really hold water. During Roman times the south of England enjoyed a good standard of literacy, yet soon afterwards Saxon regions seem, on the non-perishable inscriptional evidence, to have descended into illiteracy. When runes first spread exclusively to Anglian and Jutish regions in the fifth century it would be natural for their practical use to have spread south to the apparently illiterate Saxons, who lacked their own runes – in much the same way that runes spread across into Frisia from eastern England. This did not happen. As indicated earlier, when the first post-Roman writing reappeared in the Saxon kingdoms later, after 650, it was in Roman script. This southern picture of Roman before and Medieval Roman after seems more consistent with the archaeological-cultural continuity in the south of England described by Pryor rather than the descent into illiterate barbarism implied by Gildas.

One of the other specific archaeological markers used to link the German Anglo-Saxon homelands with England in the fifth and sixth centuries is a form of personal adornment known as a cruciform brooch (Figure 9.3). In Germany, they are typically found in both the Angeln and Old Saxony homelands at the base of the Cimbrian (Danish) Peninsula, although not in Lower Saxony (Niedersachsen, see Figure 8.1d) in the hinterland south of Frisia. In England, they are found in nearly exactly the same limited Anglo-Jutish distribution as are the early runes, and are absent from the English Saxon kingdoms.[30] Again, this void seems to de-link those kingdoms from the Old Saxony homeland over the same period.

Even in the Anglian regions, in which they are found, cruciform brooches, like early runes are also far more limited in distribution than the great political spread of the Anglians by AD 600, which is suggested in most standard texts (Figure 9.3). As mentioned, this first spread of runes was in Kent (Jutes), East Anglia (East Angles), East Midlands (Middle Angles), and a couple of sites in Northumbria (Deira and Lindsey). The centre of gravity of later runes (after 650) moved north to north Mercia, missing most of south Mercia. They also spread throughout Northumbria, including Bernicia and Lindisfarne. The later runes featured the double-barred H, Ⱶ (Figure 9.2b). The density of English runic finds is not nearly as intense as in the Angeln homeland at any period. This muted and tardy distribution pattern could be argued as just a limited elite expansion, but then it would not support the population replacement that many argue for. Furthermore, the slow wave of cultural movement implied by the runes might suggest receptive acculturation as the elite invaders moved up the eastern rivers, rather than rapid conquest and genocide.

Summary

In the last three chapters I have discussed genetic, linguistic, literary and archaeological evidence for a foreign, probably Germanic-speaking presence in Britain from before the fifth century AD, and even before the Roman invasion. The genetic evidence points towards north-west European influence on the east and south coasts of Britain, going back to the Neolithic, and a Scandinavian influence, affecting north-east Britain in particular. I have cited evidence for a deeper division between the first Old English dialects with a stronger Norse influence in the north. I have suggested that the colonizations of Anglian and Saxon parts of England had complex but different timescales and arose from several Continental sources at different times.

In the next chapter I ask how the Angles and Saxons identified themselves: who they thought they were, their origins and cultural identities. Unlike the Celts of the classical period, the Angles and Saxons wrote quite a bit about themselves.

10

OLD ENGLISH PERCEPTIONS OF ETHNICITY: SCANDINAVIAN OR LOW SAXON?

Ethnicity: how we would like to be identified

Using language and culture to trace and label previous migrations is an age-old practice, but the problem is that such identifications may or may not correlate with modern genetic evidence for migration. In the past, the word 'race' was employed quite freely and confidently, even in academic texts, on the assumption that the reader would understand exactly what the writer meant by 'race', in particular and in general. Since the 1950s, two other words have crept into the vocabulary of anthropologists and others: 'ethnic group' and 'ethnicity'. Although there is no more clarity between the user and hearer, these terms carry rather less of the spurious implications of genetic difference and purity implied by 'race'. In anthropological usage, 'ethnicity' stresses an individual's own perception of identity and group membership in addition to the externally imposed cultural or genetic definition. An example of the former would be 'I am "Welsh" because I see myself as Welsh'. Unfortunately, when the perception of the person making this statement is different from those of people hearing it, particularly in the context of recent immigration, it may cause misunderstanding.

The external observer's view of group affiliation, although more readily available in the works of historians, is likely to be biased and unreliable when the source of information is remote in time or place — one cannot take ancient assessments at face value. Philologists studying the Roman and Early Medieval periods (Latin, celtic and Old English) use the words 'British' and 'English' in a similar sense to Bede and Procopius, to indicate Brythonic celtic as opposed to Anglian and other non-celtic languages. Those early Medieval historians also used the Latin term *gens* ('people' or 'tribe') to indicate that they meant nationhood or tribal grouping in this context. An example is the title of Bede's great text: *Historia ecclesiastica gentis Anglorum*, meaning the 'Ecclesiastical History of the Anglian Nation', which specifically excludes not only Britons but also Saxons from front-stage.

By contrast, Gildas was, in Bede's view, the 'Britons' own historian' and demonstrated immaculate credentials of subjectivity. When he used the term 'Britain' he passionately identified with an island that, in his view, had belonged entirely to celtic-speaking Britons. Unfortunately this detracts from his historical credibility when it comes to the sort of details Bede was so keen to nail down in order to clarify his history of the Anglian church: who invaded which area, where and when?

The subjective view of group affiliation may still help us to see how speakers of Old English, the new Anglo-Saxon rulers, saw themselves ethnically. Is there any information which can help us to determine whether they regarded themselves as Scandinavian[1] or as Continental 'Western Germanic' folk? There is one readily available source of information, but one which is usually derided as creative, legendary and even 'mythological' — namely royal family trees.

The king lists and Woden

It is said that we can all choose our friends but not our relatives. With *king lists*, written lists of kingly succession which feature in many countries' ancient literature,[2] there is the possibility of breaking this biological rule. By adding illustrious ancestors, kings may hope to enhance themselves and their origins. The motivation for such risky creativity has to be legitimacy. One purpose that all royal genealogies share is to establish legitimacy through the identification of lines of descent. In common with many other royal trees, the Germanic king lists are indeed somewhat creative and legendary, not to say mythical, often stretching back to Biblical names such as Noah. Scythia and Asia Minor are sometimes claimed as ultimate homelands, and the Scandinavian god Thor as a prime ancestor, sometimes said to have descended from Priam of Troy. While these Medieval eccentricities might be thought to cast doubt on the veracity of more recent branches of the trees, they emphasize the overall motive of the process: to establish legitimacy via a bloodline going back to great leaders of the past.

The more recent branches of ancient Germanic royal trees from Germany, parts of Scandinavia, Iceland and England show increasing concordance with one another in their deeper roots, tending to move the Anglian and even the Saxon identity away from Lower Saxony to Scandinavia. Some of the concordance results from later Medieval scribes copying from the same older sources. This potentially circular practice can, to a certain extent, be checked for detail to ensure independence. Some of the earliest documents about Germanic-speaking peoples are from England, and royal genealogy forms a major part of their text. Luckily for the purpose of checking the distortions of later Medieval writers, the earliest such text by an English person is Bede's sober *Ecclesiastical History*, written in Latin.[3] About 150 years after Bede, in the late ninth century, came the first of a series of versions of the *Anglo-Saxon Chronicle*. This first one was commissioned in Old English by King Ælfred the Great of Wessex, but subsequent editions were added to by generations of compilers up until the middle of the twelfth century, by which time the language had matured into Middle English.

The *Anglo-Saxon Chronicle* is extensively about genealogies and dates, and draws on earlier texts. The one feature that stands out in this and later documents is that all the half-dozen Germanic royal houses in England (Saxon, Anglian and Jute) claimed descent from one recent non-Saxon legendary ancestor: a person named 'Woden'. The only partial exception was the line of Essex kings, who did still claim descent from Woden by the marriage in the sixth century of Sledd to Ricula, sister to King Ethelbert of Kent, but whose male line of descent went back, otherwise and uniquely for England, to the legendary Saxon ancestor Seaxnet.

We still honour Woden's name in variants of the word-stem *Wednes-* in English place-names and one of the days of our week. So, we read in Book 1 of the *Anglo-Saxon Chronicle*: 'From this Woden arose all our royal kindred…'.[4] This text, from one of our oldest surviving English histories, was no routine description of an accident of descent but a deliberate statement. The gap between each of the main English founders and Woden was but a few generations. For instance, the Jutish leader Wihtgils, whose name is attached to the Isle of Wight and was the father of the fifth-century and earliest semi-legendary invaders Hengist and Horsa, was claimed to be Woden's great-grandson through Wecta and then Witta.

England was not the only Germanic-speaking country whose royalty claimed Woden as their ancestor. We read from the prologue of the Icelandic bard Snorri Sturluson's *Prose Edda* that the royal houses of Denmark, Norway and Sweden all descended from various other sons of Woden. Not only that, but Snorri apparently supplied three additional sons to Woden for his royal offspring in north-west Europe, namely the 'houses' of East Saxony, Westphalia and the Volsungs Kingdom of the Franks.

However, two of Snorri's Continental Woden lines can, with slightly different spellings and details, still be recognized in the pre-invasion royal lines of the English given in the *Anglo-Saxon Chronicle*: Snorri's 'House of East Saxony' confusingly gives rise to both Jutes and Northumbrians, while that of 'Westphalia' is very close to King Ælfred's Wessex line for over ten generations. Even the three male generations preceding Woden are also very similar in the two texts. Thor Heyerdahl, who picked up on this paradox in his book *The Search for Odin*, argues on textual grounds that this particular parallelism was unlikely to be the simple result of copying.[5]

Was Woden Scandinavian?

So, if the god-king Woden was a common Germanic ancestor figure, does that help us to differentiate Dark Ages Germanic regal ethnicities? Well it might, if we can identify from relevant Medieval Germanic literature where he is supposed to have come from and when and how his 'sons' dispersed.

The first thing to notice is that the connection between Anglians and Jutes of the Cimbrian Peninsula (i.e. modern Jutland in Denmark and Angeln in Schleswig-Holstein) on the one hand and Snorri's House of East Saxony on the other is less likely to be a confusion of homelands and more likely to reflect a rapid geographical extension of elite control. The latter is consistent with territorial changes known to have taken place in the first millennium AD. During Emperor Augustus' reign 2,000 years ago, the tribal land of the 'Saxones' lay right next to Angeln at the base of the Cimbrian Peninsula, on the east side of the River Albis (Elbe) (Figures 9.3 and 8.1d).[6] A little over a hundred years later, Ptolemy again refers to the Saxons as occupying the neck of the Cimbrian Peninsula. This small region was possibly Bede's Old Saxony. By the fifth century AD, however, 'Saxony' and Saxon geographical identity had moved a long way from this region, south-west into the hinterland behind Frisia, now known as Niedersachsen (Lower Saxony).[7] Whatever the explanation for the move, these references all seem to confirm the suspicion that the word 'Saxon' was used rather loosely in Medieval histories.

Westphalia and the Frankish kingdoms, described by Snorri as ruled by the descendants of Woden, were progressively to the south and west of Lower Saxony. In other words, these two Low German royal families, supposedly twinned with the invasion of England, were originally Scandinavian and may well have started out in the Cimbrian Peninsula (i.e. southern Scandinavia) at least 2,000 years ago during early Roman Imperial times. If the genealogies are anything to go by, Woden's descendants ended up ruling a large part of north-west Europe. How much of this movement and change reflected an extension of elite-family political control and how much was real people movement is not clear. There certainly is historical evidence for south-western migrations on the Continent, but the best of it relates to the effects of flooding of large parts of Frisia and the Netherlands around AD 500, which does not tell us anything directly about England.

Saxo Grammaticus

There is another, much larger Late Medieval document which attempts to chart the history of the royal families of southern Scandinavia in great detail and has much to say about Woden or Odin, mostly uncomplimentary. This is the *Gesta Danorum* or 'Danish History', written in Latin by the Danish historian Saxo Grammaticus ('Saxo the Lettered') during the late twelfth or early thirteenth century.[8] Saxo was well aware of Bede's earlier work and admired him, but was passionately overinclusive compared with Bede and may not have had quite the same dry concern for accurate recording and cross-checking of sources. He probably got his information from a mixed bag of manuscripts, myths, legends and sagas. His agenda were nationalist – he was a Dane – and, almost certainly, Christian – his patron was a bishop.

Saxo starts his first of sixteen books by defining the two brothers Dan and Angul as the stock of Danes and immediately identifies the

'Anglian race' which took possession of Britain as descendants of Angul. So there is little doubt on which side of the north/south Germanic family divide Saxo places the Angles and the English. As far as he is concerned, the royal 'Saxon-named' lines, interact with but are not part of the mainstream of Danish, Swedish, Norwegian and English history. (But remember that 'Saxony' had a different location by the time of his writing.)

What is perhaps most interesting in terms of *perceived* English ancestry among Saxo's tangle of genealogical detail is his corroboration of some of Beowulf's Danish relations (see below) and a mass of description, albeit unfavourable, on the sexual and socio-political antics of Odin, his wife Frigga, and their origins and descendants.

As a Christian, Saxo adopts the same practice used by the Christian compilers of the *Anglo-Saxon Chronicle* in referring to Odin/Woden as a historical person. However, unlike them he makes this designation explicit by such comments as 'At this time there was one Odin, who was credited over all Europe with the honour, which was false, of godhead, but used more continually to sojourn at Up[p]sala.'[9] By this and other references, we gather that Saxo regarded Odin as no paragon or deity but as a particularly successful Swedish chieftain-warlord who had managed to induce a godhead cult in his own name.

This all seems to point to Woden and southern Scandinavia respectively as the preferred ancestor and homeland to which all the Anglo-Saxon kings wanted to be related. Of course, it could still be that Woden or Odin was just a non-specific Germanic godhead figure with an affiliation and ancestry that varied according to bard or author. There is, for instance, an isolated, dubious eighteenth-century genealogy which dates 'Wotan's' reign to the third century AD and makes him descended from a long line of prior Saxon kings, all with non-Scandinavian Low German names.[10]

There is some evidence from around the time of Bede for an early split in Woden's cult to form a new mainland branch. This is indicated in a curious 'formal response' given in an eighth-century

Formel of Pagan Renouncement intended for those taking up Christianity – voluntarily or otherwise! Written in Old Frankish, one line reads: *Ec forsacho Thunaer ende Woden ende Saxe Ote* – 'I forsake Thor and Woden and the Saxon Odin'.[11] Presumably, it was felt that if both regional cults of Odin were not covered, the convert might be insincere, in a sense crossing his fingers behind his back.

This evidence of a cultural-linguistic split in royal-sacred genealogy leads us to ask again just where the English pagan affiliations lay: in Scandinavia or in Low Saxony. Since Woden's genealogies in the *Anglo-Saxon Chronicle* were all consistent with the Scandinavian names and lines given in Snorri's *Prose Edda* and in Saxo's 'Danish History', we may still infer that the Anglo-Saxon kings had all made an informed choice of *Scandinavian* ethnic ancestor. This amounts to a clear declaration of perceived Norse ancestry, irrespective of Odin or Woden's actual existence and the veracity of his putative family trees.

Beowulf the Geat

Other, more direct evidence for English–Scandinavian affiliation rests in *Beowulf*, the first jewel of Old English literature. *Beowulf* is one of the earliest and most famous of English poems. There is doubt about its date of writing, though probably in the early eighth century, and about where it was composed, probably in Northumbria. (A later date, and in Mercia, has also been suggested on textual grounds.[12]) Irrespective of such details, *Beowulf* is written in Anglian and is essentially a Scandinavian saga. The main dramatis personae are southern Scandinavian, and with the exception of excursions to Frisia, the action takes place in Zeeland (part of Denmark) and the land of the Geats in southern Sweden. Why on earth should the English be writing a saga praising Danes around the time of the Viking raids? One answer is that the original was composed before the Viking raids, and the Norsemen might not have been strangers to the place where it was composed.

Written in a grand heroic alliterative split-line couplet style, the poem tells how Beowulf, a Geat hero of royal descent, sails to help Danish King Hrothgar deal with two monsters (Plate 20), Grendel and his mother. Having succeeded, he sails back to his home some- where in southern Sweden and is acclaimed by Hygelac, King of the Geats, whom he helps in wars with the Swedes, eventually succeeding him as King. Beowulf kills one more dragon in his old age, dying of his wounds afterwards. The dragon had been guarding an ancient treasure buried in a mound. Beowulf himself is buried in a mound with great pomp.

As mentioned, the Scandinavian genealogy described in *Beowulf*, both Swedish and Danish, is extensively corroborated in the Danish History. Hrothgar, King of the Danes, may have lived in the fifth century.[13] One key Geat character in *Beowulf*, namely his lord Hygelac, is mentioned in many independent historical sources, and we can even read of the date of his disastrous raid on the Frisians, in 521.[14] Beowulf himself has a more shadowy provenance in other sources. However, if we are looking for an answer to the common question of the heroic relevance of Beowulf, from southern Sweden, to the English rulers' sense of identity, there is a clue in the Winchester MS, the oldest version of the *Anglo-Saxon Chronicle*. Here it states that five generations before Woden, the ultimate ancestor of English Kings, was Geata from eastern Sweden.[15]

Another king-list argument can be made to link England with southern Sweden and possibly with Beowulf. The name of one of the East Anglian kings, Wuffa, appears as a stem in several place- names of the region and in the name of the royal family of East Anglia, known as the Wuffingas. Bede tells us that 'King Redwald was ... son of Tytilus, whose father was Uuffa, from whom the kings of the East Angles are called Uuffings.'[16] Wuffa ruled from around 570. The Wuffingas may have been an offshoot of the Scylfings, the royal house of Uppsala.[17]

Sam Newton, East Anglian scholar, author and enthusiast, has taken this link much further and argues that:

> *Wuffa* seems best explained as a diminutive of *Wulf* ... The patronymic *Wuffingas* seems to be a variant of *Wulfingas* or *Wylfingas* ... meaning children of the wolf ... The Wuffings, in other words, may have been descended from the Wulfings ... given the possibility that the East Anglian Wuffings may have been descended from the Wulfings, *Beowulf*'s Queen Wealhþeow may have been regarded as a Wuffing ancestor.[18]

I have done Dr Newton a disservice in condensing his convincing and lucid text, which contains numerous parallel lines of textual, place-name and archaeological evidence that link the kings of East Anglia with southern Sweden. He reveals such a tide of corroborative material in *Beowulf* that it is tempting to search further afield for allusions to the Wulfingas, such as Gildas' comment on the Saxons sending for reinforcements: 'Their mother-land, finding her first brood thus successful, sends forth a larger company of her wolfish offspring, which sailing over, join themselves to their bastard-born comrades.'[19] But then Gildas was talking about Saxons, not Angles, and had already described the first ones as lion cubs in the same chapter, and who knows – if he had heard of crocodiles, he would probably have called them spawn of crocodiles too.

Sutton Hoo: the finest boat burial in all Scandinavia

There are more links between the early East Anglians and the Swedes; but they are more in the realm of archaeology, which unlike king lists and histories is usually more even-handed, illuminating the lives of ordinary people as well as royalty. There is one English archaeological treasure, however, which concerns an actual Anglian king who not only has all the trappings of Norse splendour claimed in *Beowulf*, but who has a place in the East Anglian Wuffing king list. The weight of evidence is that the king buried in a ship in Mound One of the royal burial ground of the Wuffings at Sutton Hoo in Suffolk was King Rædwald, grandson of Wuffa. Rædwald, who died around 625, became High King of Britain, or Bretwald, after defeating Æthelfrith, King of Northumbria, at the Battle of River Idle, in about 617.[20] As there is no inscribed tombstone, there is still justifiable doubt among academics as to whether this formal burial mound held Rædwald or some other closely related contemporary, or indeed an actual corpse. That does not affect the point of the story, and, I shall call him Rædwald.

Like many, I have enjoyed the very accessible privilege of viewing the shimmering geometric inlaid gold designs, drinking horns with silver-gilt mounts and other fine pieces of Rædwald's burial treasure at the British Museum in London, and on more than one occasion. The first time was when I went to speak at an archaeological conference there (Plates 14 and 15). I also took advantage of a coach trip and guided tour of Sutton Hoo arranged for delegates. The brilliant tomb reconstructions and cherry-picked jewellery displays in the National Trust's museum there are a few minutes walk from the burial site, and screened by some trees. So when we came upon a collection of huge grassy mounds on the edge of a bleak plain looking down on Woodbridge and the River Deben, the extraordinary atmosphere we all felt was unalloyed by tourist trappings.

Ever since its excavation, the fact that Mound One and its neighbours contained buried, high-prowed longships has excited comparisons with southern Scandinavia (Plates 17, 18 and 19). There are a couple of other similar ship burials in this region of England, for instance at Snape, but ship burials are a rare and unique cultural practice, more generally associated with Scandinavia of the same period, in particular the boat-grave cemeteries at Vendel and Valsgärde in Sweden.[21] Although many of the beautiful grave goods and coins in Sutton Hoo came from all over Europe, including Byzantium, a significant proportion pointed to Scandinavia. The style of the king's helmet and shield seemed to confirm the Scandinavian connection[22] though whether they were imported or made locally is not clear.[23] The grave goods also included a range of ornaments, including clasps and square-headed brooches, giving evidence for links between Anglia and Scandinavia during the 'migration period' of the fifth and sixth centuries.[24] Multiple ship burials are also found in the Oslo fjord in south-east Norway.[25]

Some of the foregoing does seem to push the cultural centre of gravity of early East Anglians from Denmark and Angeln north across the water to Norway and Sweden – but these are not the only cultural links. Neither were East Anglian kings the only migrants to East Anglia. I have already mentioned the cruciform brooches, which point to links across the North Sea more to Angeln in Schleswig-Holstein than farther north (Figure 9.3). Other cultural links, including ceramics, ornaments and the practice of horse cremation, seem to link Norfolk and the areas of earliest English rune finds with Schleswig-Holstein and the neck of the Cimbrian Peninsula.[26]

The first Domesday document

Before moving away from English king lists and royal claims of descent, I should point out that the ethnicity and regional identity claimed by various early self-proclaimed 'kings' in England was not necessarily the same as was perceived by their 'subjects', nor even how other English 'kings' viewed them. There is a famous and mysterious Old English sheet document thought to have been composed originally in the seventh or eighth century called the *Tribal Hidage*.[27] In the spirit of the Domesday Book, but compiled four hundred years before, the Tribal Hidage appears to be a summary regional administrative assessment of land size intended for use in exacting tribute or tax. It may have been drawn up in either the Kingdom of Mercia or the Kingdom of East Anglia, depending on who was top dog at the time, and listed thirty-four political units south of the Humber, with an accountant's roundings of land size in terms of *hide* (family) numbers or *hidage*.

The relevant point of interest in this document and its copies, apart from their intrinsic historical value, is that although the recognized half-dozen kingdoms of southern Old England are included as obviously big players, they are in a numerical minority: just six out of thirty-four. The kings of West Sexena, Sussexena, East Sexena, Cantwarena (Kent), East Engle and Mercia are accorded no special political status over the small fry. If that degree of political complexity was still present in seventh-century England, over two hundred years after the invasion, who were all those others, and how much more politically complex was England in the fifth century? This is hardly the sort of fragmentation expected if ethnic cleansing followed the 'invasion', as implied by the texts. Surely, the big kings should have finished their merger game and divided up the cake by that time. Were the small players all descended from other invaders, or was there, as with the Domesday Book, a great degree of administrative inheritance from the old Romano-British order?

Good claims of Norse ancestry for the Anglian elite; how about Saxons?

The cultural evidence set out in this chapter, combined with the runes and distribution of cruciform brooches, argues much more strongly for Bede's story of the elite Anglian invasion from Angeln, than for Gildas' version, the Saxon Advent. Even the king lists of the Saxon regions of southern England, although Scandinavian in intent, are all non-Saxon, with the exception of Essex's. This deep division between Angle and Saxon will crop up again with the Vikings and the Danelaw line, but just how much of this division was an accident of a fifth-century carve-up, and how much was determined by pre-Roman relationships and long-term cultural continuity? Some archaeologists argue for continuity and others for wipeout, as we shall see next.

11

ENGLISH CONTINUITY OR
REPLACEMENT?

Archaeological evidence of continuity / replacement in England after the Romans

The Anglo-Saxon question I want to address in this book is how much population replacement occurred. It is clear that a dramatic cultural change got under way in England starting in the fifth century, and that north-west Europe is the likely source of that change. Was Gildas telling the truth? Was it a mass Anglo-Saxon invasion with slaughter and replacement of Britons, as formerly claimed by archaeologists such as John Myres and Edward Leeds? Interestingly, some geneticists still believe in complete replacement, although few archaeologists do.[1] Or was it an elite cultural and political takeover, with a degree of cultural consent and local apathy, as argued by others such as Nick Higham of the University of Manchester?[2] Perhaps it was somewhere in between, with some population migration and some indigenous cultural developments set in motion by a small nucleus of Germanic overlords. But what a gap it would leave in English history never to be sure how much!

In a recent collection of essays on migration and invasion in archaeology, edited by John Chapman and Helena Hamerow, several authors attempt to get a handle on archaeological evidence that could help. The starting point is expressed by Sally Crawford: 'Suffice it to say that few archaeologists today would agree with Freeman's declaration [1888] that the Anglo-Saxons wiped out the native British in the south-east of the country, linguistic evidence notwithstanding...'.[3] Hamerow, who specializes in this transition in history, reviews the divide and argues that the polarization of *opinion* between archaeologists still holding the increasingly unpopular migrationist view[4] and those holding an essentially indigenous acculturation and local developmental view has 'at times [amounted] to a veritable identity crisis'.[5] She believes that to resolve this question of migration or acculturation requires a more systematic methodology for measuring and comparing detailed studies of individual communities and their change of land use over time.

It might be thought that we could dispense with archaeology and just concentrate on genetics, but that would be unwise – not least because the opinions of geneticists are highly polarized as well. One discipline cannot have all the answers; personally, I think that history and archaeology should and do offer some solutions, particularly for and against the extremes of opinion. History, naturally, includes the study of administrative documents, and use of the historical demographer's skill, and so has much to tell us. In the thousand-year period that brackets the Anglo-Saxon 'invasion' with its sparse documentation, there were four serious invasions of Britain, two of which even affected Ireland. Three of these invasions – by the Romans, Vikings and Normans – were all much better documented than the Anglo-Saxon one. For these three there is consensus among historians and archaeologists that in spite of the profound visible cultural change they each brought, there was substantial continuity in the indigenous population, and that some

17 (*above inset*): The buried long-ship at Sutton Hoo during its original excavation. Very similar contemporary longboat burials have been found in Sweden and boat burials go back to 350 BC in Scandinavia. 18 (*main image*): One of the finest examples of a long-boat is Norway's Oseberg Ship, from the Viking period.

19 and 20 (*above*): Scenes from the movie *Beowulf and Grendel*, including (*top inset*) Beowulf entering the court of Hrothgar. The original Old English saga provides evidence of English-Scandinavian affiliation. 21 (*right*): 'Up Helly Aa' festival, Norse traditions are enthusiastically maintained in Lerwick, Shetland.

22: Tacitus made three stereotypical ethnic comments about the inhabitants of the British Isles, saying, in précis, that the Scots were red-haired like the Scandinavians, the Welsh like the Spanish and the English like the Belgians. We all still fall for labelling habits today, but can we really tell? Faces of our time: (*above, left to right*) a Scotsman, and Irish mother, (*below, left to right*) an English woman, and a Welsh choral singer.

displaced 'native' practices, one would expect the latter to have vanished completely. They did not. If people were not moving around in great waves of migration, how and why did the archaeological changes of post-Roman times happen? What was going on in Britain at the time?[8]

Helena Hamerow has reviewed both sides of the archaeological debate and its history, looking in particular at the cultural practices and material culture, to see what might help determine the impact of the Continent on England from the mid-fifth to the mid-seventh century. This is not an easy task, for a number of reasons. Styles of burial, building and ornaments, while clearly indicating profound change and outside influence both just before and after the Roman exit, do not give clear-cut answers. They suggest, rather than the extremes of all or nothing, a major degree of modification and hybridization and indigenous switching of cultural affiliation. One exception she cites to this in-between world of similar yet different is the absence from England of the so-called Anglo-Saxon longhouse:

> [N]owhere in England is the main type of farmhouse found throughout the continental homelands of the Anglo-Saxons to be found: the longhouse, in which the cattle byre and living room lay under one massive roof, supported on rows of internal posts.[9]

One possible reason for cultural discrepancies between 'Saxon homelands' and Anglo-Saxon England is in the historian's choice of homeland; Lower Saxony is possibly the wrong place to make such comparisons, and the base of the Cimbrian Peninsula might be more appropriate. However, in the passage quoted above, Hamerow is actually referring to the latter (i.e. the Elbe/Weser region). Another cultural domain, with less self-conscious ethnic marking, is animal husbandry and land use — which, she argues, do 'suggest a substantial degree of continuity'.[10]

Before moving on to the genetic story, perhaps I might allow myself an observer's comment on archaeological continuity and change in Dark Ages eastern Britain. The one thing on which all archaeologists seem to be agreed is that there were profound changes in English cultural styles, in the broadest sense, over the fifth to seventh centuries AD. It would be easier to accept that these could occur without complete replacement if there were pre-existing cultural relationships between eastern Britain and Scandinavia/north-west Europe. The possible elements of these relationships – trade, a common language, prior alliances with high-status marriages – would all contribute to the speed of acculturation and even facilitate a passive acceptance of inward migration and domination. As discussed earlier, archaeological evidence for such relationships goes back to the Neolithic. It was indeed the main complaint of Gildas in the west that certain 'tyrants' actually invited the 'Saxon' wolves. Germanic continuity in England would also explain the numerous linguistic anomalies I have been discussing throughout this book, which could be summed-up thus: a lack of celtic-language continuity in England.

Two parts of England in particular seem to have been less troubled by the Saxon Advent, and even seemed to have preserved some political continuity, while being ultimately as Germanic, culturally and linguistically, as the rest of Britain: these were Lindsey and Wessex. Lindsey preserved its Roman fort at Lincoln and its Roman name. Roman artefacts persisted until a late date in conjunction with very early Saxon materials in Roman forts at Horcastle and Caistor-on-the-Wolds.[11] Was this evidence just of Germanic mercenaries, or of pre-existing Germanic settlements? Wessex has only patchy archaeological evidence of early Germanic settlers, with complete absence in south Hampshire. John Myres, a traditional invasionist, suggested that an early Wessex king, Cerdic, was the head of a partly British noble family with blood ties to existing Saxon or Jute settlers who had been entrusted with the 'defence of Wessex in the last days of British sub-Roman authority.[12] Yet, by the time their first literature appears, with his descendant King Ælfred, Wessex had its own local Germanic dialect, with no trace of any celtic substratum.

In pursuing the theme of invasions from north-west Europe, it would be more logical for me to go straight to the last two waves, of Vikings and Normans, and include those with Anglo-Saxons before looking at the genetic evidence together. They all happened within six hundred years of one another and were, after all, only part of a series of such events which have occurred since the Neolithic. But I shall break here because there are several subjective reasons why the Anglo-Saxon 'invasions' should be dealt with on their own. With its dramatic cultural-linguistic schism from Roman Britain, the Anglo-Saxon Advent has led historians and geneticists to cross a speculative chasm on a bridge of flimsy evidence. The resulting extreme view of the slaughter of 'Celts' has fulfilled and reinforced Gildas' doom-laden prophecy and his anti-Saxon agenda. This hypothetical holocaust has some psychological significance for the English. The subsequent invasions of the Vikings and Normans, and the occasional subsequent bloody intrusions into remaining non-English-speaking regions by their political descendants, have had perhaps an even greater impact in terms of alienation for the other surviving tribes of the British Isles.

Anglo-Saxon invasions: old genetics

The genetic picture has some colourful history to it as well, for it has been known for almost as long as genetic studies have been possible that there is a genetic fault line between the English and the Welsh that roughly follows the line of Offa's Dyke. The big questions, as I have tried to stress in this book, are not about whom, but about when, and over what periods, this fault line was created.

Blood groups were one of the first available genetic marking systems that physical anthropologists used to try to categorize the human 'races'. As it happens, their lack of of variety and their presence, at different proportions, in all populations make them generally useless for looking in a detailed way at differences and migrations. One of the earliest studies, conducted over eighty years ago, foundered in a morass of absurd relationships, such as Russians linked with Madagascans. Blood groups are still used in population genetic studies, but only as an adjunct to many other marker systems.[13]

The genetic difference between England and Wales, however, is so clear that it can be detected by many crude markers, including blood groups. Arthur Mourant, one of the founders of modern physical anthropology, wrote extensively on the biological anthropology of blood groups in the middle of the last century; his autobiography (1995) was aptly entitled *Blood and Stones*. He was not the first to note the higher rates of blood group O, as opposed to A, in Basques, Bretons, Welsh, Irish and Scots, but in 1952, with Welsh colleague Morgan Watkin, he made a reasonable crack at suggesting why:

> [T]here appear to us reasonable grounds for the belief that, prior to the advent of Celtic-speaking immigrants, the British Isles were inhabited by a people whose domain had at one time extended over a considerable part of Europe and North Africa but who under ever increasing pressure from the east had been driven from their homelands. Some, no doubt, found refuge in the more isolated mountain regions, but the remainder were gradually driven westwards and finally came to occupy a limited area near the Atlantic seaboard of Europe.[14]

Watkin later added:

> One wonders, therefore, whether a large part of Britain's very early population did not arrive by the western sea routes and whether Celtic speech was acquired from later invaders ...[15]

If we put the language interpretation to one side for the moment, we could say, so far so good: these very general statements, without specifying dates or putting numbers to differences between regions, could be consistent with some of the mitochondrial and Y-chromosome results discussed in this book. However, this blood group evidence has been cited more recently and taken very much further than is warranted. I should like to look at this type of approach in some detail, since it still has lessons for the genetic debate on the 'Anglo-Saxon replacement' hypothesis today.

The German linguist Wolfgang Viereck of the University of Bamberg (where blood groups were first discovered) asks: 'Is there a connection between blood group and membership in a tribe or race and between blood group and language, even between dialectal differences within individual languages?'[16] He then proceeds to build a castle of cards based on correlating minor variations in English dialect with subtle variations in blood group O and A frequencies in Britain. While the dialect variations are interesting, the genetic–linguistic correlations will inevitably end on the heap. However, the use of genetics in this example allows us to look at the use and misuse of genetics and interpretations of how gene frequencies in England could have come to be measurably distinct from those of Wales, Scotland and Ireland.

The basis of Viereck's geographical comparisons is a map of relative frequencies of blood groups A and O in Britain (Figure 11.1). This frequency map is derived from a standard text on blood groups in the United Kingdom.[17] However, the map is not only confusing, it is also a statistically misleading presentation. In all regions except southern England, Norfolk, Lincoln and Nottinghamshire and the two south-western peninsular regions of Britain, blood group O is shown as more common than blood group A. The percentages given in the figure reflect the degree to which one of these two groups exceeds the other, and are intended to indicate the degree of genetic intrusion by A. What they actually do

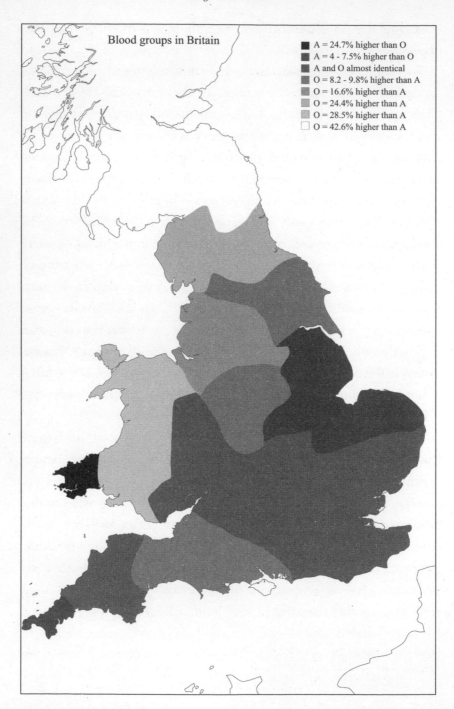

Blood groups in Britain

A = 24.7% higher than O
A = 4 - 7.5% higher than O
A and O almost identical
O = 8.2 - 9.8% higher than A
O = 16.6% higher than A
O = 24.4% higher than A
O = 28.5% higher than A
O = 42.6% higher than A

is exaggerate the differences as a result of misunderstanding the genetics of blood groups.

The impression this diagram gives is that in England, blood group A – which in Viereck's words is 'said to be the younger Germanic blood group'[18] – has partially replaced blood group O, which he regards as older and 'Celtic'. This story is wrong, for a number of reasons. For a start, you cannot date blood groups. Group A cannot be specifically 'Germanic', nor 'O' specifically 'Celtic', whatever those designations mean. The gene coding for blood group A is found throughout the world and is no more common in Germany (25–30% allele, or gene frequency) than any other region in mainland Western Europe. The highest frequencies of the A gene type are in fact found among the Saami (35–40%) and to a lesser extent among Scandinavians (30–35%). The group O gene, conversely, is the most common blood group gene type in the world, and outnumbers group A genes nearly everywhere. In Western Europe, the group O gene dominates at 60–70%, and at slightly higher frequencies of 70–80% in the Atlantic fringe. The O gene type in fact greatly outnumbers the A throughout Britain, including England.[19]

We might ask how Viereck's map could get it so wrong and over-state blood group A so much in England (Figure 11.1). The answer is simply that the map is based on *blood group* frequencies, not on *blood group gene* frequencies. Our expression of blood groups in the blood typing test depends on the genes we get – one from each parent (our *genotype*). Blood group A is *dominant*, which means that the result of our blood test (our *phenotype*) is group A, whether we get one or two A genes from our parents. Group O is *recessive*, which means that we must receive both our parental genes as O to

Figure 11.1   Viereck's map of British blood groups. Drawn from blood bank data, this map purports to show massive 'younger Germanic blood group A' invasion into 'older Celtic blood group O' in eastern England. The fallacies are: exaggeration of 'A' gene frequency, false racial labelling, and no allowance for similar previous colonization history of England and the nearby Continent.

have an O blood group phenotype. If one parent gives us an O and the other gives us an A, we will test as A. Therefore analyses of blood group frequency will automatically overstate A gene frequencies if this very basic fact is overlooked. Since B, although dominant as well, is quite uncommon in Western Europe, the division of simple blood group frequencies between A and O is a two-horse race, in which the dominant A has the advantage. Small increases in the A gene frequency will be reflected in dramatic relative falls in the O group frequency.

So, where Viereck's map misleads is in presenting blood group data (phenotypes) rather than gene frequencies (genotypes). Even in England, where blood groups A and O appear to be neck and neck, the group O gene actually still outnumbers A in frequency by up to $2:1$. Although the differences in gene frequency probably do represent relative differences in migration history, for the reasons given above they overstate the case and cannot be taken as evidence of recent massive replacement of O by A. This argument would still be a statistical fallacy, even if the A gene were specifically Germanic or the most common gene in Germany were A — neither of which is true.

My main reason for this digression is to show that we can use the simple blood group data to illustrate a major flaw in efforts to determine the size of the Anglo-Saxon or any other Germanic migration or replacement genetically. In a nutshell, the hypothetical Germanic 'source regions' were already made up of mixed populations, derived from at least two remote Ice Age refuges shared with the ancestors of people destined for England. This problem also affects analysis of the mtDNA and Y-chromosome data. Let us say (for argument's sake only) that high rates of the O allele really did represent early post-LGM migrations into Northern Europe from the Basque refuge, and that higher rates of the blood group A allele had arrived in north-west Europe by sailing up the Danube from the Balkan refuge, or at least from Eastern Europe. In this situation, what mix of these two genes would we expect to find in, say, north-west Germany or Frisia?

It is very likely that north-west Europeans had mixed Ice Age origins, but it is more difficult to know exactly what mix there was at the beginning. However, if we use the analogy of the Y-chromosome evidence, with Ruisko originating from the Basque country and Rostov, Ian and Ingert from the Balkans rather later (see Chapters 4 and 5), we could estimate a mixture of around 60% of the O gene and 40% of the A gene in Frisia/Saxony during the Neolithic. Allowing for the presence of the group B gene (5–10%), this is not far off the actual rate. But then we would also expect to see nearly the same relative frequencies in England, on the basis that England had a rather similar parallel post-glacial recolonization history to Frisia. Group A genes in England would, however, be somewhat less frequent than in Frisia, say between 25% and 30%, but higher than Wales and Ireland, which would have only around 5–10%. Again, these two regional relative frequencies would more or less fit the real blood group data, which are that blood group A gene rates are the same either side of the North Sea, at 25–30%. So on these figures, a common parallel ancient history of colonization either side of the North Sea could be a more likely reason for similarity than a recent 'Germanic' invasion and replacement.

But how could we tell from these predicted Neolithic figures what the true history of subsequent interchange was between Frisia and England, or between Germany and England, if the gene frequencies were already so similar on either side of the North Sea? Several scenarios could have produced the same outcome. First, following one of the main historical myths in this book, we could suggest a model of the holocaust/*Lebensraum* type, with 100% replacement by Anglo-Saxons (in this case Frisians and/or Germans) of 'indigenous' Britons. A second possibility is of recurrent invasions from Frisia/Germany to England over the whole post-glacial period, leading gradually to similar mixes. Third, there may have been parallel, long-term, independent colonization of England and Frisia/north-west Germany from the same two remote sources (the Balkans and the Basque Country); there need not necessarily have been any later local invasion either way across the North Sea. And there could, of course, have been some combination of the three.

Anglo-Saxon invasions: new genetics

The three models outlined above allow us to look at the current genetic literature on the Anglo-Saxon invasion in some perspective. As should be clear by now, the highest-definition genetic information available for the British Isles is from the Y chromosome. The relevant half-dozen investigations of the past five years created the systematically sampled datasets that I have combined to make the composite, Y dataset used for analysis throughout this book, but they each come to quite different conclusions. The common dataset thus helps make it possible to work out how they came to such different answers.

100% replacement

A key paper published in 2002 by Michael Weale of University College London and colleagues focused on the Anglo-Saxon question. To address the question of whether there is any difference between Wales and England, they used seven sample populations strung in an east–west 'transect line' from North Walsham in Norfolk west to Llangefni in Anglesey, north Wales (all shown in Figure 11.2a). For comparison, they also used samples from Friesland and Norway 'to look for evidence of male immigration from the continent'.[20] Their British samples were carefully selected to represent stable populations from at least the time of the Domesday Book (1086).[21]

In his analysis, Weale wanted to explore three different population processes: simple splitting with subsequent divergence, single mass migration (analogous to the first model described above) and

Figure 11.2a   Weale's British transect line and Continental 'homelands'. The aims of Weale's study were to genetically sample seven ancient market towns in a line from Norfolk to north Wales, and to determine whether the line of Offa's Dyke formed a genetic boundary – and, if so, why.

continuous background migration (analogous, but not identical, to the third model, since no other previous sources of migration are considered). Since none of the available mathematical methods allowed these three processes to be examined simultaneously, in their own words they 'developed an alternative inference method that allowed [them] to explore more flexible models under a range of historical scenarios involving both background [migration] and mass migration in the presence of population splitting and growth.'

Their results were displayed as a virtual genetic distance map. This was constructed by doing the same Principal Components Analysis (PCA) on gene frequencies I described in Chapter 6, and then plotting the value of the First Principal Component against the Second Principal Component for each population sample (Figure 11.2b). This kind of two-dimensional plot is commonly

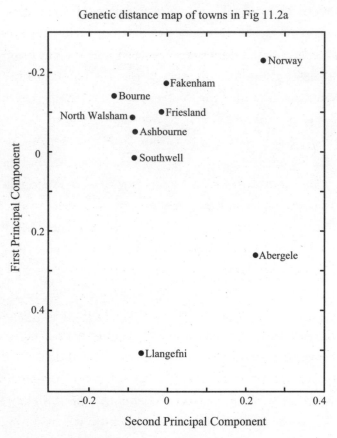

Genetic distance map of towns in Fig 11.2a

Figure 11.2b Weale's genetic distance map. Massive male Anglo-Saxon invasion was inferred from the observation that five English towns group with one another and with Friesland rather than with Wales. Problems of interpretation: the effect is not specific to Friesland, and there are better matches with Belgium and France (Figure 11.4b); plus there are other explanations, such as common previous regional colonization history. (Two-dimensional genetic distance map generated using the First and Second Principal Components in genetic analysis.)

used to simplify and display results for multiple gene types. Using two axes helps to spread out the dots more. It sounds arcane, but the results are clear on the map and can be summarized as follows.[22] The Central English towns were genetically as like Friesland as peas in a pod, and all very different from Wales, but less so from Norway. The main genetic *reasons* for the similarities between Central England and Friesland and their differences from Wales were, they argued, the similar relative frequencies of their two male groups (which are nearly coincident with Ruisko and Ivan in this book). Ruisko dominates Western Europe, but more particularly Wales and the Atlantic Fringe, like the blood group O gene, and Ivan appears, like the blood group A gene, as a minority intruder at rates of 25–30% in north-west Europe.

From this summary of identity and difference, Weale and his colleagues proceeded to computer models of migration and splitting to determine the most likely scenarios and 'to evaluate whether or not a large Anglo-Saxon migration event is needed to explain the extremely high Central English–Frisian affinity'. They then disposed of all but one of their alternative models as 'straw men' and mathematically unlikely, leaving the mass Anglo-Saxon migration with 100% replacement as the most likely event by default.

We can already see a few problems with this conclusion, apart from the size and number of Welsh sample sites. The 'statistical significance' tests required to make the inferences of 'extremely high affinity' rested on calculations based on only seven gene *groups* rather than the many gene *types* available to them.[23] While seven is nearly twice the four possible ABO blood groups, as we have seen, only two Y gene groups are really important in determining population mix throughout England, Wales and Frisia: Ruisko and Ivan. This is not much better than the old blood group story. Furthermore, as mentioned in Chapter 6, PCA analysis gives no dates or direction to gene flow. The study effectively increases the inherent risk of making prehistoric judgements of migration based on genetic distance rather than using a phylogeographic approach, as I have done in this book.

Also, Weale and colleagues effectively set up Norway and Friesland as alternative primary sources of migration from the Continent rather than exploring the possibility they were already mixed populations themselves (my first model). As we have seen in this book, the main Basque refuge group, R1b (Ruisko) dominated Western Europe because he was the first colonizer to arrive after the Ice Age, coming up from Iberia and the Basque Country. Ivan did not arrive in north-west Europe until much later, during the Mesolithic or Neolithic, and seems to have spread as a relative minority rather evenly across north-west Europe, including Frisia and into eastern Britain.

In other words, Frisia, Saxony and England could each have received rather similar secondary admixtures of specific Neolithic Ivan intruders, and then retained the same mix ever since. This 'staged parallel mixing' scenario (my first model) seemed to have occurred to Watkin and Mourant over fifty years ago, but was not tested for by Weale. As they say, 'We note, however, that our data do not allow us to distinguish an event that simply added to the indigenous Central English male gene pool from one where indigenous males were displaced elsewhere or one where indigenous males were reduced in number,' although it is not clear what timescale of 'events' is implied in their comment.

Subsequent papers written by members of Weale's group appear to have diverged somewhat in their adherence to the 2002 conclusions. One of them has taken the idea of total replacement as 'given' and proceeded to work out how Frisian Y chromosomes could so effectively have replaced British ones, for instance postulating a male apartheid.[24] Others have taken a second look at the conclusions by extending the dataset massively to include samples obtained system-atically from throughout the British Isles, and further samples from Scandinavia and the putative Anglo-Saxon homelands of Schleswig-Holstein in the Cimbrian Peninsula and part of north-west Germany at its base. This latter approach, adopted by Italian geneticist Cristian Capelli and colleagues,[25] working at University College London at the time, has incidentally supplied the bulk of the public domain British Y-chromosome data I have used in this book.

Less than 50% replacement

In his introductory paragraph Capelli acknowledges conclusions of an earlier paper published by Orcadian geneticist Jim Wilson and other members of the same group,[26] who used the similarity of Basque and Celtic Y chromosomes to argue for genetic continuity in British indigenous populations from the Upper Palaeolithic to the present. This preamble introduces Capelli's view of a substantial genetic retention of Basque gene types in 'the indigenous population of the British Isles [including England]'.

In the analysis, Capelli represents the indigenous or aboriginal pole (meaning the extreme indigenous part of the genetic distribution on the genetic distance map – see upper left hand corner in Figure 11.3a) by a sample from Castlerea, 'a site in central Ireland that has had no known history of contact with Anglo-Saxon or Viking invaders'. He did not assume that Frisia was the putative Anglo-Saxon homeland, instead using a combination of samples from Denmark, Schleswig-Holstein and north-west Germany (referred to below as NGD, for northern Germany and Denmark) as representative of the invaders.[27] They included an enlarged sample from Norway and marginally improved the resolution of their analysis by adding three STR clusters to the existing eleven gene groups, thus making fourteen.[28]

Like Weale and colleagues, Capelli's group also used Principal Components Analysis to produce a genetic distance map of the populations sampled in their study (Figure 11.3a).[29] Capelli and colleagues note two other poles of distribution in their genetic distance plot: one in the bottom middle of the plot, representing NGD, and the other at the top right, occupied by 'Norway'. This plot gives a rather different perspective on the English, and specifically of

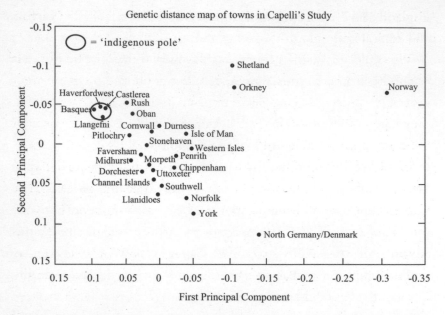

Figure 11.3a Capelli's genetic distance map. The degree of Anglo-Saxon male invasion or (vice versa) 'indigenous' survival, inferred from relative genetic position between northern Germany/Denmark and Ireland. Problems of interpretation: the effect is not specific to northern Germany/Denmark, and there are better matches with Belgium, France and Friesland (Figure 11.4b); plus there are other explanations, such as common previous regional colonization history. (Two-dimensional genetic distance map generated using the First and Second Principal Components in genetic analysis.)

their relationship with Continental 'sources', from that in the Weale paper. In their words, it

> provides significant evidence that there has not been complete population replacement anywhere in the British Isles ... Perhaps the most surprising conclusion is the limited continental input in southern England, which appears to be predominantly indigenous and, by some analyses, no more influenced by the continental invaders than is mainland Scotland.[30]

Overall, using NGD as their source region for migration, they estimate the degree of Continental intrusion (admixture) to

England at 37%, although this figure rises to over 70% in Norfolk and York (Figure 11.3b).[31]

We could well ask – I certainly did – why Capelli rejected Frisia as an alternative Continental source, and even as a point on their plot, if it was so much more like England in the Weale paper. In fact, they had considered using Frisian populations but did not consider them significantly different from NGD populations.[32]

In the Weale paper, the *lack of significant genetic difference* between Frisians and English was used specifically to argue for holocaust. The Capelli paper, on the other hand, uses the *degree of difference* between the NGD sample and the 'indigenous' group to estimate Anglo-Saxon immigration. If groups of researchers with overlapping memberships decided to use Frisia as the source for migration in one study and obtained wipeout, and then used NGD as the source in another study and found less than 50% replacement, we may well be left wondering just why they chose different sources and inference methods and why they obtained different outcomes.

Naturally, both the Weale and Capelli reports provide some of the answers to these questions of method and definition. Weale and colleagues state that: 'Friesland is thought to be one of the source locations for Anglo-Saxon immigration both because of its geographical location and because Frisian is considered to be the closest extant language to Old English.' The Capelli group explain: 'We also note that some historians view the Anglo-Saxons themselves as Germanic invaders from what is now North Germany/Denmark.' So, apart from the linguistic link, both studies appeal to our notorious historians, the former to Procopius of Caesarea and the latter presumably to Gildas, as primary source.

Now, while it is important to try to test different historical migration views genetically, there are clearly several problems of method here. Apart from the choice of replacement estimation, the historical views of both baseline homeland assumptions are disputed, as are historical views of the replacement outcome.

Figure 11.3b   Degree of north German and Danish male 'introgression' into the British Isles estimated by Capelli. According to these estimates, the Anglo-Saxon invasion penetrated much farther and was more massive than any historian's claim. (Contour map generated from Capelli's analysis.)

Furthermore, neither of these papers tested whether, as I have suggested, there are other, older reasons for the gross similarities of gene group frequency between the Low Countries of the Continent and those of eastern England.

The Capelli group does acknowledge some of these problems: 'it should be emphasized that our analyses assume that we have correctly identified the source populations. If, for example, the real continental invaders had a composition more similar to the indigenous British than our candidate sample set, our results would systematically underestimate the continental input.' This last comment, however unconsciously, underlines the incompatibility in mathematical approach of the two papers.

The broader picture of genetic similarity in Northern Europe

There are broader-based sources of information on the male gene group genetic mix of Western Europe which allow us to put the Low Countries of Saxony, Frisia, the Netherlands and Belgium in their place in a larger European picture. Perhaps the most comprehensive of these, which also allows us to compare European populations with the detailed information available on the British Isles, is a massive study published in 2000 by Zoë Rosser and numerous collaborators.[33]

Their study creates a similar genetic distance plot of European male gene groups, using the first and second principal components. This plot clearly shows a simple triangular distribution of populations, with three extreme poles. Not surprisingly, given the discussion in the first half of this book, one of these poles is Basque and the other two are Balkan and northern Baltic (Figure 11.4a). This pattern fits with the male genetic influence from the two former Ice Age refuges of Iberia, from the Black Sea/Balkan region with the third pole representing later gene flow to the Scandinavian Peninsula, the Baltic and north-east Europe, ultimately from the Ukrainian refuge.

Pan-European genetic distance map

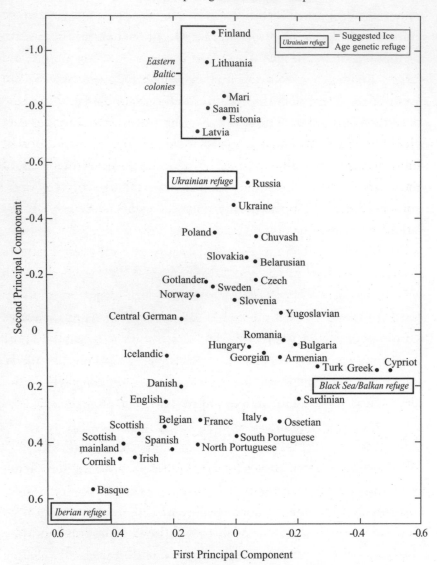

Figure 11.4a Rosser's genetic distance map of all of Western Europe, showing three main sources of male gene flow. Ice age refuges in Iberia, the Balkans / Black Sea and the Ukraine regions form the main poles of the pan-European genetic distribution map, with the eastern Baltic region as an outlier to the Ukrainian pole. Locations in the British Isles, including England, group near the Iberian pole, along with France and Belgium. (Two-dimensional genetic distance map generated using the First and Second Principal Components in genetic analysis.)

The West European genetic landscape thus seems informative and rather uncomplicated at the level of the male gene groups used in the Rosser study. In this book I have argued for there being three sources of post-LGM north-west European colonization, the oldest and largest input being from the Iberian refuge, followed by a Balkan input up the Danube during the Mesolithic/Neolithic, and then the Scandinavian/Baltic influence coming in mainly during the Neolithic. There are other interpretations, of course, with different timescales, but it is difficult to see how they could affect the details of the Basque pole, which shows the extraordinarily tight genetic conservatism of Western Europe. In Rosser's work, the closest population to the Basques is in Cornwall, followed closely by Ireland, Scotland, Spain, Belgium, Portugal, East Anglia and then northern France. At this point the triangle starts to spread, with Mediterranean countries on the lower border as we move towards the Balkan pole in the south-east, and Germanic-speaking countries on the upper-left border as we move towards the Baltic pole. The northern Scandinavian countries cluster predictably on the upper border, roughly equidistant between the Basque and northern Baltic. Germany is also far up that line, and near northern Scandinavia.

East Anglia is similar to all the Low Countries in gene group mix!

When we look more closely at the countries bordering the North Sea on Rosser's plot, we find that East Anglia is no more 'Germanic' than the Netherlands or northern France, although more so than Belgium. Given the juxtaposition of these countries, and the absence of Frisia from this study and its crucial importance for the Weale argument, I decided to re-analyse the English dataset to look more closely at the mix in the Low Countries. To do this I included not only Frisia, Norway and NGD, but all the other countries bordering the North Sea from Rosser's study, using gene group markers common to all studies. For better detail in this analysis, I excluded Central and Eastern Europe (including the Balkans).

The results were interesting (Figure 11.4b). As found previously, Frisia linked very closely with the samples from Norfolk, and as expected also with Rosser's East Anglia sample. This was a closer similarity than for the 'Anglo-Saxon homeland' of Schleswig-Holstein and north-west Germany and much closer than for Denmark, which, in contrast with Capelli's results, grouped closer to southern Norway than to their north-west German sample. However, the Continental 'Anglo-Saxon homeland' defined in this way, without Denmark, is still considerably closer to Norfolk than to Denmark.

Were Caesar and Tacitus right about the Belgae?

The real surprise was to find Belgium moving farther west into England than Frisia or the rest, and nestling into a group consisting of the three central English towns of Weale's transect line – as if to bear out Caesar's claim that some English regions were more Belgian than British. Northern France and the Netherlands, although bordering the English samples, were all much closer to England than Denmark or Rosser's central German sample.

So many independent indications of Continental similarity with England is what we should expect if the Low Countries of north-west Europe all had similar colonization histories. This view of parallel regional development makes much more sense than Weale's interpretation of 'cleansing' by similarity which, if extended, would conclude that each and any of the Low Countries could have wiped out Britons, as was claimed for Frisia.

So, it is essential to check for similarity by common descent before opting for genocide. What the story shows so far is that correlations of gene frequency mean nothing without a basic model (and knowledge) of the genetic prehistory of the regions in question. Weale's and Capelli's studies generated certain models based on historical perspectives of Procopius and Gildas and then selected the genetic populations against which to test those models.

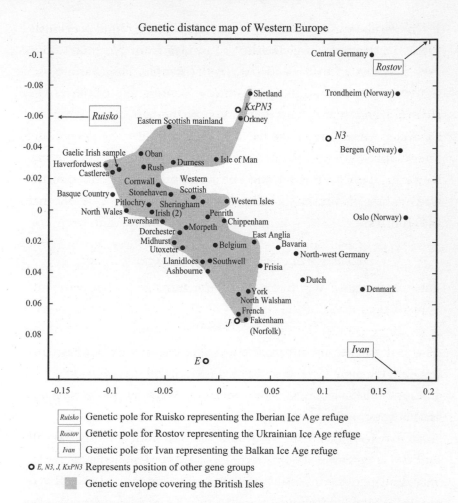

Genetic distance map of Western Europe

Figure 11.4b We are all family' – or are Belgians more English than the English? Put in the context of the whole of Western Europe, the mix in the south-eastern male British genetic distribution (lower part of the shaded area) fits best with Belgium, followed by France, Frisia and Bavaria, rather than north-west Germany and Denmark. The simplest explanation for this regional similarity is a common regional prehistory of colonization, rather than recent invasions. (Two-dimensional genetic distance map generated using the First and Second Principal Components in genetic analysis. North-west European data in Figure 11.4a extended to include dataset used in this book.)

If circularity is to be avoided, effective models need to be derived primarily from genetic, climatic and geographical information, not driven by narratives drawn from recent history or archaeology.

We cannot assume the similarities or differences of the various British tribes from their neighbours to be the consequences of controversial historical events of the recent past if we do not have some idea of what went on before. Patterns of gene group frequencies on their own cannot be given a date, but are usually the final result of a number of different events that took place at different times. Even splitting up the gene groups into numerous gene types in an attempt to define regions more precisely brings its own problems if no attempt is made to produce genetic trees and date expansions and movements of branches.

Reconstruction

The phylogeographic approach to genetic prehistory that I have used in this book aims to reconstruct a genetic history of individual gene lines and their movement from source regions to target regions. This means first examining the representation of the branches of each line in different regions in order to establish the source and target. The second part is dating the branch points and the gene line movements. Increased resolution of the Y-chromosome gene tree using STR gene markers makes it possible not only to get a more informative tree, but also to get date estimates on the branches.

To summarize, the phylogeographic approach establishes three broad aspects of West European and British colonization in the past 16,000 years which have a bearing on the Anglo-Saxon question. First, all but a few per cent of male and female gene lines appear to have arrived in the British Isles *before* the historical period (i.e. before the Anglo-Saxons). Second, most British colonizers, including about two-thirds of *English* ancestors, came from the Iberian refuge soon after deglaciation, or at least during the Mesolithic. And third, the *subsequent* colonization of the British Isles during the Neolithic and the Bronze Age was complex in time and space, but mainly came from the other side of the North Sea.

Eastern England received some input from the Balkans via the Danube and the North European Plain during the Mesolithic, and considerably more during the Neolithic from the same region and also increasingly from southern Scandinavia. The Bronze Age saw further development of this trend in eastern Britain, and Wessex and the south coast began to receive more gene flow from north-west Europe. This sequence of dated, incoming, prehistoric male lines does not leave much room for the culturally dramatic invasions of the past two thousand years.

A fresh approach to time and mixing

Clearly, not all geneticists may agree with my critique of the Weale and Capelli research or with my summary of the complex pre-historic interplay between the now Germanic-speaking areas of Europe and Britain, but my findings ought to be independently replicable. In summer 2005 I presented a paper at a Cambridge symposium on methods of simulating and estimating migrations, and was amazed to hear a paper given by Cambridge geneticist Bill Amos which had come to some similar conclusions to mine but used a completely different mathematical approach and dating methods. Amos and his colleagues clearly felt that simplistic models like Weale's were unsatisfactory, and that knowing and understanding the history of previous population mixes in the north-west European region was an essential prerequisite, and that this had not been addressed by Weale or Capelli.[34]

I shall not go into detail of their complex, novel yet robust approach to dating multiple admixtures between the Continent and Britain, but merely quote from their discussion, as they pose the problem so clearly. I preface this quote by explaining that Amos's definition of Anglo-Saxons, for the purpose of his analysis (and to be open-ended about dates), were people who had come over from

north-west Europe carrying male genes at *any* time in the past ten
thousand years, rather than the more conventional Gildas concept
of sons of Saxon wolves who ravaged England in the fourth and fifth
centuries AD. Also, when he refers to a 'time slice' he means a
reconstruction of previous admixture and affinity between populations at various specified times in the past.

> An important question raised by this analysis concerns when Anglo-
> Saxons came to England ... was it through one major influx following
> the collapse of the Roman empire or were there earlier incursions?
> Overall, English towns are tightly linked to NGD and Friesland,
> samples collected from the Anglo-Saxon homelands, but this affinity
> tells us nothing about when the Anglo-Saxons came across nor how
> long the settlement took. However, there are two reasons for
> suspecting it was earlier rather than later. First, in all but the deepest
> time slice, NGD and Friesland tend to cluster as a pair, suggesting
> stronger links between these two than between either one and
> England. Such a pattern is most consistent with NGD and Friesland
> being continental neighbours with greater exchange between each
> other than across the channel with England. The second point relates
> to the deepest time slice, where NGD joins to Europe rather than
> Britain, suggesting reduced links. At this point, Friesland is indistin-
> guishable from a number of towns that together form a very shallow
> English clade. Such a pattern could arise either if mixing with
> Friesland began earlier than with NGD or, in view of the extremely
> shallow English clade, if at this point the English and Frisians were one
> and the same. Taken together, therefore, our analysis appears to favour
> a history in which Anglo-Saxons arrive in Britain considerably before
> 4–6th centuries, possibly as a group of Friesians who moved across
> 5–10,000 years ago (depending on the time calibration).[35]

To précis and oversimplify the final point, Amos is saying that the
affinity and migration between Friesland and England is likely to be
much more ancient than that with the 'Anglo-Saxon' homelands – at

least going back to the Neolithic Age. Amos was not able to resolve any more recent gene flow, given the gaps between his time slices, but we are now collaborating to look at that question.

Only three British male founder lines in the historical period

Back to my own study. After estimating ages for all the constituent clusters of the main British founding haplogroups, there were only three founding clusters dating to the historical period of the past two thousand years. Two of these originated in Scandinavia, one from the south and one from the north. None of the three, however, could be identified with the putative homelands in north-west Germany (Saxon) or Schleswig-Holstein (Angeln) as *sources*, or with Norfolk in particular as a *target* (Figure 12.3). This does not necessarily mean that there was no invasion from those homelands, Anglo-Saxon or otherwise; merely that I could not detect it by looking for and dating specific recent genetic founding events.

There are several possible reasons for this lack of evidence, the main one being the short time period available to create detectable founding lineage clusters in Britain by random mutation.[36] The existence in England of members of a common founder line originating on the Continent does not necessarily tell us when it came, unless there are unique new mutations in England that can be used for dating. This would be even more difficult if the founding size was relatively small and recent.

Matching exact gene types to source and destination

Having, so to speak, scraped the male barrel looking for a genetic founding event to match the Anglo-Saxon invasion, I felt that there was one other possible way to exclude significant specific migration. That was to look for exact gene type matches between the samples from the British Isles and each of the various putative Continental sources (Iberia, the Anglo-Saxon homeland, Norway,

Denmark and Frisia). This is in a sense taking a leaf out of the archaeologist's book, of matching cultural items in different places, but instead using unique male gene markers. To identify the most likely Continental sources of particular gene types, I was able to refer to my earlier network phylogeographic analysis of lineages, described in this book and its Appendices.

This approach has its pros and cons. The main advantages are that full use can be made of the very detailed information provided by the STR gene types, and recent mass migration would be expected to produce such exact matches. The main disadvantage of this quantitative matching approach is that it cannot make full use of available information on time depth. There is also likely to be some gene-type overlap between the different Continental source regions (Figures 11.5 and 12.4).[37]

I can honestly say that I did not hold out much hope for this approach, reminiscent as it is of the 'happy families' card game, and given the sampling issues. So it was a surprise when there were very specific results. I shall come back to the Norwegian results (Figure 12.4) a little later, since their distribution seems generally more appropriate to locations of Norwegian Neolithic and Viking raids, but any specific matches found for Frisia, Denmark, Schleswig-Holstein or north-west Germany could theoretically be relevant as possible Anglo-Saxon intrusions. As it turns out, there were none for Frisia (see below), so that is not shown as a figure.

Since Denmark was clearly different on the genetic distance map (Figure 11.4b), and had a different mix of haplotypes from Schleswig-Holstein (Angeln) and north-west Germany, I treated it separately

11.5 Specific male intrusions from Europe into the British Isles (shown as three genetic frequency contour maps based on the percentage of exact matching STR gene types between source and target areas; arrows indicate direction of gene flow based on gene trees and geography).

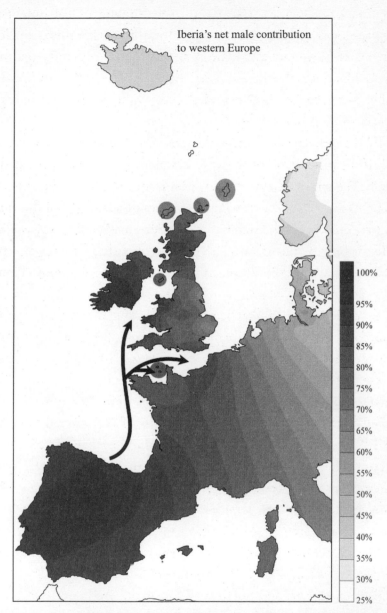

Figure 11.5a Intrusions from an Iberian source. By far the majority of male gene types in the British Isles derive from Iberia, ranging from a low of 59% in Fakenham, Norfolk to highs of 96% in Llangefni, north Wales and 93% in Castlerea, Ireland. On average only 30% of gene types in England derive from north-west Europe. Even without dating the earlier waves of north-west European immigration, this invalidates the Anglo-Saxon wipeout theory.

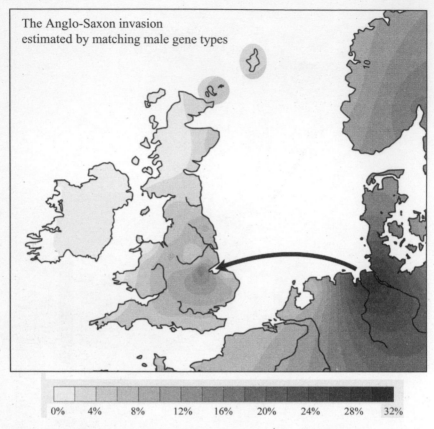

Figure 11.5.b Intrusions from the traditional Anglo-Saxon homelands of Schleswig-Holstein (Angeln) and north-west Germany (Old Saxony). Only an average of 3.8% British male gene types have matches in the Anglo-Saxon homeland region. This rises to an average 5.5% in England and 9–15% in parts of Norfolk. Frisia has no similar degree of matching, indicating that the Anglo-Saxon gene flow event was real, but very modest.

from these two, which I combined in this analysis as the 'Anglo-Saxon homeland'. They also fit the putative Anglo-Saxon homeland on cultural grounds (Figure 9.3). Figures 11.5 and 12.4 show the results of gene type matching from the south-west European source (Figure 11.5a) and the various North European sources: 'Anglo-Saxon homeland' (Figure 11.5b), Denmark (Figure 11.5c) and Norway (Figure 12.4). As might be expected, there is some degree of overlap

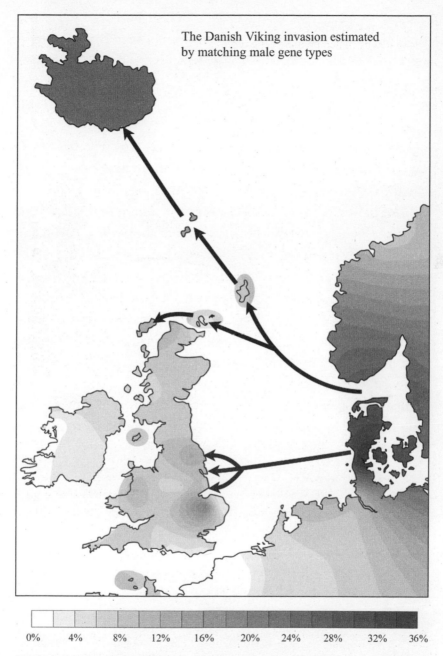

The Danish Viking invasion estimated by matching male gene types

Figure 11.5c Intrusions from Denmark. These matches are found in areas consistent with both Anglian and Danish Viking invasions (Figures 9.2a and 12.2), namely the Fens (8%), Fakenham, Norfolk (19%) and York (11%), but also other areas consistent only with Vikings, for instance Iceland.

of matched gene types between the different European sources of gene flow. The greatest degree of overlap is between Danish matches and Oslo ones in southern Norway, which is to be expected from geography and history.

75–95% of British Isles matches derive from Iberia

Figure 11.5 tells us a lot about specific gene type matches across the North Sea, but before describing the more interesting Anglo-Saxon stuff, I should mention several things which serve to validate the method. The most important source region for West European matches overall is, as expected, Iberia (Figure 11.5a). Even Trondheim in northern Norway has over 25% from the southern source. Ireland, coastal Wales, and central and west-coast Scotland are almost entirely made up from Iberian founders, while the rest of the non-English parts of the British Isles have similarly high rates. England has rather lower rates of Iberian types with marked heterogeneity, but no English sample has less than 58% of Iberian types, or what Capelli and colleagues might call 'indigenous' gene types. Looking at this picture the other way round, overall male intrusion from Northern Europe into England from any time since the last Ice Age varies from 15% to 42% (average 30%) in different samples. This conservative Atlantic coastal picture for the British Isles is completely consistent with the balance of sources of gene flow into the British Isles discussed earlier in this book.

How does my range of 15– 42% of intrusive North European male gene type markers to England compare with Capelli's or Weale's estimates? Although the dataset is nearly the same, the methods are different, so strict comparisons cannot be made. But my average figure of 30% is obviously closer to Capelli's estimate of about 40% intrusion than to Weale's 100% wipeout.[38] However, given that the cross-Channel matches inevitably include a much larger flow of founder gene types from the earlier Mesolithic and

Neolithic periods, the overall figure of 30% invasion into England from Northern Europe since the Ice Age is still most likely to be a gross overestimate of the recent Anglo-Saxon invasion, and even more so when we focus just on matches with the 'Anglo-Saxon homeland'.

There really was a (small) invasion from Angeln and northern Germany

When we do begin to look at the effects and variation in rates of specific 'Anglo-Saxon' gene types throughout Britain, several patterns emerge. First, the 30% intrusive figure falls sharply to 5.5% in England, and an average of 3.8% over all of the British Isles. This still seems to point to a real, though small, Anglo-Saxon invasion of eastern Britain and England (Figure 11.5b). Exact gene type matches from the putative Anglo-Saxon homelands are found at frequencies of 5–10% throughout England.[39] Within England the highest rates of intrusion, 9–15%, are seen in parts of Norfolk, in the Fen country around the Wash and, notably, in the Mercian and Anglian English towns of Weale's transect (shown in Figure 11.2a).

My figure of 5.5% for genetic intrusion within England, rising to a maximum 9–15% in eastern England, can be compared with archaeologists' estimates of the 'Anglo-Saxon' invasion. Presently, the boldest of these, given by Reading University archaeologist Heinrich Härke, is of an invasion of about 250,000 people, into a British population of 1–2 million. Translated back into population proportions, in his own words:

> Archaeological and skeletal data suggest an immigrant–native proportion of 1:3 to 1:5 in the Anglo-Saxon heartlands of southern and eastern England ... but a much smaller proportion of Anglo-Saxons (1:10 or less) [in] south-west, northern and north-west England.'[40]

These burial-based figures of 10–33% immigration give a range about two to three times higher than mine, but still lower than Capelli or Weale's genetic estimates.

Outside England, similar but slightly lower rates of intrusion are seen only in the Channel Islands, the Isle of Man, Cornwall and farther north on the east coast of Scotland and in Orkney and Shetland. Elsewhere outside England, 'Anglo-Saxon' intrusions are uniformly low, with Scotland at 1.7%, the Scottish Isles at 3%, Wales at 1.5% and Ireland at 0.8%. The exception to this picture is the Llanidloes sample (5.3%) which, we have seen elsewhere, always tends to group more with England and holds all the Welsh Anglo-Saxon matches (Figure 11.5b).

Less in New Saxony

Curiously, however, the counties along the 'Saxon coast' of southern England have a consistently lower rate of 'Anglo-Saxon' matches than the Anglian regions: 5%, which is close to the baseline background level for the rest of England. This lower 'Anglo-Saxon' signal for the English south coast is more consistent with Bede's fifth-century invasion of Angles than with Gildas' claimed dramatic *Adventus Saxonum* of the same date. Combined with genetic evidence for earlier Neolithic and Bronze Age intrusions into Wessex and the south coast (see Chapter 5), this would also be more consistent with a longer-term presence of north-west Europeans in the Saxon counties.

Matched Danish gene types in the British Isles, although to some extent overlapping, also differ sharply from the 'Anglo-Saxon' ones in that they are found both within and outside England in a characteristic coastal distribution, geographically suggestive of historically recorded Viking raids (Figure 11.5c). Their different distribution but similar approximate dating suggests the Vikings as more likely culprits than Angles, and I shall return to this point later.

Absence of Frisian-specific matches validates 'Anglo-Saxon' homeland

Perhaps the best validation of my matching approach to *specific* gene flow into Britain from southern Scandinavia and the Cimbrian Peninsula is the nearly complete absence from the British Isles of the numerous gene types specific to Frisia. Frisia is so much nearer geographically to eastern Britain, and so much closer in language and gene-group mix, that, on the basis of neighbourly affinities it would be expected to have more valid matches than the Cimbrian Peninsula – but it has virtually *none*. Needless to say, I have not shown the Frisian map for this non-migration.

Overall, 4% of Anglo-Saxon male intrusion into the British Isles (maximum 9–15% in those areas of eastern England which from the archaeology would have been expected to bear the brunt) seems more reasonable than the wipeout theory. Assuming that it is a true reflection, 4% overall should still not be regarded as a minor event. That is a higher immigration figure than my estimate of 3% (see Chapter 5) for the entire Bronze and Iron Ages put together, and would represent ancestors for more than a couple of million of today's population.

Anglo-Saxons and mitochondrial DNA

So far I have discussed only male markers in connection with the Anglo-Saxon invasions. This is mainly because that is where the most geographically specific, and indeed most of the published, genetic information is found. The mtDNA maternal evidence, such as it is, tends to support my story.

The Dark Age historians say rather little about women crossing from Europe, except for Procopius' second-hand report in which he mentions annual cross-channel family holidays (see Chapter 9): 'So great apparently is the multitude of these peoples that every year in large groups they migrate from there with their women and children and go to the Franks.'[41] Gildas' report spoke only of warriors, although he is obviously talking about Saxons settling. We cannot, however, assume by default of accurate history that the *Adventus Saxonum* would have been purely a male elite, wherever it came from. Part of the independent body of evidence which suggests mass movement of peoples relates to the extensive contemporary flooding of the Continental coastline, leading to land hunger, particularly in Frisia.

As we have seen, some historians and archaeologists have seen these invasions as massive, with whole communities moving in from Germany and sweeping across a defenceless and largely depopulated England.

The English maternal genetic record in mtDNA undermines this story with complementary evidence from two key maternal lines (see also Chapter 5). One of these, J1a, specifically links Germanic speaking areas of Europe with England and is not found in other parts of the British Isles except for Lowland Scotland. Cambridge geneticist Peter Forster, however, argues that this line is most likely to have entered Britain in the Neolithic, not the Dark Ages.[42] Apart from the observation that the overall European expansion date for this line is around 5,000 years ago, Forster has what is perhaps a more telling, piece of negative evidence: that English females have a low rate of a specific Saxon mtDNA marker.[43] Instead, their main cross-channel maternal links are considerably older. So, if only a few Saxon males came during the Dark Ages, they would have been matched by similarly few females.

Summary

In this chapter, I first presented three studies which used genetic similarity or distance between modern populations as means of addressing the question of Anglo-Saxon wipeout vs continuity theories. Their results were wildly at variance with one another and greatly overstated even the most aggressive modern archaeological view of that invasion. Apart from criticisms of their assumptions, and the lack of dating and resolution in their methods, my main worries were: first, that they implicitly regarded potential migration source areas such as Frisia, northern Germany and Schleswig-Holstein as regions lacking their own genetic prehistory of intrusion running parallel to England; and second, that they did not adequately consider a deeper timescale of North European migrations to the British Isles.

A Pacific proverb runs thus: 'To know where we are going, we have to know where we are; to know that we have to know where we came from.' I would add to that '… and when'. Although the phylogeographic method I have used in this book demonstrated substantial migrations from Northern Europe dating to the Neolithic and the Bronze Age, it failed to identify any specific founding male gene clusters to match the Anglo-Saxon invasion, probably reflecting the recent nature of the event and possibly its relatively small size. Exact STR matching did, however, suggest a specific immigration sourced from the putative Anglo-Saxon homeland, but amounting to only 5.5% of ancestors for modern English people.

In the next chapter I go on to use both these methods to trace evidence of Viking invasions, for which the literary and archaeological evidence is much more abundant. In the event, both the methods I used produce results consistent with each other and with the known distribution of Viking settlements.

12

THE VIKINGS

The genetic trail still has more mileage in the few hundred years after the Anglo-Saxon hegemony. For five hundred years the Saxons made their mark, if not as large as previously thought in the genetic heritage of Britain, then certainly in culture and in clerical and historical records, not to forget the shire names. But those records increasingly complain of fresh invaders from across the water. These were raiders from Denmark and Norway.

History repeats itself in patterns of invasion

This time there were more, and better, historians to record 'awful events' than the prophet St Gildas. Although the new incursions were much better documented than the Anglo-Saxon invasions, the English chronicles are still biased towards the more gory details. English historians identified these intrusive Danes and Norwegians non-specifically as 'Vikings', while the Carolingians called them 'Northmen'. The Danish historian Saxo Grammaticus, gives the raiders no special label, just their geographical origin, in his *Gesta Danorum* – which is perhaps not surprising, considering his perspective. Presumably, they were more or less the same old Scandinavians, with the same seafaring skills possessed by their ancestors of the previous few hundred years, just behaving badly overseas.

Obviously the extraordinary extent of the pan-European Viking spread, and the reasons for and ferocity of their seaborne attacks in the eighth to tenth centuries, were unique, and validate the continuing use of a special 'Viking' label rather than 'Oh God! Not more Scandinavian raiders!' But they were not the first people of Scandinavian origin in sea-going longboats to leave their mark on the British Isles, nor were they the last such Continentals to invade Britain, possibly carrying similar male genes.

Scandinavian longboats had been constructed, had travelled abroad, and had been under sail, far longer than the 'Vikings'. As we have seen, Gildas described the invading 'Saxon' boats as high-prowed, sailing longboats. In the last chapter I linked the longboat burial at Sutton Hoo with its contemporaries, the Scandinavian longboat burials in Schleswig-Holstein, Denmark, Sweden and Norway (Figure 9.3, Plates 17 and 18). Among other common cultural clues, the Sutton Hoo burial lends the Anglo-Saxon invasions of eastern and northern England a distinct Scandinavian tint. Gildas gives us one of the earliest written statements that these ships could move under sail. As Barry Cunliffe points out, 'Sleek, elegant vessels of clinker-built construction, with planks sewn together, are known [from Nydam, Denmark] as early as the fourth century BC. By the fourth century AD the overlapping edges of the planks were clasped by iron nails, and paddles had been replaced by rowing oars'.[1] The same high-prowed sailing longboats appear on the famous Bayeux Tapestry as used not only by the Norman invaders, but also by Saxon King Harold before his defeat and death.

Scandinavians honed their sailing skills over many centuries as vital for communication in a land of high mountains strung along fjord-scored coastlines. It may have been improvements in sailing techniques that widened the range of raiding. A complex of other reasons have been suggested to explain the Viking era.[2] Knowledge of the pickings for plunder south and west must have been a spur, but there was also trade and increasing land-hunger in the narrow coastal strip backing onto the high mountains of the hinterland.

Earlier in this book, I discussed the genetic evidence for significant Scandinavian intrusions into Britain, especially the north, during the Neolithic Period. I shall come back to the genetics of the Vikings a little later; meanwhile, it is important to realize that attention-grabbing television documentary titles such as 'Blood of the Vikings' should carry a health warning in the small print: not all Scandinavian gene markers need have been imported to the British Isles by Vikings.

Although historical contact[3] between Norway and north-west Britain began in the seventh century AD, serious raiding started from the last decade of the eighth century and built to a terrible crescendo in the following seventy years. Norwegian Vikings exten-sively attacked the western half of the British Isles (Figure 12.1) almost exclusively affecting what are regarded today as Celtic lands in northern and western Scotland, Wales, Cornwall and Ireland, extending down the Atlantic coast to Brittany and Spain. Cunliffe suggests that the early Viking colonization of north-west Britain was largely completed during the eighth century, providing a spring-board for more juicy prizes in the monasteries of Ireland.

That England was initially avoided by both Danes and Norwegians is in itself curious: possibly it was because previous alliances between the Anglian kings of England and their cousins in Denmark held from the sixth century, as implied by the context in *Beowulf* (and by mention there of the names Offa, Finn and Hengist),[4] were initially still active. That perspective would tend to support the view that the *Beowulf* oral saga was originally created before 793 (the year of the sack of Lindisfarne) rather than after 850, when a new order was established in the Danelaw.

Somewhat later than the Norwegian onslaught, starting from the beginning of the ninth century, the Danes threw out cousinly values in the pursuit of booty and glory and adopted the same tactics as the

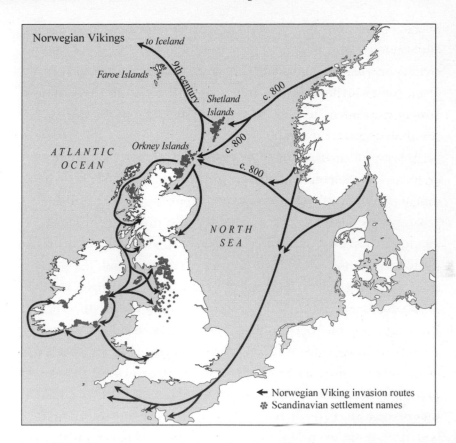

Figure 12.1 Viking invasions in the west. Norwegian Viking invasions and Scandinavian settlement names concentrated on Shetland, Orkney and the western side of the British Isles, particularly the Western Isles, Cumbria, Lancashire and Ireland.

Norwegians. Commencing by attacking their immediate neighbours in Saxony, they moved on to Frisia, then to the Frankish kingdoms, and on to England and France. All hell was let loose, and Viking raids became a free-for-all, the longships sailing out from both Norway and Denmark to scour the coastal regions of Europe for plunder.

Danish attacks on eastern and north-east England and up the River Thames intensified from 830 onwards, developing into a territorial repeat of the Anglian invasions that began four centuries before. In 865, a large Danish army invaded East Anglia and over-wintered there. Consolidating their base of operations, they went on, over the next five years, to conquer and settle East Anglia, Northumberland and northern Mercia, much as their former neighbours the Angles had done (Figure 12.2). And similarly, apart from forays up the Thames and Severn, the Danish Vikings were unable to gain any foothold in the non-Anglian, traditionally Saxon areas of southern England.

This repeat demarcation – or palimpsest, to borrow a term used earlier – of settlements between former Anglian lands and Saxon territory, the latter defended by revitalized fortifications of the old 'Saxon shore', seems quite a cultural and historical coincidence. It cannot be simply explained by King Ælfred's stout defence of the South, since the Danes never settled there at all. Essex is a geographical exception that could prove this rule. Although Essex was a vulnerable and isolated Saxon pocket north-east of the Thames bordered by East Anglia, unlike Norfolk and Suffolk it was never settled by Vikings (Figure 12.2). This was in spite of the fact that it was ultimately included under the Danes' control, north of the agreed Danelaw line. That may tell us something about the effect of common culture on ease of conquest and settlement. Geographically, Angeln is a stone's throw from Denmark, and it was the Anglian rather than Saxon territories that fell so quickly to the Danes.

'Blood of the Vikings'

As we have seen, there tends to be some overlap between genetic markers held by the Danes and those of other Germanic-speaking peoples of north-west Europe, but the northern Norwegian markers are more clearly recognizable. However, none of these male markers have labels such as 'Neolithic Scandinavian' or 'Iron Age Viking' written on them in ink. So again it becomes important to try to identify when such markers arrived and from where.

As I mentioned above, by using the phylogeographic approach I have managed to identify just three founding clusters that date to the historical period.[5] Two of these are of Scandinavian origin and account

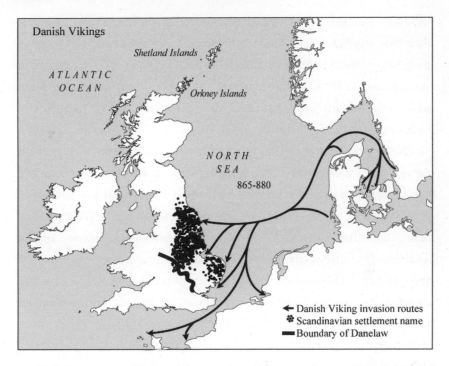

Figure 12.2 Viking invasions in the east. Danish Viking invasions concentrated in East Anglia and north-east England, particularly York. The Danelaw Line is shown as well as Scandinavian settlement names in eastern Britain. The latter are much more numerous than Viking graves, which have a more coastal distribution (not shown).

for 5% of extant male lines in the British Isles (Figures 12.3a and 12.3b). Given the stringent criteria applied in the identification of these most recent founder clusters, this figure is likely to be an underestimate, but it is somewhat greater than for the Anglo-Saxon intrusion, given that I could not identify any clear Anglo-Saxon founding male clusters. Gene type matches reinforce this measure of Viking influence – as we shall see.

Danes

Since there was already a geographical overlap between the Anglian territories of eastern Britain (discussed earlier) and those invaded by the Danes, I shall consider the Danish Vikings first. Danish gene type matches account for 4.4% of the entire British sample, which is marginally greater than the similar figure of 3.8% for Anglo-Saxon matches.[6] Of the three British founding clusters of the historical period, one, I1a-3, was identifiable as a founding gene line from Denmark and its derivative British cluster. Although found elsewhere in Britain, particularly on the east coast, I1a-3 points to a clear founding event in Fakenham, west Norfolk (Figure 12.3a).[7] This cluster dates to around 1,200 years ago.[8] It is found throughout the areas known to have been raided by Danish Vikings, and also to some extent in Norwegian Viking colonies on the west coast of Britain. The latter may be explained by overlap between Denmark and nearby southern Norway in the distribution of the founding gene type for this cluster.[9] This is reflected in the map (Figure 11.5c) by the presence of 'Danish matches' in Scotland, its islands, and west-coast Britain traditionally associated with Norwegian Vikings.[10]

It might be thought that, with so many different inputs from related north-west European groups, any differentiation between sources of gene flow would have been lost. Yet some pattern persists. In line with overwhelming historical and place-name evidence (Figures 12.1 and 12.2), male intrusive gene types in York and Norfolk are more characteristic of northern Germany and Denmark than of Norway (ratio of NGD to Norway 27 : 5; compare Figures 11.5b, 11.5c and 12.4).

If we look at shared gene type matches between Denmark and the British Isles, we can see this pattern of distribution in more detail (Figure 11.5c). Danish-linked gene types are found in several parts of England, but in a patchy distribution rather different to that of the 'Anglo-Saxon' matches. They are nearly absent, for instance, from the parts of East Anglia (North Walsham) and Sussex coasts previously identified by the Romans as the 'Saxon shore'. Conversely, they are most common in York and around the Wash. Their highest rate anywhere is in a single pocket in the west Norfolk town of Fakenham, making up 19% of the sample.[11] Most of these

12.3   Three British male genetic founding clusters of the historic period, none of which clearly relate to the Anglo-Saxon Advent.

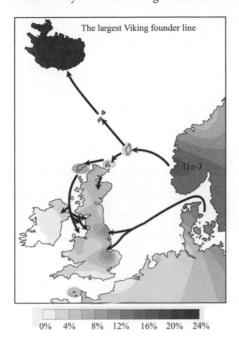

Figure 12.3a   The largest Viking gene cluster, I1a-3, arose in Bergen, Norway in the Bronze Age. The British founding I1a-3 cluster dates to 1,200 years ago and is found in all regions visited by the Vikings in Britain, including York (11%), Norfolk (11%), the Western Isles (7%) and the west coast. Additionally there was a strong founder effect in Iceland (22%).

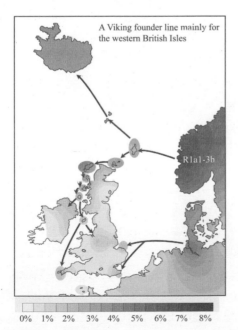

Figure 12.3b   Gene cluster R1a1–3b also arose in Bergen in the Bronze Age and arrived in the British Isles within the last 2,000 years, settling mainly in the Western Isles (5%), and to a lesser extent in Shetland (2%), Orkney (3%) and other parts of the western British Isles.

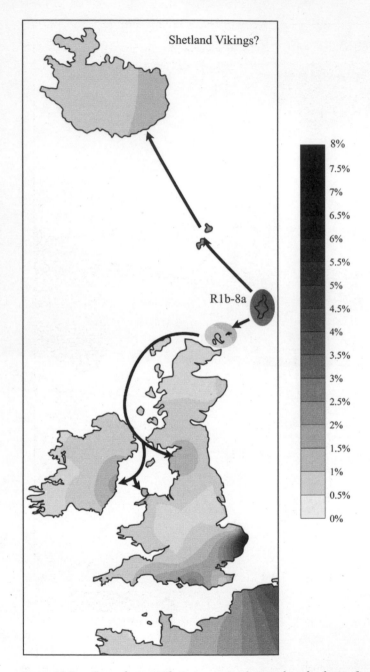

Figure 12.3c   Gene cluster R1b-8a is a uniquely British male cluster found in Shetland and Norfolk and does not derive from northern Europe. He appears to have expanded down the west coast of Britain and also ultimately to Iceland, as if he were a Viking.

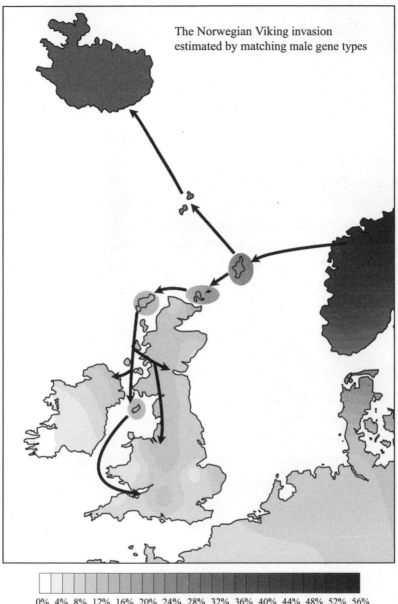

The Norwegian Viking invasion
estimated by matching male gene types

0% 4% 8% 12% 16% 20% 24% 28% 32% 36% 40% 44% 48% 52% 56%

Gene frequency

Figure 12.4 Matching male gene type intrusions from Norway. These are found in typical Norwegian Viking locations, in Shetland (20%), Orkney (17%), and the Isle of Man (10%), and at rates of 7–8% down western Britain and in Northern Ireland. 33% of Icelandic male genes have matches in Norway.

latter belong to the same gene type and constitute part of the Danish founding cluster I1a-3. The frequency of Danish gene type matches in the areas under the Danelaw in the late ninth century is consistent with the density of archaeologically visible Viking settlements in those areas.

Norwegians

Julian Richards, the archaeological presenter for the BBC documentary 'Blood of the Vikings', eloquently described the general difficulty of finding enough physical traces of Vikings to match history's horror stories and estimate the impact experienced by the British. As he pointed out, part of the reason was that they left little in the way of non-perishable remains, although there were many settlements. But in spite of the physical silence, place-name and linguistic evidence is abundant (Figures 12.1 and 12.2). The Norwegians, like the Danes, did leave their genetic mark with its strongest traces in the most vulnerable of the western British Isles and in the northern islands of Scotland.

The male genetic group markers most clearly linked to Norway, in particular Trondheim in the north, belong to the more eastern European gene group R1a1, or Rostov (Figure 4.10). Five of Rostov's descendent sub-clusters date to the Bronze and Neolithic Ages and earlier, and are represented in eastern Britain, East Anglia, the Isle of Man, and Scotland and its northern islands (the Western Isles, Shetland and Orkney). The two Norwegian Bronze Age clusters in this group focus more specifically, one in Shetland and Orkney and the other on the Isle of Man[12] (Figure 5.14a, Chapter 5, and Appendix C). Only one Rostov cluster, R1a1-3b, dates to anywhere near the historical Viking period.[13] Coincident with known locations of Norwegian Viking raids, R1a1-3b is found most in the Western Isles and in Shetland and Orkney, as well as locations in Western Scotland, the Isle of Man and north-east Ireland (Figure 12.3b). Based on this division in dates between the Bronze Age and Neolithic on the one hand, and the time of the Viking raids on the other, only 17.5% of the R1a1 lines (1% of extant British lines) would have arrived in the British Isles during the Viking period.[14] Although this may be an underestimate, and the true figure may be as high as 2%, it underlines how easy it is to overestimate high-profile historical migrations from the Continent when there were multiple migrations from the same source at earlier dates.

Indigenous Shetland Vikings?

Having dealt with the first two of the three British male genetic cluster expansions from the historical period, I should mention that the third Dark Ages founding cluster is the biggest surprise. It cannot in any way be identified with the nearby Continent across the North Sea, or with a recent founding event from Iberia. In the context of the Dark Ages, it was actually an indigenous re-expansion and should not really be called a founding cluster in the British context. This is cluster R1b-8a, which is a sub-cluster of Rob, who arrived in northern and southern Britain during the Mesolithic (Chapter 4) and re-expanded during the Neolithic (Figure 5.5c) and ultimately derives from the Mesolithic Iberian male R1b-10 cluster (Ruy). The R1b-8a cluster is uniquely 'off-Continent' and characteristic of the far north of Scotland. It appears to have expanded from Shetland to Iceland and down the western British Isles and nowhere else (Figure 12.3c).[15] In other words, these are 'indigenous' British male types spreading genetically as if they were Vikings, and maybe acting in the same way (Plate 21). The western British locations seem typical for the Viking spread and include Penrith in Cumbria, Llangefni in north Wales, Rush on the east coast of Ireland and also Dorchester and Midhurst on the south coast of England.

The implications of Shetland harbouring its own 'indigenous British-Viking raiders' are interesting. Of course, there are many Scandinavian male gene types in Shetland; this is well known, and Shetland is acknowledged to have been a major Viking outpost.

However, in spite of the strong Norwegian influence, 64% of Shetland males have Ruisko types with a very local flavour, and 72% of all Shetland male gene types are 'indigenous' in the sense that they have no matches anywhere in north-west Europe.[16] Given the bellicose, male-dominated political structure of Medieval Scandinavian society, it is hardly likely that a local 'Pictish' chieftain would have been able to spread his seed so widely in the Scandinavian Atlantic maritime network of the north unless he was an equal member of the Viking culture.

Shetland and Orkney were the springboards for the Norwegian Viking colonization of Iceland, but it has always been somewhat puzzling why there should be so many British male and female gene types in Iceland. Icelandic geneticist Agnar Helgason and colleagues at Oxford University have carried out much of the work on this. Although one might expect Viking gene flow to be male-dominated, a complex male–female interchange occurred between the Vikings and the British Isles. It includes a large contemporary contribution from British Atlantic coastal females carried to Iceland and *outnumbering* the Norwegian maternal component there by a ratio of 2 : 1 today. The male contribution to Iceland from the British Isles was less and in a minority (1 : 4, British male lines : Scandinavian male lines). Yet even a 25% contribution by British males to the Icelandic genetic heritage still needs explaining.[17]

The overwhelmingly female matches between British and Icelanders have usually been put down to male Vikings acquiring British women, either with their consent or as slaves. This view is based on mentions of slaves in early Icelandic records, though there are no exact records of numbers. Male British migrations to Iceland are another matter. It is difficult to use the male slavery story to explain the contemporary expansion of the Shetland R1b-8a male cluster down the west coast of Britain – by analogy, rather like taking coals to Newcastle. Iceland is supposed to have been colonized by northern Norwegians mainly from the Trondheim region.

Yet on the database I am using, Iceland's Scandinavian component has a gene group profile nearer to southern Norway (Bergen or Oslo) and Denmark, as a result of dilution with the historic founding cluster I1a-3 and the Ruisko group. Of Icelandic males, 37% have the Atlantic coastal Ruisko gene group.[18] This is significantly higher than Trondheim, which has only 29%.[19]

In Iceland, the high preponderance of the more southerly male Viking gene cluster I1a-3 has several possible interpretations, the simplest of which is that there was a much larger southern Scandinavian component in these northern islands than has previously been assumed. For I1a-3 this means southern Norway, since there are hardly any exact Danish gene type matches in Shetland and Orkney (only 3%), while there are plenty from Norway.[20]

But that still does not explain the Ruisko contribution. I tested the additional possibility of British male influence through indigenous R1b (Ruisko) clusters. I found eighteen (13%) exact matches in Iceland for unique British indigenous gene types.[21] These matches, although they include the types I identified for the Shetland 'indigenous Viking' expansion, were not all derived from Shetland and Scotland, but from different locations around the coasts of Britain. My figure of 13% matches is likely to be an underestimate of British influence; Oxford geneticist Sara Goodacre and colleagues, who included Agnar Helgason, have recently estimated a 25% British/Irish male contribution to Iceland.[22]

The indigenous Shetland male R1b-8a cluster expanded both down the British west coast and to Iceland around the time of the Vikings (Figure 12.3c). And as I have shown in earlier chapters, there were several Neolithic Scandinavian intrusions into Shetland, Orkney and Scotland. The implications of a significant British male involvement in the ninth-century colonization of Iceland from Shetland are therefore that, rather than being the unfortunate enslaved or slaughtered victims of a major Viking incursion, Shetland Islanders may already have been an established part of the show. Politically and culturally, they could have been vigorous participants in the Scandinavian world long before the Vikings set sail in the seventh century. There is circumstantial evidence suggesting that the advent of the Vikings could have been low-key 'business as usual'. Cunliffe paints a picture of quiet continuity in the Shetlands during the Viking Age:

> no centralized trading port has been identified [and] what exchanges there were (... timber was imported from Norway) would have been organized on a small-scale local basis, with the merchants visiting individual coastal settlements ... Norse settlement was much like that of the early first millennium settlements it replaced. In some cases like Jarlshof on Shetland and Udal on North Uist, the same settlement sites simply continued in use. On the island of Rousay Viking burials were found in a cemetery which had been established centuries earlier, the Viking graves carefully avoiding the earlier burials that had been marked by boulders.[23]

Consistent with the view of these islands as a stable part of the Scandinavian world, Orkney was also formerly a base for regular Viking raiding parties, as we read in the lifestyle of Svein Asleifarson recalled in the *Orkneyinga Saga*:

> In the spring he had more than enough to occupy him, with a great deal of seed to sow which he saw to carefully himself. Then when the job was done, he would go off plundering in the Hebrides and in Ireland on what he called his 'spring-trip', then back home just after midsummer where he stayed till the cornfields had been reaped and the grain was safely in. After that he would go off raiding again, and never come back till the first month of winter was ended. This he would call his 'autumn trip'.[24]

The genetic picture in Shetland, Orkney and the Western Isles is consistent with the substantial Scandinavian Neolithic intrusions of male and female gene types, but at the same time a retention of around 60% of indigenous Ruisko male lines. The Ruisko lines have an identity and diversity which is different from the nearby mainland, denying the conventional story of recolonization following the 'Viking slaughter'.[25] Sara Goodacre's recent study specifically argues for a family-based Scandinavian settlement of Shetland, Orkney and the Scottish north-west coast, based on equal Scandinavian male and female intrusions of around 44%. She contrasts this with the lesser Norse influence in Skye and the Western Isles, where male Scandinavian intrusions are twice those of females (22.5% vs 11%), and may reflect more lone Scandinavian males settling.[26]

All this goes against the Viking genocide story that some are keen to promote.[27] And I should not forget to mention the Shetland and Highland ponies, which arrived long before the Vikings, but whose matching mitochondrial DNA has been identified in Viking burials in Scandinavia (see Chapter 5).

Would the Shetland and Orkney islanders have been speaking Norn, the extinct Norse dialect of Caithness and the Islands, or some celtic-Pictish language? There is no clear evidence of what was spoken before the Vikings in Shetland and Orkney, and all the place-names there are Norse or later. There are some Ogham inscriptions in both Shetland and Orkney – thought to be celtic, and one at least Gaelic – from between the sixth and the tenth century.[28] The location is too far east for indigenous Gaelic, so these may derive from Gaelic-speaking monks from Ireland or Argyll. The rest of northern Scotland, particularly the Western Isles – another Scandinavian stronghold – is practically devoid of celtic inscriptions of any period, pre- or post-Viking Age. So, we are still left wondering what that other Pictish language in northern Scotland and the islands was.

Norwegian exact gene type matching

When I turned to the level of male gene type matching in carefully sampled populations of small towns throughout the rest of the

British Isles, I found a general 'Norwegian' admixture of varying rates (Figure 12.4). There was a higher male Norwegian admixture in island and west-coastal regions but, with the exception of Wales and west central Ireland, Shetland and Orkney show the highest Norwegian intrusion, followed by the Western Isles, the Isle of Man and the Channel Islands.[29] The Scottish west-coast ports of Oban and Durness also show over 10% Scandinavian intrusion.[30] Further inland and on the east and south coasts, the Scandinavian male influence is less, but is still present (Figures 11.5c and 12.4).

Overall, 6.2% of male gene type markers in the British Isles are identifiable as Norwegian in origin.[31] While their presence on the Western Isles, Orkney and Shetland, the east coast of Ireland and the Isle of Man is to be expected, their distribution elsewhere is surprisingly widespread and different in this respect from the more localized Danish matches. This, together with the high overall intrusion rate of 6.2%, further supports my view that Norwegian influence in northern Britain is older than generally assumed.

Given that only about one out of six of these intrusive Norwegian clusters date to the last 2,000 years, an overall figure of about 1.1% for Norwegian Viking intrusion to the British Isles would be more realistic. Combined with the figure of 4.4% that I found for Danish Vikings, this would give about 5.5% for the overall Viking intrusion, which is slightly higher than figures of 2–4% estimated by archaeologists.[32]

Normans

The Normans, being the most recent successful invaders of Britain a thousand years ago (AD 1066), had a profound effect, which is still felt today, on culture, language, class structure and even our names. Dorset writer Thomas Hardy in *Tess of the D'Urbervilles* wove the last point into a moving but tragic tale of class by modifying the surname of Tess, the working-class poor relation, to Derbyfield. The physical, cultural and linguistic impact of that invasion is much better documented than the Dark Ages and Viking ones. Perhaps as a result the various estimates of the ratio of incoming Normans to indigenous people are relatively small and in the same modest range as my genetic estimates for Vikings and Anglo-Saxons.

The overall English population at the time of the Domesday Book has been estimated variously at 1.1–2.6 million. This is about half of estimates for the population of Roman England, and four to six times estimates given for the Dark Ages and the time of the 'tribal hidage' (see Chapter 10 and discussion in Chapter 11).[33] Total Norman immigration is unlikely to have been in six figures. Heinrich Härke suggests that it was more likely 'in the low 10,000s'. An immigration on that scale would not have amounted to any more than 1–2% of the native population of England.[34] A rather larger but unreferenced measure is confidently stated in Melvyn Bragg's delightful biography of the English language: 'It has been estimated that in the beginning the Norman French accounted for no more than three or five percent of the population.'[35]

From my own genetic analysis of Anglo-Saxons and Vikings in this book, 1–5% seems to be the going rate for a successful Medieval invasion, but the economic effects of the Norman invasion were profound, with the disinheritance of the entire Anglo-Saxon landowning class:

> That conquest resulted in an unparalleled enrichment of an alien aristocracy ... Their numbers were small ... [A]t the time of the Domesday survey over 50 per cent of the recorded landed wealth of England was in the hands of less than 190 lay tenants-in-chief, but nearly half of this wealth was held by just eleven men.[36]

As I have suggested for earlier invasions, pre-existing cultural links were a great advantage to prospective elite invaders. Whatever the rights or wrongs of Duke William of Normandy's claims to legitimacy, his pretext was a statement of familial and cultural relationship. A study

of the Domesday Book, now available in cheap popular form,[37] illustrates a number of points of cultural continuity. Not only did the
Saxons determine the English place-name landscape for the Norman
clerks, but also the bulk of those names have survived until today, with
relatively much less vigorous Norman influence. The measure of land,
a *hide*, remained in the Norman Domesday Book as in the previous
Anglo-Saxon 'tribal hidage'. The Domesday Book, informative as it is,
was not, however, a national census as we understand the term today.[38]

It might be thought that personal and family names would be
useful in determining the male Norman contribution to modern
British populations. The answer is, only up to a point. Many of the
French family names that persist today and festoon *Debrett's Peerage*
can be traced back to the Norman conquest, and titles such as baron
and marquess are from Norman times. But such titles continued to
be bestowed by later monarchs, thus diluting measurable numbers.

We might also look at non-titled family names, which in England
are overwhelmingly Anglo-Saxon in derivation rather than French.
There are several problems with this approach to detecting a small
French admixture. First, the practice of surname use started in
England in the century following the aftermath of the conquest itself,
tending to confound attempts to use surnames as an ethnic marker.
Then many French names became anglicized among English-speaking
communities, such as 'Carteret' to 'Cartwright' and – fictionally, in
Hardy's novel – 'D'Urberville' to 'Derbyfield'.

Ethnic relabelling is worldwide. One of my father's cousins changed
his family name from Oppenheimer to his mother's name, Newton, to
avoid anti-German feeling in the First World War. The original name
referred to Jews living in the Medieval town of Oppenheim who
could, if they paid their taxes, obtain this as a valid census name. My
near neighbour, an economist who carries the same name – and is
almost certainly not closely related – told me this titbit. The practices
of relabelling and name imposition also create difficulties in using
personal names to determine celtic affiliation in Roman England.

English surnames, as elsewhere in Europe, are of four main types: *place names* (John London), *patronyms* (John Robertson or Williams), *occupational* names (John Smith) and *descriptive names* (John Little). Patronyms ought to be the most useful, but even in former celtic-speaking parts of the British Isles, where they have longer and more consistent usage, they still account for only 70% of the total. In any case, when the Late Medieval English revolution in the use of surnames started, patronyms would not have meant much. In the twelfth and thirteenth centuries, the increasing use of patronyms may have been partly in response to the poverty of first names. For instance, at the time half of all male first names were of Norman regal derivation, namely Robert, William, Henry and Richard.

The Normans had a devastating initial effect on written English. Documents tended to be in French or Latin, although this bias towards the landed literate classes did not mean that English died out as a written language, only that it was not generally used for official purposes. The masking effect lasted until its re-appearance as Middle English in the fourteenth century. Nearly 30% of the 2,650 words in the epic English poem *Sir Gawaine and the Green Knight* are of French origin. Later, although the French component steadily decayed, romance words in the shape of Latin borrowings were still seeping into the language. Melvyn Bragg charts the changes in a very accessible way.[39]

My favourite piece of linguistic trivia is evidence of a Norman pidgin surviving today in the English words for live domestic animals and French ones for dead meat. Presumably, the Norman lord would ask his steward to arrange roasts of *boeuf*, *porc* and *poulet*, and the latter would round up cows, swine and hens from the Saxon serfs. A thousand years later, we now have beef, pork and poultry as meat on the butcher's tray and the Saxon animals still alive in the fields.

Genetic evidence for the Norman invasion

Unfortunately, I am not in the same position as with previous invasions to use knowledge of source markers to determine the genetic

Epilogue

The most important message of my genetic story is that three-quarters of British ancestors arrived long before the first farmers.[1] This applies to 88% of Irish, 81% of Welsh, 79% of Cornish, 70% of the people of Scotland and its associated islands and 68% (over two-thirds) of the English and their politically associated islands.[2] These figures dwarf any perception of Celtic or Anglo-Saxon ethnicity based on concepts of more recent, massive invasions. There were later invasions, and less violent immigrations; each left a genetic signal, but no individual event contributed even a tenth of our modern genetic mix.

There certainly is a deep genetic division between peoples of the west and east coasts of the British Isles, particularly between the English and the Welsh, but this does not merely reflect the Anglo-Saxon, Viking and Norman invasions. These were only the most recent of a succession of waves of cultural and genetic influx from north-west Europe, going back to the first farmers and before. Even the first settlers to come up from the Basque Ice Age refuge left different genetic traces on the east and west coasts of Britain. That difference was merely added to by subsequent migrations across the North Sea.

West Side Story

The second main issue, which should be resolved before any further discussion of British heritage, relates to how Celts are and were defined in terms of history, language, culture and genetic heritage. While most people of the British Isles hold an unambiguous, community-derived cultural identity, there are a number of different popular modern perspectives on Celtic national identity. Popular history books and the media now have more effect on these views than do orally transmitted folk memory and original written sources. Fashions among archaeologists and historians in the past couple of hundred years have in turn informed the media. Fashions change, and the growing 'Celto-sceptic' view among archaeologists and others argues for a rejection of the term 'Celtic' as hopelessly corrupted and too vague in classical sources to be meaningful.

My view, argued in this book, is that Celtic ethnicity is a valid concept, both modern and ancient. I think the sceptics go too far and risk losing the baby with the bathwater. I do agree that the term 'Celtic' has become corrupted, but only to the extent of rejecting the archaeological orthodoxy that 'Celts' as a people arose from somewhere in Central Europe, and during the Iron Age swept en masse across Western Europe and into the British Isles. There is no genetic or convincing historical evidence for this conviction, and it is the main source of corruption of Celtic perceptions. The picture of 'Iron Age Bohemians' came from the archaeologists – not, as the sceptics seem to imply, from early antiquarian philologists. The latter correctly linked the insular-celtic languages of Brittany and the British Isles to languages spoken during Caesar's time among Gauls living south of the Seine. Such celtic languages were later found inscribed on metal and stones in Italy, France and Spain from the late first millennium BC.

To the earliest historians of classical times, Celts were a real, defined Continental nation, not just 'western barbarians'. I disagree with the Celto-sceptic view that the classical historians were collectively so vague and contradictory on Celts to the point of having nothing useful to say to us. They certainly differed in their degree of inclusion and exclusion, in the same way as many British do today when using such broad terms as 'Asian'. But we can find sufficient cross-checks in classical sources spanning a thousand years to derive definitions better than those found, for example, in the older Webster dictionaries.

Celts in the British Isles have real cultural and linguistic connections to former Continental Celts. There is sufficient corroborated evidence in the classical texts to place Celts in south-west rather than Central Europe at an early stage – specifically, in France south of the Seine, in Iberia and Italy. Gaulish, Celtiberian and Lepontic inscriptions have been found in these same regions, dating from a few centuries BC and showing a clear linguistic relationship with insular-celtic languages. In short, the inferred location and languages spoken by people called 'Celts' by the ancients need not be arcane or confusing, so long as the Central European Celtic homeland is seen for what it is – a modern archaeological myth.

Languages geographically associated with the ancient Celts and modern insular-celtic languages all have a common south-west European origin. I have used the recent literature in several disciplines, and my own re-analysis, to ask when celtic languages moved from the European mainland to the British Isles, with which culture, and carried by how many people. My answers are (1) Neolithic, (2) Neolithic and (3) not many.

Barry Cunliffe suggests that celtic language developed along the Atlantic fringe during the first four millennia of maritime trade, spanning the Neolithic and the Bronze Age, and was carried north into Ireland and Wales with Maritime Bell Beakers, by early metal prospectors from the south from 4,400 years ago. This suggestion has support from the material culture as shown by archaeology and in extrapolated dates for the legendary Irish king lists. Also, there appears to be at least one clear genetic colonization event in

Abergele, north Wales, to match the archaeological evidence of an early copper mining colony there from 3,700 years ago. However, the genetic evidence, including Abergele, is more consistent with a Neolithic date for 'Celtic' arrival.

Following the Mesolithic, the gene flow up the Atlantic fringe to the British south coast dates mainly to the Neolithic, nearer the *beginning* of Cunliffe's long period of maritime contact. Following new attempts to date the break-up of insular from Continental celtic, Peter Forster's speculative estimate of 5,200 years for the fragmentation of Gaulish, Goidelic, and Brythonic from their most recent common ancestor is also in the Neolithic. Other date estimates from New Zealand argue strongly for the Neolithic being the driving force behind the spread of the entire Indo-European language family, which includes celtic. Given this evidence stretching back to the first farmers, sceptical archaeologists should not be so ready to pull the plug on the idea of a common heritage between those ancient Celtic areas of south-west Europe and the celtic-speaking western side of the British Isles.

Gildas' Dark Ages horror story of the Saxon invasion has generated the view that Celts were somehow *the* aboriginal population of the British Isles. The view that languages in Britain were 100% celtic when the Romans invaded is part of this false assumption. If 'aboriginal' means Neolithic immigrants speaking celtic languages who had replaced the former inhabitants of Britain, then nothing could be further from the truth. The genetic evidence does not support this at all. There was no Celtic replacement, any more than there was an Anglo-Saxon replacement.

The arrival of celtic languages and associated gene flow could hardly be classed as evidence for the establishment of a Celtic replacement of a former unknown British population on genetic grounds. The highest single rate of Neolithic intrusion from the Mediterranean route in the British Isles was in Abergele at 33%. But in Ireland, such Neolithic intrusion was only around 4%, while it was

2% in Cornwall, 6–9% in the two Welsh peninsulas, and 8–11% in the Channel Islands and southern England (Figure 5.4a). For England and the Channel Islands, the Neolithic contribution from the East via the northern route, just across the North Sea, was the same or greater than for the Atlantic coastal source (Figure 5.4b).

In other words, Ireland and the Welsh peninsulas – which, on the basis of recent history and language, might be thought to be Celtic bastions – have less evidence of Neolithic genetic intrusions, let alone from the Bronze or Iron Age, than anywhere else in the British Isles. Of course, the flip side of this is that their descendants are *truly* aboriginal and genetically represent the most conservative parts of the British Isles, retaining respectively 88% and 89% of their pre-Neolithic founding lineages (Figure 11.5a). And where do those founding lineages come from? They come from the same part of Europe, the south-west, but more specifically they match the equally conservative region of the Basque Country.

Ultimately ancestors for the modern Irish population, male and female did come from the same region as those ancient celtic inscriptions, but thousands of years before celtic languages. But then every other sample in the British Isles shows at least 60% retention of those pre-Neolithic aboriginal male founders, reflecting the very conservative nature of the British Isles after the Last Glacial Maximum.

Translating all this back to question the assumption that 'Celts', however defined, were the aboriginal peoples of the British Isles, we can see new perspectives, which depend on how that definition is applied. First, if Celts were to be defined by their languages, the small proportion of associated gene flow would make them an invading cultural elite with no stronger claims to aboriginal status than the Anglo-Saxons. If we focused more specifically on those 2–10% of immigrating southern Neolithic, Bronze or Iron Age genes as identifying people rather than language, they would be even less 'aboriginal' in Ireland and Wales than in the rest of the British Isles.

I think we should take Cunliffe's gradualist concept of the *Longue Durée* of the Atlantic cultural network as a paradigm for the genetics, as Irish geneticists Brian McEvoy and Dan Bradley of Trinity College Dublin, with English colleagues Martin Richards and Peter Forster, have done. Rather than being on the fringe of a celtic-speaking Neolithic revolution, the Atlantic fringe countries of Ireland and peninsular Wales then become the genetic aboriginal strongholds of post-LGM and Mesolithic gene flow from the Iberian glacial refuge, now best represented in south-west Europe by the equally conservative genetic profile of the Basque Country. The rest of Britain and the northern isles off Scotland then become more or less aboriginal with rates varying from 60% to 80% of 'indigenous' male markers (Figure 11.5a). In a sense, this is similar to the position taken on the Y gene group markers of 'the indigenous population of the British Isles' by geneticist Cristian Capelli (see Chapter 11), only my estimates for indigenous survival are much higher.

I feel that the genetic picture, both male and female, best reflects the broader picture of cultural and migration influences repeatedly moving up along the Atlantic coast to the western British Isles since the last Ice Age, so well described by Barry Cunliffe. Celtic languages need not be left out as irrelevant latecomers from this deep view of British ancestry, since they are part of the spray of repeated cultural waves along the Atlantic coast. We can also see that even if celtic languages did arrive on the western coasts of the British Isles as long ago as the Neolithic, there is no reason to assume that they spread universally in Britain, let alone remained universal during the Neolithic. What about languages arriving from across the North Sea before the Dark Ages?

East Side Story

The third question I have posed in this book is how old are the genetic and cultural divisions between eastern and western Britain. An even greater flow of Neolithic-dated genes (10–19% of modern gene lines) were impinging on eastern and south-east Britain from Scandinavia and north-west Germany, from across the North Sea, over the same Neolithic period (Figure 5.4b). The male genetic evidence shows that even before the advent of farmers, the first pioneers were making a choice of east or west when they arrived at the mouth of the English Channel. Where this differs from the Atlantic coastal tin trade in the west is that we have no cultural clues with which to argue that the people in eastern Britain were, by default, speaking celtic languages like those in the west. Rather the opposite, since Neolithic cultural influences were arriving on the east and south coasts of Britain from north-west Europe, along with genetic markers of clearly north-west European rather than Iberian origin. The fact that Britain was one island is of no help in arguing for one language, since the highways of influence were maritime, not land-based.

Was there an Anglo-Saxon genocide?

The key historical source that has led to the conviction that the English originated as recently as the Dark Ages is Gildas' tract 'On the Ruin of Britain'. The gory embellishments of this latter-day Job have led to the entrenched view that Angles and Saxons came over from the Continent, slaughtered the Celts in England and became the 'English'. Few of his core claims hold water. Even Gildas' Saxon Advent is contradicted by Bede, who claims it was Anglian. There is little evidence for genocide, but it remains in schoolbooks. Genocide means the deliberate extermination of a nation. Now, if that means the death of over 50% of the people, I am certain, after studying the genetic story, that there was no genocide in Dark Ages England.

There is specific evidence of an invasion from the region of Schleswig-Holstein at the base of the Danish Peninsula, but on my estimation this amounts to only 4% of male gene types in the British Isles. This does not give enough genetic evidence for even a 10% cull (literally, a decimation), except in parts of Norfolk and the Fens, which reached about that level of intrusion (Figure 11.5b). This means there was not just substantial continuity of population, but a survival of around 95% of the indigenous lines. Even the Vikings achieved a higher estimated overall level of genetic invasion. Increasingly, this lack of 'wipeout' is what many archaeologists are inferring from their detailed cultural and burial evidence. In Francis Pryor's words, 'massive war graves, settlement dislocation and "knock-on" impacts ... have not been found'. Rather, there is evidence of continuity in spite of cultural change: 'if Anglo-Saxon people and culture displaced "native" practices, one would expect the latter to have vanished completely. They did not.'[3]

One of the fascinating results of this matching of source gene types from the base of the Cimbrian (Danish) Peninsula to England is in the target distribution. As would be expected from Bede and all the Anglian cultural matches, including runes and cruciform brooches, the higher rates of intrusion (9 –17%) fell in Anglian regions of England. These were in Norfolk, the Fens and, to a lesser extent, north-eastern parts of Mercia and Lincoln (Figure 11.5b). By contrast, Saxon England in the South could only muster the background English rate of 5% invasion; and, unlike further north, the south has no evidence for any specific genetic founding event dated to this period. So, why did Gildas, but not Bede, tell us that Saxons committed the slaughter? As may be clear by now, I think he had his own nation's agenda.

I would go much further than doubting Gildas on the genetic and cultural evidence. The so-called Anglo-Saxons were not even the first English nation. They did not all arrive at the same time. The Saxons, in particular, were already in residence during Roman times. The Angles were not genetically, culturally or linguistically close to the Saxons, nor for that matter to Frisians. The Angles and Jutes were more Scandinavian culturally and linguistically, with clear genetic and archaeological matches to the Danish Peninsula and Sweden. They were not even our first Scandinavian visitors, nor the last.

Two Englands

Nowadays, southerners and northerners talk about the north/south cultural, political and economic divide, much in the same terms as the Welsh and Scottish borders. One can find recent historical-economic reasons for the perception, but there is also a cultural-historical divide. Having paternal relatives living in the north and having lived on both sides, I am sure of it.

Angles and Saxons are usually hyphenated together with reference to early English history, but the disagreement between Bede and Gildas over which nation invaded our east coast is just one aspect of the many differences in the cultural and political history of the two English regions. The stronger genetic and archaeological trail for Anglians and Jutes supports Bede's view that Saxons were not the main invaders, and raises the question of whether Saxons were already in residence before the Romans left. Francis Pryor points to clear cultural evidence that this was the case, for instance in Mucking, Essex (see p. 314). He tends to discount the theory that these British Saxons were just visiting mercenaries, and questions the currently accepted view of what was implied by the Roman references to the 'Saxon shore', from Norfolk to Hampshire. A fresh view of the Saxon coast, unbiased by Gildas, might be that this term meant the shore of English Saxony – rather like New England in North America, but before the Dark Ages. In other words, these were not the shores being defended against the Saxons, but the shores of established Saxon colonies.

The Viking raiders, who contributed at least as much genetically as the 'Anglo-Saxons', also seemed to realize, in their choice of settlements, that there were two Englands. From the perspective of previous Neolithic gene flow, we can see that they settled specifi-cally in their respective former haunts. Norwegian Vikings, like their Neolithic forebears, first concentrated in the far north of Scotland, and on Orkney, Shetland and the Western Isles. Shortly after, the Vikings of the Cimbrian Peninsula (i.e. Danes from Jutland

and others from Schleswig-Holstein) invaded England. However, they avoided Saxon England and settled extensively and exclusively in those north-eastern regions that their recent ancestors, the Jutes and Angles, had invaded a few hundred years before. This exclusive pattern was replicated right down to the border between Anglian Suffolk and Saxon Essex.

Can these north/south divisions be traced further back in English 'prehistory', before the Dark Ages, in the same way as the east-coast/west-coast division? They certainly can be, genetically. As I described in Part II of this book, the separate evolution of the genetic character of east-coast Britain from the west coast goes right back through the Neolithic to the Mesolithic and even the Late Upper Palaeolithic. It is paralleled repeatedly in two geographical sources and inputs of cultural influence from the Continent, one in the south the other in the north. England north-east of the Danelaw line, however, has a much greater overall genetic intrusion from the nearby Continent, at around 25–40%, than the south, at 15–18%. Furthermore, the majority of the gene lines making up this impact on eastern Britain arrived long before the Romans, mostly during the Neolithic, and some even before that.

Looking at the distribution and source of the eastern intrusive Neolithic lines, we can see a north/south separation. The extreme north of Scotland and its neighbouring islands, Orkney, Shetland and the Western Isles, show both male and female Neolithic genetic connections with Norway (Figures 5.3a, 5.3b and 5.6a). Orkney even seems to have acted as a source of 'British Viking' raiders. Eastern England, on the other hand, has strong Neolithic genetic connections that point more to southern Scandinavia and Denmark (Figure 5.6b).

How old is English?

What about the possibility that English was spoken as some form closer to Norse or even a separate Germanic branch in one or both of these two Englands before the Anglo-Saxons arrived? I am sure this will be the most contentious aspect of my argument, but that does not deter me from suggesting it. The various academics I have quoted or cited on this issue are united on one aspect of the oldest recorded English, whether written in runes or Roman script. This is the strong, unexplained Norse influence, both culturally and linguistically, before the Vikings came on the scene. But the evidence against a Dark Ages root of English goes deeper than that. In terms of vocabulary, English is nowhere near any of the West Germanic languages it has traditionally been associated with. It actually roots closer to Scandinavian than to *Beowulf*, the earliest 'Old English' poem and probably written in the elite court of the Swedish Wuffing dynasty of East Anglia. One study suggests that, on this lexical evidence, English forms a fourth Germanic branch dating 'to before AD 350 and probably after 3600 BC'.[4]

There is inadequate evidence for the current view that celtic languages were spoken *universally* in England in Roman times. Caesar, corroborated by Tacitus, tells us that an aboriginal population occupied the south-east English hinterland, while extensive Belgic settlements replaced them on the coast, with languages shared across the Channel. Place-name evidence confirms this. Caesar tells us that most of the Belgae were more closely related to the Germani, a view again supported by place-names. There is also much negative evidence against the orthodoxy of universal celtic languages in Roman England, for instance the near-total absence of Celtic inscriptions at any time.

English identity

Why should the question of when people started speaking Germanic tongues in Britain matter to more than a few linguists? They would have been unintelligible today, in any case. I think that, along with the rest of Gildas' poison and the genocide myth, the historical perspective matters to the Welsh and Cornish. But English history should matter to the English as well. A *real* issue about

English ethnicity is found in their own ideas of historic identity. I was talking to two television executives when one of them, a psychologist, casually asked, 'Who is really concerned about ethnic identity here in the United Kingdom?' My jaw dropped, and I ventured that, at the very least, people in Wales, Scotland and Northern Ireland, among others, might be. He acknowledged that with a nod, and then without a pause went on, 'Well, yes of course it is something we should put to the regional networks.' I realized it would have been a mistake for me to pick up on this and ask whether he felt that English and British identity were one and the same thing. Maybe he thought that the various British minorities, including the Celtic nations were, well, just quaint – different and 'not like us'.

Historical context

There is no doubt in my mind that the English, individually and collectively, do have a sense of identity even though they may need to be nudged just a little. You only have to open a newspaper to realize that. One problem, as suggested by another television figure, Jeremy Paxman, in his book *English: A Portrait of a People*, may be that in the post-Empire era they are a little unsure of their identity. Maybe, they have lost some of their sense of cultural superiority. Which implies that, whether at the family or community level, identity depends on historical context as well as the rest of the cultural and linguistic baggage.

I was idly watching a 'reality' television show recently, which brought this point home to me. Well-known personalities were being helped to trace their ancestry. The process began with nearest relatives, then to older and more distant ones; the producers used standard documentary methods and civil records. Jeremy Paxman could hardly be regarded as a typical victim. His genealogical trail went back through comfortable middle-class settings to grinding poverty and indentured labour transports, and to an ancestral couple who had both died of tuberculosis. Paxman was completely unaware of these ancestors until he saw the certificates. The female death certificate recorded death 'from consumption and exhaustion'.

Then something small but surprising happened. The steely anchorman, who could face down a prime minister or worse, any day of the week in front of millions, sat staring at the scraps of paper in the council office, for a while repeating the cause of death to himself. Then he subsided into tears. By the end of the programme, Paxo was back to his old self, chiding the producer for asking a final, fatuous question. The cynic might say that pressing such buttons was the 'reality' pay-off for the programme, the equivalent of fisticuffs on Jerry Springer, and this man of peace was a willing victim. That may well be, but what large buttons they were. Not being Paxman or a psychologist, I can only assume that the emotion was about family, identity and bleak, anonymous tragedy, but the trigger was unambiguous *historical context*.

My two-pennyworth for another problem with English identity is that, thanks to Gildas, our English ancestry was orphaned and stripped of any historical context beyond the Dark Ages threshold. According to Gildas, there were no English before the Saxons arrived in their three keels. The genocide of the fifth and sixth centuries as inferred from his account validates Gildas' view of the Saxons in England as 'bastard-born'. Six hundred years after his 'Saxon Advent', new invasions of Vikings and Normans had dispossessed first the Anglians in the north, and then the Saxons in the south, and for a time even English stopped being written.

The British Isles today

Unlike with all earlier migrations, we cannot determine the impact of the last successful invasion of Britain by the Normans from the genetic trail, mainly because the genetic make-up of those founding noblemen is as yet unknown, and it is impossible to measure impact without a source. Documentary evidence suggests perhaps a 3% intrusion, although this figure may still be biased towards the landowning class if the French names in *Debrett's Peerage* mean anything.

England has, however, benefited from considerably more recent immigration. As transport improved, passenger ships began to cross the oceans between continents, with vessels from the *Empire Windrush* onwards bringing immigrants from Jamaica. With the advent of regular intercontinental air travel, immigrants now fly to the British Isles from all over the world. These islands have as on numerous occasions before, changed from a multicultural to an even more multicultural society.

But in this book I have avoided reference to the more recent period, simply because that is not what it is about. The DNA sample frameworks used in the analysis specifically exclude more-recent immigration, even within Britain. This exclusion is not intended as any jingoistic statement about perceived Britishness, it is simply a practical necessity for looking further back in our history.

Statistics, founder events and damned lies

I have given a number of percentages relating to individual migration events in this book. The largest figures come from our first hunter-gatherer founders, while figures for more recent arrivals, such as Anglo-Saxons and Vikings, rarely top 10% locally and 6% overall. Perusing our national census for more recent minority immigration, those figures seem no more dramatic. Recent migrations into Britain have been proportionately minor compared with the pioneer events, so minor that the biggest 'increases' are attributable merely to ways of measuring. Without going into the complex analysis of immigration and census, I found a short government statement which was of interest, both for its comment on the statistical effect of changes in categorization and for its relevance to my own family make-up:

> [2001] Comparisons with the 1991 Census show: The proportion of minority ethnic groups in England rose from six per cent to nine per cent – partly as a result of the addition of [the category of] Mixed ethnic groups in 2001.[5]

In spite of my German Jewish name, which might predict a Near Eastern Y chromosome, less than 13% of my ancestry derives from that source. The rest is a collage of more or less 'British' ancestors, including a large dose of Scots on both sides, smaller doses of Mancunian and Brummy, and even the possibility of Flemish (Huygens through my mother's maiden name).

This sort of mobility in marriage is a recent but rising trend. I guess I have taken it a step further. My kids have two cultural and genetic backgrounds, English and Malaysian Chinese, thus finding themselves in the 'Mixed ethnic groups' – 2001 census category. Occasionally, well-meaning persons ask me in a concerned way whether my children have problems of cultural identification. I pass the question on to them and receive amused responses. They both feel enriched, having not one but two different cultural and culinary resources to relate to; in other words, the confusion of identification remains in the eye of the observer. I would personally prefer ethnic identity to be more a self-chosen smorgasbord than anything that might be imposed by others.

This lucky example is not smugly intended to underplay the problems of belonging to a minority in the United Kingdom. However, an increase in immigration over the past half-century has not, despite the efforts of political rabble-rousers, resulted in Enoch Powell's prophecy of 'rivers of blood'. As with St Gildas, another British prophet before him, there is the faintest suspicion of an agendum of wish-fulfilment in that prophecy. Speaking of the aftermath of the supposed Anglo-Saxon invasion, Gildas tells us of:

> fragments of human bodies, covered with livid clots of coagulated blood, looking as if they had been squeezed together in a press

and exhorts the surviving and unrighteous British kings to

> seek for the rule of right judgment [on] the proud, murderers, the combined and adulterers, enemies of God, who ought to be utterly destroyed and their names forgotten.[6]

Should we lightly wish such a fate on our worst enemies? After all, Celts, Angles, Jutes, Saxons, Vikings, Normans and others, we are all minorities compared with the first unnamed pioneers, who ventured into the empty, chilly lands so recently vacated by the great ice sheets.

Appendix A

INTRODUCTION TO GENETIC TRACKING

In the main text I make little reference to the methods used for the genetic tracking that figures prominently in this book. The real revolution in understanding human genetic prehistory covers the last 200,000 years, the most recent 20,000 years of which concerns us here. For a large part of the period before the first farmers, the new genetics has shone a bright light onto a contentious field previously dominated by collections of European and African stone tools and a few poorly dated skeletal remains. But even for the Neolithic and later periods, archaeology tells us more about cultural spread than human migrations. Before turning to details of genetic tracking, it may help to look at some of the ideas behind genetic inheritance and how they have evolved. The concepts are mainly simple, being related to our own everyday understanding of and preoccupation with inheritance, but are often misrepresented, either for reasons of hype or because they are veiled in jargon.

Within each of the cells of our bodies we all have incredibly long strings of DNA. It is the stuff of the genes. It stores, replicates and passes on all our unique characteristics – our genetic inheritance. These DNA strings hold the template codes for proteins, the building blocks of our bodies. The codes are 'written' in combinations of just four different chemicals known as *nucleotide bases* (adenine, guanine, cytosine and thymine, represented by the letters A, G, C and T), which provide all the instructions for making our bodies. We inherit DNA from each of our parents, and because we receive a unique mixture from both, each of us has slightly different DNA strings from everyone else. Our own DNA is like a molecular fingerprint.

During human reproduction, the parents' DNA is copied and transmitted in equal proportions. It is important to know that although most of the DNA from each parent is segregated during reproduction, small bits of their respective contributions are shuffled and mixed at each generation. The mixing here is not that of mass random allocation of genes brilliantly inferred by Gregor Mendel, but tiny crossovers, duplications, deletions and swaps between maternal and paternal DNA contributions. This is known technically as *recombination*. Luckily for genetic researchers, there are two small portions of our DNA that do not recombine. Non-recombining DNA is easier to trace back through previous generations since the information is uncorrupted during transmission from one generation to the next. The two portions are known as mitochondrial DNA (mtDNA) and the non-recombining part of the Y chromosome (NRY).

Mitochondrial DNA: the Eve gene

To say that we get exactly half of our DNA from our father and half from our mother is not quite true. One tiny piece of our DNA is inherited *only* down the female line. That piece is called mitochondrial DNA because it is held as a unique circular strand in small tubular packets known as mitochondria which function rather like batteries within the cell. Some molecular biologists believe that, aeons ago, the mitochondrion was a free-living organism with its own DNA, and possessed the secret of generating lots of energy. It invaded single-celled nucleated organisms and has stayed on ever since, outside the cell nucleus, dividing, like yeast, by simple binary fission with its own DNA. Males, although they receive and use their mother's mitochondrial DNA, cannot pass it on to their children. The sperm has its own mitochondria to power the long journey from the vagina to the ovum, but when it enters the ovum the male mitochondria wither and die. It is as if the man has had to leave his guns at the door.

So each of us inherits our mtDNA from our mother, who inherited her mtDNA intact from her mother, and so on back through the generations – hence mtDNA's popular name, 'the Eve gene'. Ultimately, every person alive today has inherited their mito-chondrial DNA from one single great-great-great-(etc.)-grand-mother, nearly 200,000 years ago. (Our descent from a single ancestral maternal line is a result of natural wastage every gener-ation and happens with all genes, so it does not literally mean that we all descend from one woman and one man who lived 200,000 years ago.) This mtDNA provides us with a rare point of stability among the shifting sands of DNA inheritance. However, if all the Eve chromosomes in the world today were an exact copy of that original Eve mtDNA, then clearly they would all be identical. That would be miraculous, but it would mean that mtDNA is incapable of telling us much about our prehistory. Just knowing that all women can be traced back to one common ancestral Eve is exciting, but it doesn't get us very far in tracing the different lives of her daughters. We need something with a bit of variety.

This is where DNA mutations come in. When mtDNA is inherited from our mother, occasionally there is a change or mutation in one or more of the 'letters' of the mtDNA code – about one mutation every thousand generations.[1] The new letter, called a *point mutation*, is then transmitted through all subsequent daughters. For the large part, these mutations are 'neutral'– that is, of no functional or health relevance. Although a new mutation is a rare event within a single family line, the overall probability of mutations is clearly increased by the number of mothers having daughters (i.e. population size). So, within one generation, a million mothers could have more than a thousand daughters with a new mutation, each different from the rest. This is why, unless we share a recent maternal ancestor over the past 10,000 years or so, we each have a slightly different code from everyone else around us.

Using mutations to build a tree

Over a period of nearly 200,000 years, then, a number of tiny random mutations have steadily accumulated on different human mtDNA molecules being passed down to daughters of Eve all around the world. For each of us this represents between seven and fifteen mutations on our own personal Eve record. Mutations are thus a cumulative dossier of our own maternal prehistory. The main task of DNA is to copy itself to each new generation. We can use these mutations to reconstruct a genetic tree of mtDNA, because each new mtDNA mutation in a prospective mother's ovum will be transferred in perpetuity to all her descendants down the female line. Each new female line is thus defined by the old mutations as well as the new ones. As a result, if we can identify all the different combinations of mutations in living females around the world, we can logically reconstruct a family tree right back to our first mother.

Although it is simple to draw a recent mtDNA tree on the back of an envelope when there are only a couple of mutations to play with, the problem becomes much more complex when we have the whole human race to deal with, sifting through thousands of combinations of mutations. So, computers are used for the reconstruction. By looking at the DNA code in a sample of people alive today, and piecing together the changes in the code that have arisen down the generations, biologists can trace the line of descent back in time to a distant shared ancestor. Because we inherit mtDNA only from our mother, this line of descent is a picture of the female genealogy of the human species.

Not only can we retrace the tree, but by taking into account where the sampled people came from we can see *where* certain mutations occurred – for example, whether in Europe, or Asia or Africa. What's more, because the changes happen at a statistically consistent (though random) rate, we can approximate *when* they happened. This has made it possible, during the late 1990s and in the new century, for us to do something that anthropologists of the past could only have dreamt of: we can now trace the migrations of modern humans around our planet. It turns out that the oldest changes in our mtDNA (i.e. the earliest in the tree) took place in Africa between 190,000 and 150,000 years ago. Then new mutations start to appear in Asia, about 80,000 to 60,000 years ago. This tells us that modern humans evolved in Africa, and that 80,000 years ago some of us began to migrate out of Africa into Asia.

It is important to realize that because of the random nature of individual mutations, the dating is only approximate. Various mathematical ways of dating population migrations were tried during the 1990s with varying degrees of success, but in 1996 a method was established which dates each branch of the gene tree by averaging the number of new mutations in daughter types of that branch.[2] This method (estimation of *rho*) has stood the test of time, and is the main approach used to calculate the mtDNA genetic dates I give in this book.

Y chromosome: the Adam gene

Analogous to the maternally transmitted mtDNA residing outside our cell nuclei, there is a set of genes packaged within the nucleus that is passed down only through the male line. This is the Y chromosome, the defining chromosome for maleness. With the exception of a small segment, the Y chromosome plays no part in the promiscuous exchange of DNA indulged in by other chromosomes. Like mtDNA, the non-recombining part of the Y chromosome thus remains uncorrupted by recombination, with each passing generation, except for random point mutations, and can be traced back in an unbroken line to our original male ancestor.

Y chromosomes have been used for reconstructing trees for less time than mtDNA has, and there are more problems with using them to estimate time depth. These methods have been considerably improved over the past few years[3]; for the Y-chromosome analysis presented in this book, I use a variant of the same method mentioned above for mtDNA.[4]

So, with improved dating, the non-recombining Y-chromosome (NRY) method may have the potential for a much greater power of geographical resolution than mtDNA, for both the recent and the distant past. This is simply because the NRY is much larger than mtDNA and consequently has more potential for variation. NRY also has much more geographical resolution in Europe. A major reason for this is thought to be that women have tended to join to the clan of their husbands, rather than the other way round (see below).

Y-chromosome analysis has already helped to chart a genetic trail parallel to the mtDNA trail. At the major geographical branch points, such as the single exit from Africa, they support the story told by mtDNA: they point to a shared ancestor in Africa for all modern humans, and a more recent ancestor in Asia for all non-Africans. In addition, because men's behaviour differs in certain key ways from women's, the story told by the Adam genes adds interesting detail. One difference is that men have shown more variation in the number of their offspring than women: a few men father considerably more children than the rest. Women, in contrast, tend to show less variation in the number of children they have. The main effect of this is that most male lines become extinct more rapidly than female lines, leaving a few dominant male genetic lines. This is an example of genetic drift, and could be another reason why the Y chromosome has more local geographical detail and specificity than mtDNA.

Another difference between the sexes is in movement from place to place. It has often been argued that because women have usually moved to their husband's village, their genes are inevitably more mobile. Paradoxically, while this may be true within any one cultural region, it results in rapid mixing and dispersal of mtDNA only within that cultural region. For travel *between* regions (or long-distance intercontinental migrations, e.g. in the Pacific) the burden of caring for children would have limited female mobility. Predatory raiding groups would also have been more commonly male-dominated, resulting in increased mobility in the Y chromosome.

A final point on the methods of genetic tracking of migrations: it is important to distinguish this new approach to tracing the history of molecules on a DNA tree, known as phylogeography (literally 'tree-geography'), from the statistical study of the history of whole human populations, which has been used for decades and is now known as classical population genetics. (An example of a classical method is Principal Components Analysis, described in Chapters 6 and 11.) The two disciplines are based on the same Mendelian biological principles, but have quite different aims and assumptions, and the difference is the source of much misunderstanding and controversy. The simplest way of explaining it is that phylogeography studies the prehistory of individual DNA molecules, while population genetics studies the prehistory of populations. Put another way, each human population contains multiple versions of any particular DNA molecule, each with its own history and different origin. Although these two approaches to human prehistory cannot represent exactly the same thing, their shared aim is to trace human migrations. Tracing the individual molecules we carry is just much easier than trying to follow whole groups.

Naming gene lines

In this book I refer to maternal or paternal clans, gene lines, lineages, genetic groups/branches and even haplogroups (gene groups and sub-groups). All these terms mean much the same thing: membership of a large group of genetic types, unambiguously sharing a common ancestor with the same identifying neutral mutation (usually through their mtDNA or Y chromosome). The size of the group is to a certain extent arbitrary and depends on how far back the base of the branch is on the genetic tree. Additionally I use the term 'gene cluster' in the context of my own analysis, which splits up the major Y gene groups in Europe into smaller clusters. This is because there is potential for ambiguity in the multiple mutations defining those younger clusters (I enlarge on this in Appendix C).

One thing that quickly becomes apparent from a study of genetic trees is the lack of uniformity in the names attached to these branches. For the Y chromosome in particular, each new scientific paper has, in the past, proposed a new system of scientific nomenclature based on 'in house' genetic markers used by a particular research laboratory. This can make comparison between different studies tedious and repeatedly tests the limits of the reader's memory. The underlying tree, however, is more or less the same from lab to lab. Recently, a consensus Y nomenclature was published which uses letters from A to R for the main genetic branches (called 'gene groups' in this book) in the tree.[5] I use this nomenclature as far as possible in my referencing and figures. The trouble is that even these eighteen letters and their location on the tree are difficult to hold in the mind – at least in mine.

Luckily, our memory is often aided by context and association. For this reason, and this reason alone, I have introduced nicknames for the major branches and for other branches I refer to frequently. For the larger and more frequently mentioned European gene groups and clusters, I have used personal names as nicknames. In each case I have chosen a name from the appropriate region of origin or highest frequency, which starts with the first letter of the technical name (i.e. from the consensus scientific nomenclature). So, for example, the male gene group R1a1, which is strongly associated with the Ukraine, I have called 'Rostov'. I have also retained some alliterative and biblical names from my previous book, *Out of Eden*, such as the out-of-Africa Y-line founder Adam and his three descendent lines, Cain, Abel and Seth. There is no intention in adopting such names to infer any deeper meaning – they are simply aides-mémoires.

The mtDNA picture is slightly easier. Many of the different labs agreed at an early stage to try to use a single nomenclature (perhaps there was less testosterone involved in the process!). For instance, there are two agreed non-African daughter and granddaughter lines relevant to the colonization of Europe and Western Eurasia in general: N, from the single out-of-Africa line L3, and her daughter R. In *Out of Eden* I called them Nasreen, in keeping with a southern Arabian origin, and Rohani, to be consistent with an Indian subcontinental origin. I have retained that practice in the genetic trees in Appendices B and C.

Appendix B

OUR MATERNAL ANCESTORS: AN OVERVIEW OF THE EUROPEAN GENE TREE

In this book I have referred extensively to genetic tracking using maternally transmitted mitochondrial DNA (mtDNA). I have mentioned a number of mtDNA gene groups as having arrived in the British Isles and elsewhere in Northern Europe variously from an Ice Age refuge around the Basque Country (Helina, haplogroup H; Vera, haplogroup V; and U5b1b), from the Near East (J, T and K) and even possibly from Scythia (N1a). I have also given estimated dates for the spread of these gene groups and their descendants after the Ice Age. Without ancestry and context, they become mere names and letters. But of course they do have ancestry: they all belong to the genetic tree of West Eurasian lineages and ultimately arrived from Africa via South Asia.

In Figure A1 I update the West Eurasian mtDNA tree from an earlier version[6] to give an overview of that context for the past 50,000 years of European prehistory.[7] How and when they arrived via the Near East and the Trans-Caucasus is described elsewhere.[8] Helina and Vera, the most important mtDNA lines in Western

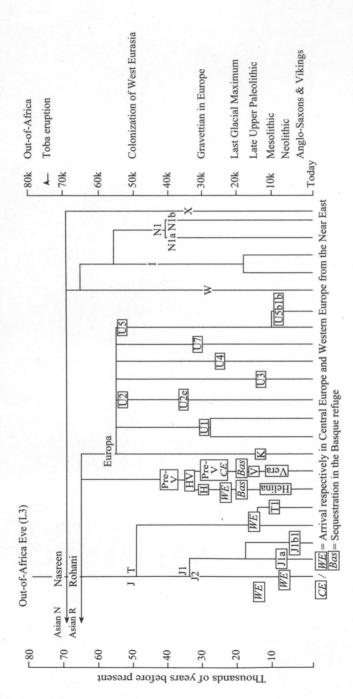

Figure A1 Tree of the main maternal gene groups (haplogroups) used in the analysis in this book in the broader context of the West Eurasian mitochondrial DNA tree. The ancestry of the major Western European group HV and of Europa go back in Europe to pre-glacial times. Helina (detail in Figure A2) and Vera diverged before the Ice Age, arriving in the Basque/Iberian refuge from farther east before the freeze. Along with minority groups such as U5b1b, they made up the major re-expansions after the LGM. Groups such as J, T and K expanded into Europe from the Near East after the Ice Age during the Mesolithic and Neolithic.

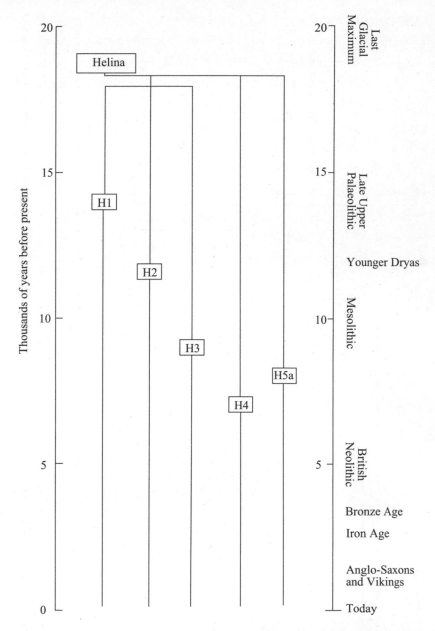

Figure A2 Detail of dated branches of Helina expanding from the Iberian refuge. H1 re-expanded during the Late Upper Palaeolithic, and H2-H5 during the Mesolithic.

Europe, had their own ancestry, arriving from farther east in Europe,[9] just before settling in for the Ice Age in the south-west Franco-Spanish refuges.

After the Ice Age, Helina expanded from those refuges by splitting into perhaps more than a dozen daughter branches or sub-groups. Several of the larger H sub-groups can be dated in Western Europe.[10] These dates are consistent with the view that Helina's expansion was divided by the Younger Dryas freeze-up into an immediate phase after the LGM and during the Late Upper Palaeolithic, and another during the Mesolithic immediately after the Younger Dryas (Figure A2).

Appendix C

PATERNAL TREES OF ANCESTRY IN EUROPE

The Y chromosome is transmitted from father to son, and, being larger than mitochondrial DNA, has potentially far greater power to chart ancient migrations in time and place. For reasons discussed in Appendix A, it has also far greater geographical specificity and resolution. In the British Isles, the mix of Y gene types varies even from county to county.

Tree of gene groups found in Western Europe

A huge collection of Y gene type datasets from the British Isles and its Continental neighbours has been published over the past few years. I have pooled the data, available in the public domain (listed in notes 39–42 and 44 to Chapter 3) into a composite dataset of 3,084 individually analysed samples. Most of the analyses in these publications share the common mutational markers, so-called unique event polymorphisms (UEPs), which allow an unambiguous genetic tree of gene groups to be constructed for combined data. By far the largest body of data in the composite British Isles dataset (85%) was collected by Christian Capelli and colleagues from stable populations in a systematic grid of the region.[11] Their analysis, by a short margin, also uses the largest number of UEPs. The tree they can construct for this is shown in Figure A3, with UEPs (e.g. M17) shown on the branches.[12]

This tree identifies eleven gene groups, of which all but one (N3) occur in the British Isles. The commonest male gene groups which have entered are, in order, Ruisko (73%: 1,511/2,082), Ivan (16%: 336/2,082),[13] Rostov (6%: 126/2,082), E3b (2%: 47/2,082) and J (2%: 38/2,082). These gene groups are likely to have entered the British Isles piecemeal at different times and by different routes. So if the overall male colonization of the British Isles is to be understood and differentiated from historical speculation of Anglo-Saxon or Celtic invasions, it is important to try to break up the larger gene groups into smaller branches, and to date those branches.

Use of STR gene types for branch building within the main gene groups

The only markers available to carry out this refining of the gene groups into further branches are the STR gene types (Single Tandem Repeat sequence haplotypes), which were established for each of the samples in the composite dataset. STR types are less reliable than UEPs for building trees because they are made up of a combination of rapidly mutating genetic sites which can mutate forward and backwards, thus introducing ambiguity. None the less, they are used for this purpose, and their rapid mutation has the advantage of facilitating analysis of shorter time periods.[14]

After dividing the entire composite dataset of 3,084 into gene groups and sub-groups, I then ranked the haplotypes within each group.[15] As found previously,[16] within each group there were several groups of related gene types clustering around a common or modal haplotype. Capelli, for instance, created three STR-defined clusters of one-step relatives around three of the commoner haplotypes. The difference is that whereas Capelli et al. have chosen only three clusters, thus adding to their eleven UEP-defined haplogroups to make fourteen groups, I have systematically broken up each of

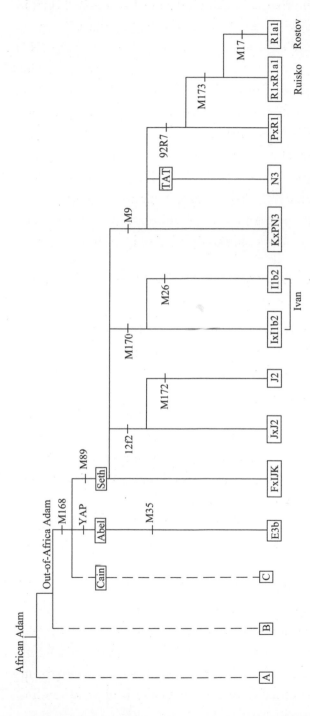

Figure A3 Tree of the main Y-chromosome gene groups (haplogroups) used in the analysis in this book. Gene group boxes for Europe are labelled in consensus nomenclature[5] qualified after Capelli,[11] with relevant nicknames Ruisko, Ivan and Rostov below. These three clusters are expanded respectively in Figures A4 to A6. The dotted lines indicate major world gene groups A, B and C not represented in Europeans; nicknames are shown for the common male ancestor and the three out-of-Africa lines.[8] (Haplogroups are based on 'unique event polymorphisms' – labelled horizontal lines along branches; branch length does not represent age.)

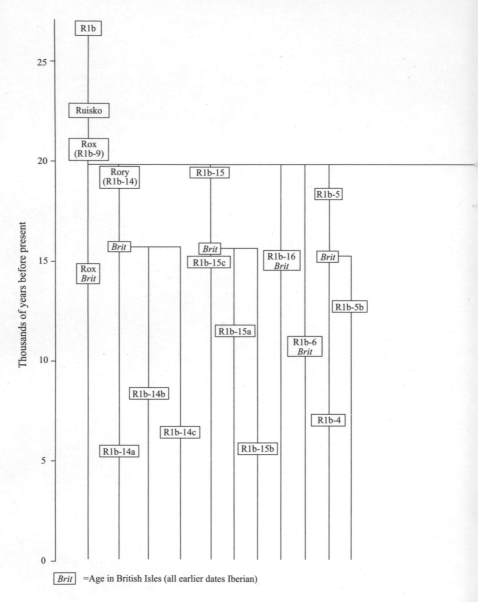

Brit = Age in British Isles (all earlier dates Iberian)

Figure A4   Tree of Ruisko gene clusters described in this book, with branch dates. Two major expansions from the Franco-Spanish refuge are evident: one after the LGM and another after the Younger Dryas. Only R1b-8a expanded in the historical period in the British Isles. (Here and in Figures A5 and A6, cluster labels are located along branches according to approximate age, and subtitled 'Brit' if the date of the founding event in the British Isles is substantially different from the overall age of the cluster in Iberia, or the cluster is unique to the British Isles – e.g. cluster R1b-7.)

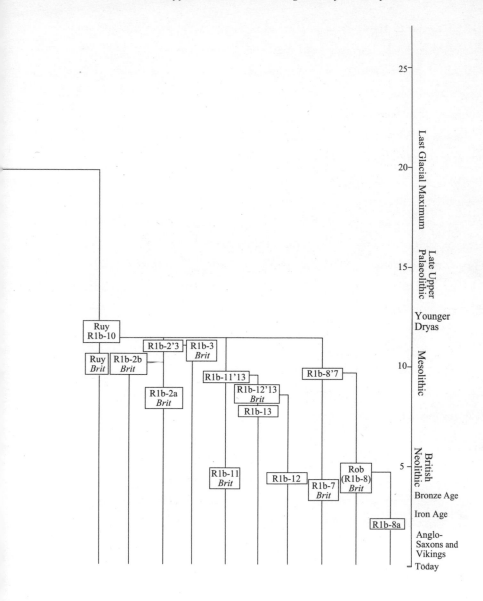

the three large haplogroups into clusters in this way, thus adding thirty clusters to eleven haplogroups, making over forty groups including all sub-clusters (Figures A4–A6). I also performed the same process on the two smaller Neolithic groups J and E3b, but have not displayed the data as figures here.

I used a standard network software package[17] to confirm the phylogenetic validity of these clusters and date them. This confirmed the integrity of the clusters and suggested a very small number of rearrangements between minor clusters. I have called these derived sub-groups 'clusters' rather than sub-groups to differentiate them in the text, and identified these clusters within a gene group by inserting the cluster number hyphenated after the gene group name (e.g. R1b-9).

I have also dated each cluster (with the same software), using Forster's calibration and dating method, which is similar in principle to that used for mitochondrial DNA.[18] The age, number in founding cluster used for dating (*n*), root haplotype and standard deviation for each gene cluster (or gene group) are given where relevant in the notes.[19] The composite dataset with haplotypes and IDs are available on request.

I have displayed spatial frequency distributions of gene groups and clusters graphically throughout the text. This appears rather like modern television displays of weather variables such as temperature, where curved 'contour lines' join up areas of similar temperature. Because such displays are easier on the eye than pie charts, they have become the graphics of choice for illustrating such genetic frequency distributions in the literature.[20]

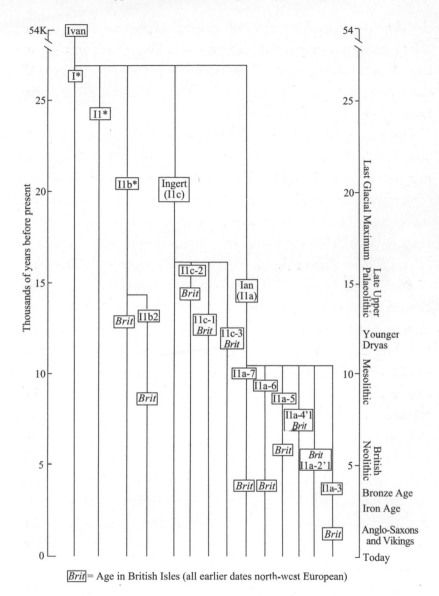

Brit = Age in British Isles (all earlier dates north-west European)

Figure A5   Tree of Ivan gene groups[12,13] and clusters described in this book, with branch dates. I1c expanded in north-west Europe during the Late Upper Palaeolithic, but founding events were staggered in Britain later, up to the Mesolithic. I1a1 expanded in north-west Europe during the Mesolithic, but founding events were staggered in Britain later into the Neolithic. Only I1a-3 expanded in the historical period into the British Isles. (Cluster labels as in Figure A4.)

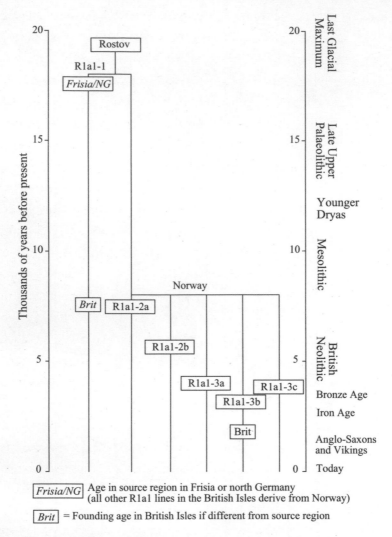

Figure A6   Tree of Rostov gene clusters described in this book, with branch dates. Rostov may have arrived on the North European Plain in the Late Upper Palaeolithic (R1a1–1); this line dates to the Mesolithic in Britain. Other branches of Rostov expanded in Norway during the Late Mesolithic, but founding events were staggered in Britain later, during the Neolithic and Bronze Age. Only R1a1–3b expanded in the historical period into the British Isles. (Cluster labels as in Figure A4.)

Glossary

A word of warning: terminology is used variably in several of the fields covered in this book. This glossary summarizes the way I have used technical terms. In several instances I have used non-jargon words, to replace jargon, for instance 'gene type' for haplotype. Such usage is clarified below. Terms in bold type within a definition have their own entries.

acculturation
The process of adopting a new culture. Shares functional meaning with the less fashionable term 'cultural diffusion' (*see* **diffusionism**).

allele
One of two or more alternative forms of a gene that occupy the same locus (area) on a chromosome.

All-over-Corded (AOC) Beaker
A type of **bell beaker** decorated all over the surface with twisted cord impressions, found in the Rhine Valley, the Netherlands and the eastern part of the British Isles from the third millennium BC.

Atlantic façade (fringe)

In the context of this book, the coastal areas of Western Europe extending from the British Isles to the Iberian Peninsula. In the British Isles, this includes Ireland and the west coast of Britain.

Atlantic Modal Haplotype (AMH)

The most common male STR **haplotype** (gene type) in Western Europe as reported by Wilson et al. (2001) (the root gene type of R1b-10, or Ruy in this book). *See* **single tandem repeat**.

bell beaker

A general style of upside-down-bell-shaped pot common to and shared between several cultures of Western Europe and North Africa, in their Late Neolithic and early metal-age phases.

Bronze Age

A period following the **Neolithic**, associated with the use of metal implements made of bronze. In Britain the Bronze Age commenced about 4,500 years ago, and gave way to the **Iron Age** around 1200–800 BC.

Brythonic

(also 'Brittonic', 'British' in some historical texts, 'P-celtic') A branch of **insular-celtic** languages, including Welsh, Cumbrian, Cornish and Breton.

calibrated dates and calendar years

The level of atmospheric radiocarbon is not strictly constant, so raw radiocarbon dates, often expressed as 'radiocarbon years before present', have to be calibrated or corrected using tables to give 'calendar years ago'. Unless otherwise stated, 'calendar years ago' are used for all radio-carbon-derived dates in this book, based on corrections already made in the primary literature sources cited.

Cardial Ware (Cardial Impressed Ware)
A style of pottery associated with the Late Mesolithic/Early Neolithic in the north-west Mediterranean region featuring patterns created by making impressions on the wet clay with the edge of a cockle (*Cardium*) shell.

Celt(ic)/celtic
(with a lower-case 'c', used for celtic languages in this book) 'Keltoi' (Greek) and 'Celtae' (Latin) were ethnic terms used in the classical period with varying inclusiveness to describe peoples in Western Europe including parts of Spain. Regions of Keltiké and Celtica were roughly coincident with Caesar's Middle Gaul, south of the Seine. The term 'Celt' fell out of use after the classical period but was resurrected in the eighteenth century in association with so-called celtic languages. The linguistic association, the modern meaning of Celt and the geographical origins of classical Celts are all the subject of controversy, and are discussed in the first part of this book.

cognate
(from Latin for 'born') Two words are cognates if they match each other to some degree in sound and meaning and both can be demonstrated to derive from a common root in an ancestral language, but do not descend one from the other.

comparative method
A method of comparative language study using systematic and regular sound changes in **cognates** from two or more languages to reconstruct an ancestral or proto-language. It is the most widely credited method used by historical linguists to reconstruct trees of relationship between languages, for example the **Indo-European** family. Several problems exist, e.g. assumption of strict tree-like branching, when in fact related languages can continue to affect each other by direct areal diffusion.

Corded Ware/Battle Axe Culture

A culture that extended across northern Europe from Russia to Germany from the Late Neolithic to the Bronze Age, named for its characteristic pottery, bearing the patterns of impressed cords, and the stone battleaxes found in single male burials.

Creswellian

A stone age culture coincident with the Late Upper Palaeolithic in Britain and characterized by trapezoidal backed blades called Cheddar and Creswell points, found at Creswell Crags in Derbyshire and elsewhere in England. It has contemporary parallels with the north-west European mainland and the **Magdalenian** of Southern Europe.

diffusionism

Cultural diffusion is the movement of cultural ideas and artefacts among societies (*see also* **acculturation**). The view that major cultural innovations in the past were initiated at a single time and place (e.g. Egypt), diffusing from one society to all others, is an extreme version of the general diffusionist view – which is that, in general, diffusion of ideas such as farming is more likely than multiple independent innovations. *Compare* **migrationism**.

founder analysis

A comparison of the **haplotypes** (gene types) in a region that has been settled with those in candidate source populations in order to identify specific migrating founder gene types. New gene types, *unique* to the target region but arising from such a founder type, are then used to date individual migration founding events.

founder effect

Frequencies of **alleles** or **haplotypes** (gene types) in founding groups are often unrepresentative of those in the source population. There are changes in frequency (to high or low) of particular gene types in a rapidly expanding population founded by a small ancestral group when one or more of the founders were, by chance, carriers of those types.

gene cluster

Term used in this book to describe a group of very closely related male STR **haplotypes** (gene types) clustering around a common root or modal type (used in the same sense as 'cluster' in Capelli et al. (2003) – technically meaning *putative clade*). Purpose here: to differentiate large clearly defined **haplogroups** (gene groups) into smaller clusters (or putative founder clades) with more local geographic specificity and ability to date founder events. *See* **single tandem repeat**.

gene flow

The movement of genes from one population to another. Gene flow can be quantified as the proportion of **haplotypes** (gene types) or **alleles** that are derived from a different population.

gene group

Term used in this book to mean haplogroup – *see* **haplogroup**.

gene lines

Term used in this book collectively to describe **haplogroups** (gene groups), **gene clusters** and **haplotypes** (gene types).

gene type

Term used in this book to mean haplotype – *see* **haplotype**.

gene tree

A branching diagram displaying the relationships between different **haplotypes** (gene types) of the same region of DNA (e.g. mitochondrial DNA, or a particular gene), using the patterns of variation in the DNA sequences in that region (locus). *See* the figures in the Appendix.

genetic drift

The process of random change of **allele** or **haplotype** (gene type) frequencies. These changes can lead to loss (or, alternatively, very high frequency) of certain alleles. The loss of certain alleles erodes genetic variability, and its effects are greatest in populations of small size.

genetic distance

An average measure of relatedness between populations, based on the frequencies of a number of different **alleles** or **haplotypes** (gene types). Genetic distances are used for understanding effects of **genetic drift** and **gene flow**, which should affect all alleles to the same extent and can be expressed graphically in *genetic distance maps*.

genotype

The genetic makeup of an individual, as determined by the **alleles** present. In the example of blood groups, the blood group phenotype A may be the result of either AO genotype or AA genotype. *Compare* **phenotype**.

glottochronology

A technique in linguistics which uses vocabulary comparison (lexico-statistics) to estimate the time of divergence of two related languages by comparing the numbers of shared **cognates**. It is analogous to the use of radiocarbon dating in that it uses decay as a measurement of time, but has been widely rejected by linguists, owing to variation in the rate of lexical decay. Recent modifications attempt to overcome this drawback.

Goidelic (Gaelic)

A branch of **insular-celtic** languages formerly widespread in Ireland (Irish Gaelic) and Scotland (Scottish Gaelic).

Gravettian

A cultural phase of the Upper Palaeolithic in Europe dating from about 28,000 to 22,000 years ago and distinguished by a variety of stone tools such as backed blades and points. Named after the archaeological site at La Gravette, in the Dordogne region of France.

Hallstatt culture
A culture of Central Europe spanning the Late Bronze to the Early Iron Age, from about 1200 to 500 BC, associated with salt-mining and intricately fashioned metal weapons and jewellery. It preceded the Iron Age **La Tène culture**; both are still widely and controversially regarded as 'Celtic' homelands in the reference literature. Named after the village of Hallstatt in Upper Austria, the site of an extensive Iron Age cemetery.

haplogroup (gene group in this book**)**
A group of **haplotypes** (gene types) that share a common genetic ancestor. An example is R1a1 (shown in Appendix C, in Figures A3 and A6).

haplotype (gene type in this book**)**
A set of closely linked **alleles** (or genes) that tend to be inherited together as a unit; in this book a haplotype is called a 'gene type' and the term is extensively used in relation to individual combinations of STR types (*see* **single tandem repeat**).

Holocene
The most recent geological epoch, also known as the post-glacial era, which followed the **Pleistocene** and commenced with the end of the last Ice Age, conventionally about 10,000 radiocarbon years ago, although much deglaciation had already occurred by then (*see* discussion of corrected dates on p. 134).

Indo-European
A language family to which belong most of the languages of Europe, the Near and Middle East, and South Asia, including such languages and language groups as Germanic (including English), celtic, Italic (including Latin), Greek, Baltic, Slavic and Sanskrit.

insular–celtic

A branch of celtic languages comprising those spoken or having originated in the British Isles, and divided into the **Goidelic** and **Brythonic** groups. Although primarily a geographical description, in some definitions it also implies the 'insular hypothesis' – that the Brythonic and Goidelic language branches evolved together in those islands, having a common ancestor more recent than any shared with the Continental celtic languages.

Iron Age

A cultural phase following the **Bronze Age**, associated with the introduction of implements made of iron. In Britain the Iron Age began about 3,000 years ago and lasted into the Roman period.

La Tène culture

A Late Iron Age culture, originally of Western and Central Europe, which developed around 500 BC out of the earlier **Hallstatt culture** and by the turn of the millennium had spread to Britain and Asia Minor. Like the Hallstatt, it was characterized by intricately patterned metalwork. It is named after the archaeological site of La Tène, on the north side of Lake Neuchatel in Switzerland.

Last Glacial Maximum (LGM)

The peak of the last Ice Age, between 22,000 and 17,000 years ago, during which time the British Isles were mostly covered in ice and were almost certainly cleared of their populations.

Late Glacial

The period, from about 16,000 years ago, between the major thaw following the last Ice Age, and the **Holocene**, during which the recolonization of Britain may have begun. It coincided with the cultural phase of the **Late Upper Palaeolithic**.

Late Upper Palaeolithic (LUP)

In Britain, the period of the Upper **Palaeolithic** lasting from the end of the Last Glacial Maximum, about 17,000 years ago, until the **Mesolithic**.

Linearbandkeramik (LBK)
An Early Neolithic culture of Central Europe, associated with a style of pottery decorated with distinctive straight or linear bands.

Magdalenian
A culture of the Late Upper Palaeolithic, starting towards the end of the last Ice Age from about 18,000 years ago, which spread through Western Europe. It was characterized by tools of flint, bone, antler and ivory, and is named after the archaeological site of La Madeleine, in the Dordogne region of France.

Mesolithic
A cultural phase following the **Palaeolithic** and associated in particular with the introduction of microliths – small flint blades. Dates given for its commencement vary from 15,000 to 10,000 years ago. In this book, it is taken to have begun at the end of the **Younger Dryas**, 11,500 years ago.

migrationism
The view that major cultural changes in the past were caused by movements of people from one region to another carrying their culture with them. It has displaced the earlier and more warlike term *invasionism*, but is itself out of archaeological favour. *Compare* **diffusionism**.

mitochondrial DNA (mtDNA)
Genetic material in the mitochondria (self-reproducing organelles inside cells, but outside the nucleus). Because mitochondria (and hence mitochondrial DNA) are inherited only from the mother, mtDNA provides an unmixed link to past female generations. It is useful for dating and recording initial colonizations and migrations of humans. *Compare* **Y chromosome**.

Neolithic
Literally 'new stone age', a period that followed the **Mesolithic** and was associated with the introduction of agriculture and pottery. In Britain the Neolithic lasted from about 6,000 to about 4,000 years ago.

Ogham

A celtic alphabetic script, probably originating in Ireland, used throughout non-English, **insular-celtic**-speaking areas of Britain from the fifth century AD, mainly on standing stones and often with celtic–Latin bilingual inscriptions. Characters consisted of one to five slash marks, vertical or diagonal, carved against a running base line.

Palaeolithic

Literally 'old stone age', a cultural phase preceding the **Mesolithic** and **Neolithic** ages. These phases occurred at different times in different cultures. In Britain the Palaeolithic commenced over half a million years ago.

palimpsest

A classical term meaning a manuscript which has been reused by scraping off the original text and writing over the top. Used in this book and in certain genetic and archaeological literature as an analogy for multiple European migrations: in this sense, it means genetic evidence of one migration route overlying an earlier one, both migrations having been channelled by the same geographical corridors and barriers.

P-celtic

Alternative name for the **Brythonic** branch of the **insular-celtic** group of languages, so called because it uses a 'P' sound where *Q-celtic* languages, in the **Goidelic** branch, use a 'Q' (or hard 'C') sound.

phenotype

The expression of the genes present in an individual. This may be directly observable (eye colour) or apparent only with specific tests (blood group). The blood group A is an example of a phenotype with several possible underlying genotypes (AO and AA), since A is dominant and O recessive. *Compare* **genotype**.

phylogeography

The study of the patterns of genetic differentiation on a **gene tree** across landscapes. The geographical distribution of gene lines is analysed with respect to their phylogenetic position on a gene tree, usually within a species, to reconstruct their origins and routes of movement.

Pleistocene

Literally 'ice age', the geological epoch that preceded the **Holocene**, lasting, by convention, from around 1.8 million to 10,000 years ago. It was a time of global cooling and increased climatic changes, long *glacial* periods alternating with much shorter, warmer *interglacial* periods.

Principal Components Analysis (PCA)

A statistical technique used to simplify the presentation of a dataset. For instance, it takes the variance of a number of different genetic markers between populations – the degree to which they vary from an average value – and parcels them into a series of components, the First Principal Component, Second Principal Component, and so on, in decreasing order of statistical importance.

Q-celtic

See **P-celtic**.

refuge

An isolated area (also called a *refugium*) where the extensive environmental changes caused by changing climate, typically glaciation, are more benign, and where plants and animals typical of a region – including humans – may survive. After a return to favourable climatic conditions, survivors may spread out from the refuge and repopulate areas from which their species had become extinct.

rune

A letter in a runic alphabet, or an inscription using such letters. Runic alphabets were used to write Germanic languages, mainly and originally in Scandinavia, but also in the British Isles and Frisia, from around the second century AD to the fifteenth.

single tandem repeat (STR) sequence

Multiply repeated DNA sequences in which the repeat units are short (typically three to seven nucleotide base pairs in length). The numbers of such repeats can be used as genetic markers. In the non-recombining Y chromosome, combinations of these numbers at different genetic sites (loci) can be combined to form unique **haplotypes** (referred to as STR *gene types* in this book). The rapidity of individual number changes over time is an advantage for studying short timescales, but a disadvantage for constructing an unambiguous gene tree or network.

Solutrean

A cultural phase of the Upper Palaeolithic in south-west Europe dating from about 22,000 to 18,000 years ago and distinguished by finely crafted flint projectile points, ornamental beads and bone pins, and evocative prehistoric art. It followed the **Gravettian** and preceded the **Magdalenian**. Named after the archaeological site at Solutré, a village in south-eastern France.

Y chromosome

The male sex chromosome and held only by the male. Its advantage for tracing the geographical movements of the genes people carry is a lack of recombination at each generation, thus allowing detailed reconstruction of male ancestry in prehistory. It tends to give a sharper geographical pattern than **mitochondrial DNA**.

Younger Dryas (Event) (YD)

A short intense period of glaciation from about 13,000 to 11,500 years ago (the Older Dryas was a less severe glaciation about 2,000 years previously). The *Dryas octopetala* is a hardy Alpine plant known from pollen deposits to have flourished during these cold spells.

NOTES

These notes are intended as a facility to readers, academic or otherwise, seeking technical clarification and sources of evidence. They contain technical terms and detail which, in the space available, cannot be explained to the same level as in the main text. Sources are cited in full in the bibliography. Explanation of the genetic analysis and dating used in this book and relevant genetic trees are found in the appendices.

Preface

1. Rhiannon Edward, 'Scots and English aulder enemies than thought', *The Scotsman*, 12 April 2004.
2. Keith Sinclair, 'English-Scots split goes back 10,000 years: genetic proof of Celts' ancient ancestry', *The Herald*, 12 April 2004.
3. Louise Gray, 'Celts and English are a breed apart? Absolutely, says Professor', *Independent*, 12 April 2004.
4. 'Scientist mulls Anglo-Scottish split: cultural differences which divide the Scots and the English date back 10,000 years before Britain was an island, a professor has suggested', <http://news.bbc.co.uk/1/hi/uk/3618613.stm>, accessed 11 April 2004.

Prologue

1. While the romance may have been created in the so-called Dark Ages ('dark' mainly for lack of written texts), its various extant versions were all written during the Late Medieval period, generally with chivalric overlay.
2. Furthermore, today these areas are Germanic-speaking rather than 'Celtic-speaking'.
3. Tacitus, *Agricola* 11.
4. Tacitus, *Agricola* 11.
5. Julius Caesar, *Gallic Wars* 1.1.

Part 1

Chapter 1

1. Yeats (1902).
2. Quoted in Renfrew (1989), p. 225.
3. James (1999), pp. 67–85.
4. Invalidation: James (1999); myth: James (1999) and Collis (2003), and see also discussion in Renfrew (1989), chapter 9.
5. James (1999), pp. 45–51.
6. James (1999), pp. 49, 124; but see Collis (2003), p. 52.
7. Pezron (1703). He was a Cistercian theologian.
8. Pezron (1706), Lhuyd (1707).
9. Pezron, incidentally, had not included Ireland in his model, although Lhuyd had included Brittany.
10. Buchanan (1582).
11. Buchanan (1582).
12. James (1999), pp. 45–50.
13. James (1999), pp. 44–51, 54–62.
14. James (1999), pp. 56–7.
15. James (1999), p. 57.
16. Herodotus, *Histories* 2.11. The identical mistake was made by Aristotle in *Meteorologica* 1.13.
17. Rufus Festus Avenius, *Ora Maritima* (4th cent. AD). Extracts as translated in Rankin (1996), pp. 4–7.
18. Livy, cited in Rankin (1996), p. 3.
19. Rankin (1996), p. 3.
20. Rankin (1996), p. 7.
21. For: Rankin (1996), Cunliffe (1988); against: Collis (2003), James (1999).
22. Cunliffe (1988); see also Cunliffe (1997).
23. Rankin (1996).
24. Collis (2003).
25. Cunliffe (2004); see also Cunliffe (2003).
26. Rankin (1996), pp. 2–9. Note that an alternative view of this part of Himilco's 'voyage' is that it was conflated by Avenius with that of Pytheas the Greek: see Cunliffe (2002).
27. Rankin, (1996), p. 5.
28. Herodotus 3.115; Diodorus, *Historical Library* 5.21,22,38; Strabo, *Geography* 2.5.15, 2.5.30, 3.5.11; Pliny the Elder, *Natural History* 4.119, 7.197. See Cunliffe (2004), pp. 302–6 for interpretations.
29. Cunliffe (2004), pp. 30, 306.
30. Cunliffe (2004), p. 306.
31. Rankin (1996), pp. 2–9.
32. Including, presumably, the Channel Islands.

33. Cunliffe (2004), p. 104.
34. Belerion, or Cornwall, according to Sicilian Diodorus Siculus writing in the first century BC: Cunliffe (2004), p. 305. Note that an alternative view of this part of Himilco's 'voyage' is that he never took it; rather it was conflated by Avenius with that of Pytheas the Greek: see Cunliffe (2002).
35. Cunliffe (2004), pp. 302–10.
36. Cunliffe (2004), pp. 306–7.
37. Cunliffe (2004), p. 317. See also Cunliffe (2003).
38. Strabo, 4.1.14; Renfrew (1989), p. 223.
39. Diodorus Siculus, 5.21; Renfrew (1989), p. 222.
40. Rankin (1996), p. 167.
41. Rankin (1996), p. 166.
42. Strabo, 4.4.6.
43. Diodorus Siculus, 5.33; see also Rankin (1996), p. 167 and Renfrew (1989), pp. 222–3.
44. Rankin (1996), pp. 81, 166.
45. Cunliffe (2004), figure 8.3.
46. Rankin (1996), pp. 38–44.
47. Rankin (1996), p. 38.
48. Rankin (1996), p. 103.
49. Cunliffe (1997), figure 15; Cunliffe (2004), figures 8.2 and 8.3. There is no dispute here about the reality of these cultural expansions from Central Europe, merely whether that was the 'Celtic homeland'.
50. Cunliffe (1997), figure 45; see also discussion in Collis (2003), pp. 115–18, where he says that the Boii had been on the Roman side of the Rhine.
51. Strabo, *Geography* 5.1.6.
52. Strabo, *Geography* 5.1.6.
53. Strabo, *Geography* 4.4.1.
54. Strabo, *Geography* 4.1.13.
55. Julius Caesar, *Gallic Wars* 1.1.
56. Cunliffe (1997), figure 45.
57. Collis (2003).
58. James (1999), p. 57.
59. Collis (2003), p. 60.
60. Collis (2003), p. 63.
61. Collis (2003), pp. 63–7.
62. Collis (2003), p. 66.
63. Collis (2003), p. 66.
64. Collis (2003), p. 66.
65. By Danish antiquarian Christian Thomsen: see Collis (2003), pp. 73–4.
66. Collis (2003), pp. 73–4.
67. Collis (2003), p. 219.

Chapter 2

1. Collis (2003), p. 224.
2. See e.g. Lambert (1994), Marichal (1988), Lejeune (1988). See also Collis (2003), pp. 130–32.
3. More radical interpretations are possible, such as an Italian-celtic-language homeland, consistent with the controversial hypothesis of an Italo-celtic linguistic common ancestor – see Cowgill (1970).
4. Cited in Collis (2003), pp. 110–12.
5. Collis (2003), pp. 110–12.
6. Lambert (1994), Marichal (1988), Lejeune (1998); see also Renfrew (1989, pp. 230–33), who has reviewed the evidence for Gaulish and other such languages.
7. Anderson (1988), Untermann (1961), Villar (1997).
8. See e.g. Diodorus Siculus, *Historical Library* 5.33; Strabo, *Geography* 3.4.12; Pliny, *Natural History* 3.1.8; Lucan, *Pharsalia* 4.1.9–10; Appian, *Hispania* 2; Isidore of Seville, *Etymologiae* 9.2.113–14. See Collis (2003), pp. 112–13.
9. Villar and Pedrero (2001).
10. Untermann (1997).
11. Collis (2003), p. 113.
12. See e.g. Parsons and Sims-Williams (2000).
13. Parsons and Sims-Williams (2000), and Hachmann et al. (1962), map 9.
14. Parsons and Sims-Williams (2000).
15. Parsons and Sims-Williams (2000), pp. 113–42; see also figures 6.1, 6.2, 6.3, 7.1, 7.2 and 11.1 in Sims-Williams (2006).
16. Renfrew (1989), p. 232.
17. Sims-Williams, P. (2006), summarized in his figure 11.1, and converted to 'contour lines' in Figure 2.1b in this book.
18. Renfrew (1989), p. 233.
19. <http://www.ucl.ac.uk/archaeology/cisp/database/maps/bigmap_all.html>. See the Celtic Inscribed Stones Project (CISP) based at the Department of History and the Institute of Archaeology, University College London: on-line database at <http://www.ucl.ac.uk/archaeology/cisp/database>.
20. Tacitus, *Agricola* 11.
21. Jackson (1955).
22. Via the Anglo-Latin word *talea* (1189) and the Anglo-French *tallie* (1321).
23. This account of Cumbric sheep-counting is based on <http://www.lakelanddialectsociety.org/counting_sheep.htm> (accessed October 2005). See also Ted Relph (ed.) (2006), 'Lakeland Dialect', *The Journal of the Lakeland Dialect Society*.
24. Sims-Williams (2006), p. 54 and figure 4.2.
25. Sims-Williams (2006), p. 54 and figure 4.2.
26. Collis (2003), p. 114.
27. Jackson (1953).
28. Jackson (1953), pp. 703–4.
29. CISP, <http://www.ucl.ac.uk/archaeology/cisp/database/maps/bigmap_all.html>.

30. Jackson (1953), Evans (1967), Rivet and Smith (1979).
31. Parsons (2000).
32. Parsons (2000).
33. Jackson (1955).
34. Adamnan, *Life of St Columba* 1.27, 2.33.
35. Bede, *Ecclesiastical History* 1.1.
36. See e.g. Saxo Grammaticus, *Gesta Danorum* Book 3: 'But the gods, whose chief seat was then at Byzantium, (Asgard)'. See also Heyerdahl and Lillieström (2001). In this book and the archaeological expedition he funded, Thor Heyerdahl was inspired by the thirteenth-century Icelandic writer Snorri Sturluson, who describes in his *Ynglinga* saga how a chief called Odin led a tribe called the Æsir in a migration, northwards through Saxland, to Fyn in Denmark, finally settling in Sweden. Heyerdahl believed that Odin may have been a real king in the first century BC in what is now southern Russia, near the Sea of Azof.
37. Bede, *Ecclesiastical History* 1.1 and 3.6.
38. Jackson (1955).
39. Jackson (1955), pp. 151–2.
40. Renfrew (1989), pp. 226–7.
41. Campbell (2001).
42. Campbell (2001).
43. Sims-Williams (2002).
44. Campbell (2001).
45. Rankin (1996), pp. 304–6.
46. *The Book of the Taking of Ireland*, compiled and edited in Middle Irish by an anonymous scholar in the eleventh century.
47. Rankin (1996), pp. 306–7, O'Rahilly (1946).
48. Rankin (1996), pp. 13–14.
49. Schmidt (1988).
50. McCone (1996).
51. McCone (1996).
52. See e.g. Sims-Williams (2003).
53. Dyen et al. (1992).
54. Gray and Atkinson (2003), Atkinson et al. (2005), McMahon and McMahon (2003, 2006).
55. See discussion in McMahon & McMahon (2006) pp. 153–60.
56. McCone (1996).
57. McMahon and McMahon (2003).
58. McMahon and McMahon (2006).
59. Gray and Atkinson (2003).
60. The calibration method used takes the following events as fixed points: Germanic tribes united against Rome, AD 1; Gothic migration to Eastern Europe, AD 180. The earliest attested North Germanic inscriptions date from the third century AD.
61. Forster and Toth (2003).
62. But see the critique in McMahon and McMahon (2006).

63. Forster and Toth (2003).
64. 3200 BC ± 1,500 years (Forster and Toth 2003).
65. 8100 BC ± 1,900 years (Forster and Toth 2003).
66. Renfrew (1989), Forster and Toth (2003).
67. Based on the Roll of the Kings of Ireland after Ireland was divided up between the invading Milesian-Gaelic lords Éber and Éremón, as given in O'Donovan (1848–51) and Macalister (1938–56).
68. Roll of the Kings of Ireland, taken from <http://www.ucc.ie/celt/online/T100005A>.
69. Roll of the Kings of Ireland, taken from <http://www.ucc.ie/celt/online/T100005A>, Annal M3266.1.
70. According to O'Rahilly (1946), the Cruithne or Priteni arrived c.700–500 BC, the Builg or Érainn c.500 BC, and the Lagin, the Domnainn and the Gálioin, c.300 BC.
71. Cunliffe (2004) – discussed in Part 2 of this book.
72. *Lebor Gabála Érenn* (Macalister 1938–56).
73. *Lebor Gabála Érenn* (Macalister 1938–56).
74. Gildas, *De excidio Britanniae*.
75. Cunliffe (2004), figure 8.3.
76. Cunliffe (1988), figure 15.
77. Cunliffe (2004), pp. 222–6, 296.
78. Cunliffe (2004), pp. 225–6.
79. Cunliffe (2004), p. 243.
80. Cunliffe (2004), p. 296.
81. Renfrew (1989).

Part 2

Introduction

1. Sims-Williams (1998a,b, 2003).
2. Gray and Atkinson (2003), Atkinson et al. (2005).
3. 5,200 (±1,500) years (Forster and Toth 2003).
4. 6,100 years using Dyen's cognate data: Gray and Atkinson (2003); 5,600 years using Ringe's cognate data: Atkinson et al. (2005). The Cambridge–Zurich group (Forster and Toth 2003) cannot estimate the date of this node (the root of celtic), but bracket it above by an estimate of the date of the break-up of Indo-European as a whole in Europe at about 10,000 years ago (10,100 ± 1,900), and below by a date of 5,200 (±1,500) years for the fragmentation of Gaulish, Goidelic and Brythonic from their most recent common ancestor.

Chapter 3

1. Oppenheimer (2003).
2. See figure 6 in Barton et al. (2003). For the British Isles at the Late Glacial see also Barton (1999).

3. Figure 6 in Barton et al. (2003); see also figure 1 in Gamble et al. (2004).

4. i.e. Solutrean and Magdalenian/Badegoulian – Gamble et al. (2004); see pp. 108–9.

5. Gamble et al. (2004), p. 247.

6. Oppenheimer (2003), pp. 248–53 and figure 6.2.

7. Oppenheimer (2003), pp. 113, 276. I qualified this observation by adding 'at least until the last five hundred years', indicating that the reason for this conservatism could be the truism that it is easier to invade unoccupied than occupied lands – unless one has guns.

8. Archaeological: Barton et al. (2003) and Gamble et al. (2004). Genetic: see pp. 111–13, the expansion is greater when estimated from mtDNA than from the Y chromosome.

9. Barton et al. (2003).

10. Barton et al. (2003).

11. Barton et al. (2003).

12. Ripoll et al. (2004). See also images at <http://www.bradshawfoundation.com/creswell.html>.

13. Oppenheimer (2003), Appendix 1, pp. 365–9.

14. Table 4 in Richards et al. (2000).

15. Present study (i.e. my own re-analysis on a pooled dataset carried out for this book: see the section 'Re-analysis of the Y-chromosome data' later in this chapter, and the Appendices).

16. 52.7% (±4.5%): table 5 in Richards et al. (2000).

17. 58.8% (±4.7%): table 5 in Richards et al. (2000).

18. Semino et al. (2000).

19. Diamond and Bellwood (2003).

20. Oppenheimer (2003), pp. 214–15.

21. Gamble et al. (2004).

22. Gamble et al. (2004).

23. Gamble ct al. (2004); for other genetic evidence see also Torroni et al. (1998) and Pereira et al. (2005).

24. i.e. during the Early Upper Palaeolithic; see Oppenheimer (2003), pp. 130–41, and Richards et al. (2000).

25. Oppenheimer (2003), pp. 144–50, and Torroni et al. (1998, 2001).

26. When H was characterized using the short-sequence (i.e. HVS1 and RFLP) tools of the 1990s, the branch was monolithic and poorly resolved.

27. Torroni et al. (1998).

28. Torroni et al (2001).

29. Although haplogroup V's highest population frequency (52%) is among the Skolt Saami of Norway, followed by 12% among the Basques, the former high rate, lacking diversity, is almost certainly a founder effect. This would have resulted from a small founding population, possibly carrying only V and a couple of others from the south-west, which then amplified V by drift and growth. Her young age among the Saami and the complex aspects of her phylogeography suggest that she may have reached the Fenno-Scandian Peninsula more recently via a circuitous route through Central Europe (Torroni et al 2001, Tambets et al. 2004).

30. Torroni et al. (2001), figure 4.

31. It should be noted, however, that while H1 seems to date to the Late Upper Palaeolithic, c.14,000 years ago, the smaller H3 group appears to be younger, possibly post-Younger Dryas, and is dated to c.9,000 years ago. H1: 14,000 years ago (SE ±4,000) using coding-region data, and c.16,000 years ago (SE ±3,500) using HVS-I; H3: c.9,000 years ago (SE ±3,000) based on coding-region data and c.11,000 years ago (SE ±3,000) using HVS-I. See Pereira et al. (2005).

32. As some measure of this, in the Founder Analysis published in 2000, Martin Richards and colleagues inferred that up to 53% of today's female lines arrived in north-west Europe (for 'north-west Europe' see Richards et al. (2000), table 5) during the Late Upper Palaeolithic, shortly after 16,000 years ago. In this estimate, they included most of H anyway, which covered a large proportion of what were at that time unresolved H types. Their figure 1 shows e.g. 38% + 3.9% = c.42% of all lineages for the whole of Europe in the LUP contributed by H, plus a further 10% addition from other gene groups (T, T2 and K); the last three groups, however, overlap with the Mesolithic.

33. Oppenheimer (2003), pp. 150–54, 250–53.

34. R1b designation in The Y Chromosome Consortium (2002). It is equivalent to Eu-18 in Semino et al. (2000).

35. Distribution of 'R1b' (Ruisko) haplogroup in Europe: present study, Oppenheimer (2003), Underhill et al. (2000), Wilson et al. (2001), Capelli et al. (2003), Semino et al. (2000).

36. Unless otherwise indicated in this note, data are collated from table 3 in Tambets et al. (2004): Spanish Basque 88.9% and Dutch 70.4% (Semino et al. 2000); French 59%; Germans 50% (as in Semino et al. (2000) – note the typographic error of '0' in published table 3, Tambets et al. (2004)); Danes 36.1%; Norwegians 27.8%; Swedes 22.0%; northern Russians 21.3%; Estonians 18.2%; Poles 16.1%; Latvians 7.0% and Finns 0%.

37. In Central Europe and the Balkans there are varying mixtures of the two European members of the R clan, R1a1 and R1b. In the west, including the British Isles, Germany, the Netherlands, Belgium and France, it is more straightforward: R1b dominates, especially south and west of the Rhine. R1a1, defined by a unique mutation known and referred to in my last book (Oppenheimer 2003) as M17, is nicknamed Rostov in this book.

38. The overall age of R1a1 (rooted on Ht. 87/R1a1–2 in database) is 19,500 years ($n = 291$, SD ±5,540) (see Appendix C). This implies post-LGM re-expansion. For pre-LGM history, note Ruslan/M17 in Oppenheimer (2003), p. 152, which is likely of great antiquity in South Asia. Data from several sources: Tambets et al. (2004), table 3; Rosser et al. (2004), Hg1 in table 1, and figures 3 and 4; Weale et al. (2002), Hg1 in tables 2 and 3; Cappelli et al. (2003), Hg R1xR1a1 in table 1; Semino et al. (2000), Hg Eu18 in table 1.

39. Age estimates for subgroup I1c from table 3 in Rootsi et al. (2004): time since sub-clade divergence 14,600 (±3,800) years, age of STR variation 13,200 (±2,700) years, time since population divergence 11,200 (±2,300) years. The second and third figures agree with my own estimate for I1c in Western Europe of 12,000–14,000 years, although I estimate sub-clade divergence as older, at 21,000 years.

40. For example, eleven haplogroups as given in Capelli et al. (2003); these are extended using STR clusters to fourteen. Relevant published Y-chromosome datasets for mainland Western

Europe include all those cited by Tambets et al. (2004), and also Rosser et al. (2000), who subdivide the dataset into twelve haplogroups, but only seven of these are relevant to the British Isles; Scozzari et al. (2001), who attempt to break up other haplogroups using STR markers, but unfortunately not the R haplogroup; Gusma et al. (2003), who look specifically at STR haplotypes within the R1b haplogroup and find geographical structure, but do not proceed to create any phylogeny; and Roewer et al. (2005), who look at STR variation geographically but do not use it to resolve the relevant haplogroups.

41. Key papers for the British as opposed to Continental data are Wilson et al. (2001), Capelli et al. (2003), Weale et al. (2002) and Hill et al. (2000). Additional indigenous Orkney and Shetland STR data courtesy Jim Wilson (from D.K. Faux and J. Wilson, unpublished) at <http://www.davidfaux.org/shetlandislandsY–DNA3>. Methods and protocols used for typing STR Y-chromosome microsatellites DYS388, 393, 392, 19, 390 and 391 can be found in Thomas et al. (1999) and Goodacre et al. (2005).

42. The main papers listing British unique event polymorphism (UEP) and STR data used in this book are as given in the previous note.

43. This was a difficult decision, since in spite of the volume of data, the information provided by STR markers, unlike the UEPs used to identify haplogroups, is ambiguous and difficult to resolve even by using network software programs. The differences between the 'bi-allelic' haplogroup markers (i.e. dichotomous unambiguous 'UEPs') and STRs is that the latter are unstable, rapidly mutating, multi-allelic and thus potentially ambiguous in interpretation. The amount of work involved in network building for such large haplogroups is horrendous, without the availability of more UEPs, which is why the field has previously stayed fallow. It was necessary not only to resolve branches but also to date them. The whole task took me about six months.

44. For source papers on the British and other Western European populations, see methods in the Appendices. For comparison I have also referred to papers giving additional STR data on Ireland and Iceland (Helgason et al. 2000a,b, 2003), other European areas including various parts of Spain (Galicia, Valencia, and additional Basque: Brion et al. (2003, 2004)); Catalan (Bosch et al. 2001), the Saami (Tambets et al. 2004), and Hb 'I' in the Balkans and the rest of Europe (Rootsi et al. 2004). Wilson et al. (2001) also have data for Friesland, Basques, Syria and Turkey. Weale et al. (2002) also include Friesland and Norway.

45. When all the information is collected together, the 1,947 samples with the R1b Y group can be split into 180 individual STR haplotypes of varying frequency. Two of these haplotypes (Ht. 157 and Ht. 155 in the present study – see Appendix C) are the most common in the northern Atlantic coastal region, and are distributed throughout Western Europe, accounting on their own for respectively 15% and 9% of all male gene types in the entire dataset that I use. One paper has even given the former founder type the grand name of 'Atlantic Modal Haplotype' (AMH) (Wilson et al. (2002); see also Capelli et al. (2003), who also use the term 'AMH'). Note the rates of two most common Atlantic founding lines: 15% (467/3,084, Ht. 157 in the present study) and 9% (272/3,084, Ht. 155 in the present study), where 3,084 is the number of individuals in the entire dataset in my analysis, and the former percentage (i.e. 15%) refers to the AMH type.

46. Unfortunately, in spite of the plethora of papers recording British male genes, there are none which report having tested for the R1b type-marker, P25. However, they have all tested for R1a1, and most have done enough other tests to identify R1b types of the British Isles to the exclusion of other R groups. See note 38 for references.

47. I have used a similar approach to that described in Martin Richards' European mitochondrial DNA paper as Founder Analysis (Richards et al. 2000).

48. For the methods, see Appendix C. In this preliminary description I have labelled these clusters R1b-1 to R1b-16 according to their rank in the sorting process. The geographical distribution of these clusters and their individual members helps a lot in tracing the process of that initial recolonization and of subsequent migrations from Iberia. My re-analysis of the Y tree confirmed the validity and integrity of these clusters as real branches, with minor misclassifications and rearrangements mainly found in clusters R1b-1 and R1b-16. In my subsequent discussion, I therefore use this 1–16 numbering convention, but with all the corrections included.

49. R1b-9 root: Ht. 155/6. R1b-9's age in the Iberian refuge: 20,600 years (rooted on Ht. 155/R1b-9, $n = 138$, SD $\pm 6,830$); overall R1b-9 age in database, but excluding R1b-10 and descendants: 19,200 years ($n = 513$, SD $\pm 5,790$); northward expansion from the refuge towards the British Isles dated to 15,600 years ago ($n = 1,513$, SD $\pm 4,800$).

50. Ht. 155 and 156 in present study (i.e. = R1b-9 = Basque Haplotype).

51. For archaeological dating of early re-expansion, see note 67.

52. 27.4% = 570/2,082 in the British database (i.e. the British part of the present study).

53. I1a–c and I* in Rootsi et al. (2004).

54. See note 39.

55. R1b-15a, age 11,539 years (rooted on Ht. 253, $n = 10$, SD $\pm 6,014$); R1b-15b, age 5,729 years (rooted on Ht. 250, $n = 46$, SD $\pm 3,037$); R1b-15c, age 14,958 years (rooted on Ht. 247, $n = 30$, SD $\pm 4,623$).

56. Hill et al. (2000). The distribution of R1b-14 correlates with that of R1b-9: regression (correlation) of R1b-14 with R1b-9 in the present study: $y = 0.4385x + 0.029$; $R = 0.376$.

57. Age of R1b-14 in the British Isles: 15,760 years ($n = 170$, SD $\pm 8,440$).

58. I1c: maximum frequency in Germany 12.5%, Netherlands 10% (table 1 and figure 1D in Rootsi et al. 2004). For eastern England, the maximum in York, for example, is 13% (present study).

59. My estimated age for I1c in my dataset as a whole (rooted on Ht. 346) is 20,800 years (SD $\pm 4,550$). See Appendix C.

60. I1c-1: 12,800 years ($n = 12$, SD $\pm 7,400$), I1c-2: 14,200 years ($n = 19$, SD $\pm 3,880$) and I1c-3: 11,700 years ($n = 25$, SD $\pm 5,470$) – see Appendix C.

61. Two-thirds of I1c types in Britain are unique. This is an unusual picture for any haplogroup in this dataset, and is a measure of its antiquity there reflected in the ages given in the previous note.

62. Barton et al. (2003), figure 1. Note that unless otherwise stated, these dates are corrected or 'calibrated' and represent calendar years.

63. Barton et al. (2003).

64. Barton et al. (2003).

65. Oppenheimer (1998), pp. 29–32.
66. Figures 1 and 4 in Gamble et al. (2004), and figure 6 in Barton et al. (2003).
67. Figure 1 in Barton et al. (2003). Note that not all these types of human evidence are present simultaneously at the same sites during these climatic phases, possibly because of changes in the locations of humans' settlement and their lifestyle.
68. Oppenheimer (2003), pp. 248–53 and figure 6.2.
69. Richards et al. (2000) and present study (see Appendix and Figure A3).
70. Oppenheimer (1998), pp. 30–48, figure 1.
71. Stuart and Lister (2001).
72. Schirrmeister et al. (2002): 'After that time [LGM, 24,000–18,000 years ago] fossil insect assemblages point to a far more continental climate with much warmer summers (Sher et al. 2001). The decrease of mammoth fossils dated to 20 kyr–15 kyr BP and their subsequent increase indicates less favorable environmental conditions for large animals during the first substage, and better conditions during the second ....'
73. Oppenheimer (1998), pp. 38, 228, 234, 255, 260.

Chapter 4

1. Oppenheimer (2003), pp. 24–7, 76–80.
2. Barton and Roberts (2004).
3. Oppenheimer (1998), pp. 29 32.
4. Cunliffe (2004), pp. 119–38.
5. Cunliffe (2004), pp. 123–6.
6. Cunliffe (2004), pp. 123–6.
7. Cunliffe (2004), pp. 126–38.
8. Cunliffe (2004), pp. 126–38.
9. Cunliffe (2004), pp. 138–9.
10. On the maternal Mesolithic contribution and Founder Analysis, see Richards et al. (2000). Note that there are problems in trying to match the Founder Analysis paper too closely to the peopling of the British Isles. First, it measures intrusion only of putative Near Eastern lineages, which ignores potential repeat Mesolithic/Neolithic expansions from the Basque refuge. Second, apportioning founding lineages between the LUP and the Mesolithic is affected not only by the foregoing (i.e. multiple expansions and/or non-Near Eastern lineages) but also by poor resolution of haplogroup H. Richards et al. (2000) acknowledge these problems; for further discussion of relative Mesolithic component see Richards (2003) and McEvoy et al. (2004).
11. For instance, a large part of the LUP component in that Founder Analysis (Richards et al. 2000) was made up of unresolved H types, so potentially recurrent migrations from the same Iberian source up the Atlantic coast (i.e. not from the Near East) could not be excluded. Such issues were mentioned in that paper, and Richards has since discussed the possible underestimation of the size of West European Mesolithic populations (Richards 2003) and, as mentioned, has subsequently resolved the large H group somewhat more (Pereira et al. 2005).

12. Pereira et al. (2005), tables 1 and 2. Ages estimated using coding-region data: H3, 9,000 years (SE ±3,000), also H2, 11,600 years (SE ±8,000), H5a, 8,000 years (SE ±4,000), but also H4, 7,000 years (SE ±4,000). In all, these four putative Mesolithic lineages account for 10.8% of Irish and 10.6% of UK maternal lines.

13. T, T2, T4 and K dates straddling the YD: Richards et al. (2000), figure 1. T, T2, T4 and K found in the Basque Country, Spain and Cornwall: Richards et al. (1996). See also discussions on the size of the Mesolithic component in Richards (2003) and McEvoy et al. (2004).

14. British 17.57% (366/2,082), Basque 19.0% (15/79); age of R1b-10 cluster and derivatives overall in British Isles: 9,800 years (*n* = 693, SD ±3,410). Present study.

15. When Principal Components Analysis (see Chapters 6 and 11) is performed, the PC plot for R1b-10 links Wales strongly with the Basque Country. Using the coordinates of each cluster, it is possible to plot the geographical region in the British Isles of highest relative frequency for each cluster.

16. 26.1% (23/88).

17. 10.5% (8/76) and 17.8% (26/146).

18. 25.6% (10/39) and 21.6% (19/58).

19. Respectively: R1b-11 and 13: 9.6% (200/2,082); R1b-7, 8 and 12: 14.9% (311/2,082); and R1b-1 to R1b-3: 2.2% (46/2,082). Ages: age of combined cluster R1b-11 to R1b-13: 9,500 years (rooted in Ht. 172, *n* = 225, SD ±4,740); R1b-2a, 8,500 years (*n* = 9, SD 7,264); R1b-2b, 10,300 years (*n* = 30, SD ±5,040); R1b-7 and R1b-8 – see notes 54–56 to Chapter 5.

20. R1b-13: frequency 3.8% (79/2,082), age 7,800 years (SD ±4,090).

21. R1b-13 in North Wales (Llangefni) 11.4% (10/88), and Cumbria (Penrith) 8.9% (8/90). Present study.

22. Age of R1b-11 to 13: 9,540 years (*n* = 225, SD ±4,740); R1b-13, 7,790 years (*n* = 79, SD ±4090); R1b-11, 4,560 years (*n* = 121, SD ±3,370); R1b-12, 4,620 years (*n* = 25, SD ±4,140). Rate of R1b-11: 5.8% (121/2,082).

23. 10.8% (225/2,082).

24. 13.0% (271/2,082).

25. Haplotype 151, in this dataset.

26. Local: R1b-7.

27. Age of R1b-4: 7,000 years (*n* = 9, SD ±3,140); R1b-14b, 8,400 years (*n* = 40, SD ±4,060); R1b-14c, 6,500 years (*n* = 53, SD ±4,600). R1b-15a: 11,500 years (*n* = 10, SD ±6,010). 112/2,082 = 5.4% of Mesolithic lines surviving until today.

28. I1b-2: 8,600 years (*n* = 6 unique descendants; SD ±4,270) dated in British expansion; total I1b-2 in British, including types shared with Iberia =12 in this study. Dates of I1b-2 in southwest Europe and Sardinia according to Rootsi et al. (2004). Time since sub-clade divergence 9,300 (±7,600) years, age of STR variation 8,000 (±4,000) years, time since population divergence 7,900 (±3,600) years.

29. R1b-6 derived from Rox, but re-expanding just after the YD: age 11,250 years (rooted on Ht. 140, *n* = 24, SD ±4,182).

30. British I1c-3 age 11,600 years ($n = 25$, SD $\pm 5,470$), total n in cluster $= 26$; other Mesolithic I lines: I* in British Isles ($n = 11$), found mainly in Ireland and the Isle of Man, not possible to date in British Isles, but dated at 23,000 ($\pm 7,700$) years in Europe (Rootsi et al. 2004); British I1b* age: 12,800 years ($n = 5$, SD $\pm 6,780$) total $n = 11$. I1b* in the Balkans: age of sub-clade divergence 10,700 years ($\pm 4,800$, Rootsi et al. 2004).

31. YD dates (calendar-corrected) as per figure 1 in Gamble et al. (2004).

32. Jonathan Adams, 'Europe during the last 150,000 years', <http://www.esd.ornl.gov/projects/qen/nercEUROPE.html>, with carbon-14 dates corrected.

33. Oppenheimer (1998), pp. 30–38.

34. J.M. Adams and H. Faure, 'Preliminary land ecosystem maps of the world since the Last Glacial Maximum', <http://www.esd.ornl.gov/projects/qen/eur8ky.gif> and <http://www.esd.ornl.gov/projects/qen/eur8k.gif> – for key to vegetation see <http://www.esd.ornl.gov/projects/qen/NEW_MAPS/europe5.gif> and also Figure 4.2 in this book.

35. Kozlowski and Bandi (1984), Nygaard (1989), Sumkin (1990).

36. 8,630 $\pm$ 85 years ago: Olofsson (2003).

37. Sumkin (1990).

38. Sumkin (1990), but see also Olofsson (2003), who gives dates but also says that the widespread lithic traditions linking northern Germany and Denmark with Norway and Sweden may have been more acculturation than migrational.

39. Although to a much lesser extent for Y chromosomes. Franco-Spanish mitochondrial haplogroup representatives include V and H, while the Y chromosome is represented by R1b.

40. i.e. higher than in the dataset used for the present study.

41. Haetta (1996).

42. Pereira et al. (2005): 'The distribution of H1, the largest sub-clade, displays two peaks, one in Iberia and another in Scandinavia (Fig. 2B). However, the Norwegian sample size is low ($n = 18$) and haplogroup H is overrepresented ($\sim 70\%$, while larger data sets for Norway point to a frequency of $\sim 50\%$: Richards et al. 2000). When we removed the Norwegian sample, the Scandinavian peak disappeared, and the picture showed only the decreasing frequency of sub-haplogroup H1 from the south-west to the north and east ... H1 has an age of $\sim 14,000$ years (SE 4000) using coding region data and $\sim 16,000$ years (SE 3500) using HVS-I.'

43. Pereira et al. (2005): H3: c.9,000 years old (SE $\pm 3,000$) based on the coding-region data and $\sim 11,000$ years old (SE $\pm 3,000$) using HVS-I.

44. Although the wide confidence intervals (the SEs given in the previous note) prevent confirmation of this.

45. 16,300 ($\pm$ 4,800) years: Torroni et al. (2001).

46. Tambets et al. (2004).

47. Tambets et al. (2004).

48. 8,500 ($\pm 2,300$) years: Torroni et al. (2001).

49. Pitkänen (1994).

50. Sather (1995).

51. Tambets et al. (2004), Achilli et al. (2005). The Saami are characterized by a unique further mutation at nucleotide 16144, creating subgroup U5b1b1, which is found at very low rates elsewhere in north-east Europe and the Volga–Ural region, but particularly among Finns (6.7%) and Karelians (6%), thus indicating the likely immediate source of this founding effect. The ratio of U5b1b1 : Vera varies from 2 : 1 in Norwegian Saami to 1 : 3 in Swedish Saami, and roughly equal proportions in Finnish Saami groups.

52. Achilli et al. (2005): 'Intriguingly, the Saami of Scandinavia and the Berbers of North Africa were found to share an extremely young branch [U5b1b], aged merely ~9,000 years [8,600 ± 2,400 years]. This unexpected finding not only confirms that the Franco-Cantabrian refuge area of south-western Europe was the source of late-glacial expansions of hunter-gatherers that repopulated northern Europe after the Last Glacial Maximum but also reveals a direct maternal link between those European hunter-gatherer populations and the Berbers.'

53. While the great majority of modern Saami maternal lines ultimately derive from the Franco-Spanish refuge, a few derive from the Asian super-group M (Tambets et al. 2004). The proportion of these eastern M lines varies from less than 3% in Norwegian and Swedish Saami to 16% in Finnish Saami. This observation suggests a later gene flow into the ancestral Saami colonies, originating ultimately from the Volga–Ural region in neighbouring Siberia to the east. For M (or 'Manju') see Oppenheimer (2003), pp. 83–4.

54. Tambets et al. (2004).

55. Rootsi (2004) argues for R1a1 as effectively a 'Slavic marker'. For suggestion of post-glacial expansion, see Semino et al. (2000).

56. 7,000 years ago: see Rosser et al. (2000) and discussion in Tambets et al. (2004).

57. R1a1-2b in Norway is rooted on Ht. 87 (this study) and dates to 5,700 years ($n = 29$, SD ±2,160). As we shall see, there is evidence that this Neolithic Norwegian cluster expanded to Shetland, where a genetic founding event dates to the same period of the Mesolithic–Neolithic transition. See Melton and Nicholson (2004).

58. My view here is a variation on Tambets et al. (2004), who draw their R1a1 migration line parallel with N3 directly through Finland. Rates for R1a1 and N3 in this and the next paragraph are mainly from table 3 in Tambets et al. (2004).

59. Tambets et al. (2004) are sceptical of any significant Siberian contribution to the N3 in Saami that entered along with the recent introduction of Uralic languages. Rather, they take the view that this lineage could equally have arrived earlier, from Eastern Europe. See also Semino et al. (2000), Wells et al. (2001), Zerjal et al. (1997) and Underhill et al. (2001).

60. After the Mesolithic era or, at least at the Mesolithic–Neolithic transition.

61. Percentages from present study.

62. Age of R1a1-2b in Norway: 5,700 years (rooted on Ht. 87, $n = 37$, SD ±2,160).

63. Olofsson (2003).

64. The preponderance of East European male lines, such as R1a1, over Basque refuge R1b types is greater than for female lines, where the balance (between both male and female and east and west) is more even.

65. British proportions of I subgroups: I1a constitutes 11.1% (233/2,082); I1b, 1% (23/2,082); and I*/I1*, 0.5% (11/2,082) of lines in the British Isles today. My dating of haplogroup I, including the entire European, Turkish and Syrian datasets, gives 54,000 years (rooted on Ht. 311, $n = 627$, SD ±16,240). Age of I*/I1*: 32,500 years (rooted on Ht. 296, $n = 19$, SD ±9,180). See also dates in the same range as the above ages of I* from the present study, given in table 3 of Rootsi et al. (2004) (their population includes Turks): I* age of STR variation: 24,000 (±7,100) years; see also Semino et al. (2000) (Eu7 and Eu8 inferred estimate c.22,000 years); and Inos (I) (33,000 years by analogy with HV) in Oppenheimer (2003), pp. 146, 151.

66. Present study: I1a age 14,940 years ($n = 458$, SD ±6,428); I1c, 20,830 years ($n = 51$, SD ±4,548). See also comparable dates in Rootsi et al. (2004), table 3: times since sub-clade divergence for I1a, 15,900 (±5,200) years; for I1c, 14,600 (±3,800) years, respectively.

67. For rates of 'I' branches, see Rootsi et al. (2004).

68. Rootsi et al. (2004), tables 1 and 2, figures 1 and 2.

69. i.e. apart from south-west France and the Basque Country: Rootsi et al. (2004).

70. Rootsi et al. (2004).

71. The four main subgroups after the LGM were I1a, I1b, I1c and I* (Rootsi et al. 2004). Note, however, that the Rootsi group argues that 'the I1a data in Scandinavia are consistent with a post-LGM recolonization of northwestern Europe from Franco-Cantabria'. Their opinion is based on diversity which is related to a later period than the separation of branches. In this book I take a different view, namely that all the post-glacial expansions of all the I subgroups arose from the Balkans (a view partly shared by Passarino et al. 2002). This variation in opinion does not really affect issues concerned with the dates, direction and effect of I1a and I1c on eastern Britain (discussed here and in Part 3), although it is critical to interpreting the post-glacial period and spread of the Balkans Neolithic.

72. Analysis in the present study of the oldest I1c sub-cluster, I1c-2 (rooted on Ht. 346), shows, for all of north-west Europe, an age of 15,700 years ($n = 50$, SD ±4,140) and for the British Isles only, 14,200 years ($n = 19$, SD ±3,880). Compare the age of I1c STR variation determined by Rootsi et al. (2004), of 13,200 (±2,700) years; time since population divergence, 11,200 (±2,300) years. Note that the Rootsi group use the new, more realistic Zhivotovsy Y-clock calibration and an appropriate date method in their estimation.

73. Age analysis in the present study of the two younger sub-clusters of I1c shows for I1c-1 a British age of 12,800 years (estimate) or less; and for I1c-3 (rooted on Ht. 363) 12,200 years ($n = 40$, SD ±4,090) from the entire dataset and a British age of 11,700 years ($n = 25$, SD ±5,470).

74. I1c follows the British pattern of I1a distribution in some ways, but not all. This older entrant to north-west Europe is less common generally than I1a, but characterizes Germany and Frisia on the mainland and is much less common in Scandinavia. However, I1c has an eastern British distribution and is again relatively common in non-Welsh, non-Irish parts of the British Isles at about half the rate at which it is found on the nearby Continent.

75. From European frequencies and data tables 1 and 3 in Rootsi et al. (2004). Looking at British dates/ages in the present study, we see rather younger (i.e. mainly Neolithic and later) dates for I1a clusters than for I1c ones (I1c clusters given in note 73): (1) I1a (whole branch rooted

on Ht. 393), 14,900 years ($n = 458$, SD $\pm 6,430$); (2) I1a-2'1 (rooted on Ht. 393), 5,800 years ($n = 75$, SD $\pm 3,020$); (3) I1a-3 (rooted on Ht. 398), 1,200 years ($n = 68$, SD ± 400); (4) I1a-4'1 (rooted on Ht. 401), 7,700 years ($n = 20$, SD $\pm 4,440$); (5) I1a-6a (rooted on Ht. 424), 3,800 years ($n = 10$, SD $\pm 2,870$); (6) I1a-6b (rooted on Ht. 430), 16,000 years ($n = 4$, SD $\pm 9,610$); (7) I1a7 (rooted on Ht. 461), 3,940 years ($n = 13$, SD $\pm 2,240$). Note that numbers used for founder calculations include only uniquely British derived haplotypes – i.e. less than total number of immigrants for each cluster (see Appendix C).

For comparison, ages estimated by Rootsi et al. (2004) for I1a: time since sub-clade divergence (i.e. whole branch), 15,900 ($\pm 5,200$) years; age of STR variation, 8,800 ($\pm 3,200$) years; time since population divergence, 6,800 ($\pm 1,900$) years.

76. Mesolithic I1a + I1c ($n = 58/2,082$).

77. McEvoy et al. (2004), Pereira et al. (2005). And there is some negative evidence for massive Continental gene flow into England in terms of the relative absence in England of a female marker – the 'Saxon' mtDNA marker (H/16189) – for the Saxon homeland (see the next chapter).

78. Present study.

79. The East European line R1a1 (Rostov), which probably arrived in Scandinavia in the Neolithic, accounts for nearly half of male lines in Trondheim in northern Norway, but has a limited and patchy distribution in the British Isles, including Ireland. The gene types and age of several Rostov clusters within Britain suggest that Scandinavians were already arriving on the east coast of Britain and Scotland during the Neolithic. As we shall see in Chapter 12, the high spots of the Rostov distribution also correspond to places visited by Norwegian Vikings from the Trondheim region much later.

By contrast, I1a, which characterizes southern Scandinavia, in particular Denmark and Oslo, and to a lesser extent extreme north-west Germany, is relatively common throughout Britain and its smaller islands (Figure 4.11a in this book). Data from present study and also table 1 and figure 1D in Rootsi et al. (2004).

Group I1a is found in the British Isles generally at about half the Continental rate, namely around 10–20%, with the notable Atlantic-fringe exceptions of north and south Wales and Ireland, where rates are only 0–4%. The highest British rates for Ian are on the east coast of England around the Wash and in York at 20–30% (Figure 4.11a). However, this does not necessarily mean that Danish Vikings flooded all over England, Cornwall and Scotland like Carlsberg beer, since Scandinavian gene lines were arriving long before the Vikings. I1a is evenly distributed throughout those parts of Britain, like Heineken beer, occupying parts that the Danes – and even the Frisians and Anglo-Saxons, who would also have brought their own I1a – never reached. Genetic dating reveals that I1a clusters in Britain are mainly Neolithic and later than those for I1c.

80. Available from relevant papers in the public domain – see the Appendices.

81. For the Anglo-Saxon replacement view, see e.g. Weale et al. (2002) and Hill et al. (2000).

82. The external Mesolithic contribution to the British Isles is 46.8% ($979/2,082$); the southern component is 95.6% ($936/979$). Source: present study.

83. This is reflected in the distribution of mitochondrial DNA lineages, as noted in a recent paper which focused on the relationships between Ireland and its neighbours: 'The recolonization of western Europe from an Iberian refugium after the retreat of the ice sheets 15,000 years ago could explain the common genetic legacy in the area. An alternative but not mutually exclusive model would place Atlantic fringe populations at the "Mesolithic" extreme of a Neolithic demic expansion into Europe from the Near East. In any event, the preservation of this signal within the Atlantic arc suggests that this region was relatively undisturbed by subsequent migrations across the continent. The identification of likely dispersal points for some Irish haplotypes in northern Spain and western France is further evidence for links between Atlantic populations' (McEvoy et al. 2004).

84. In one of the only attempts to date R1b along the Atlantic coast, a group led by Dan Bradley of Trinity College, Dublin acknowledge the possibility of an 'agriculturally facilitated population expansion, which, at the fringe of Europe, may have taken place in an insular Mesolithic population of hg 1 [R1b] genotype' (Hill et al. 2000).

Chapter 5

1. Reviewed in Richards (2003).
2. The traditional view of the Near Eastern Neolithic is described by e.g. Piggott (1965), but slower and earlier origins are now argued: see e.g. Pringle (1998). For the earliest British Neolithic see Cunliffe (2004), pp.153–4.
3. Cunliffe (2004), p.153.
4. Cunliffe (2004), pp.153–4.
5. Cunliffe (2004), p. 154.
6. Mesolithic: R1b-14b age 8,440 years (n =40, SD ±4,060). Neolithic: R1b-14c age 6,541 years (n = 53, SD ±4,614); R1b-14a age 5,449 years (n = 83, SD ±2,090).
7. Zohary and Hopf (2000), p. 86. This theme of an early Far Eastern Neolithic is developed in Oppenheimer (1998), pp. 78–112.
8. Gronenborn (2003).
9. Gronenborn (2003).
10. Cunliffe (2004), figures 4.23 and 4.25; see also Gronenborn (2003).
11. Gronenborn (2003).
12. Cunliffe (2004), pp. 140–44.
13. Cunliffe (2004), pp. 140–44.
14. Cunliffe (2004), pp. 140–44.
15. Gronenborn (2003).
16. Oppenheimer (2003), pp. 203–4.
17. Cohen (1989), Hoppa and FitzGerald (1999), Ulijaszek (2000), Cohen and Armelagos (1984), Hladik et al. (1993).
18. Bailey et al. (1989).
19. Cunliffe (2004), p. 141.

20. Richards (2003).

21. As I mentioned earlier, the most careful and extensive analysis of maternal migrations into Europe from the Near East is undoubtedly the one carried out by Leeds geneticist Martin Richards and collaborators, using data from many sources. In 2002 they stated further on this: 'Neolithic lines: The mtDNA founder analysis can also be drawn upon to gloss the PC analysis further. The founder analysis suggested that the main Neolithic founder haplotypes were members of mtDNA haplogroups J, T1, and U3. None of these haplogroups contribute substantially to the first PC of mtDNAs in Europe. Rather, the first PC is mainly shaped by haplogroups H, pre-V, and U5, which the founder analysis suggests either originated in Europe or spread into Europe during the Upper Paleolithic period. The haplogroup (pre-HV)1, by contrast, may have spread along the Mediterranean either during the Neolithic period or in more recent times or both. Thus, we seem to be witnessing, in the mtDNA data (and perhaps in the autosomal and Y chromosome data as well), the results of a palimpsest of processes, some possibly more recent than the Neolithic period and some much more ancient' (Richards et al. 2002).

22. Richards et al. (2000).

23. Richards et al. (2000).

24. See e.g. Adams and Otte (1999).

25. J1a (J-16231): as the 'LBK line', see Richards (2003); as the 'Germanic line', see Forster et al. (2004).

26. 'J1a'/J-16231 age: 5,000 years, SE ±3,000 (7,000 ± 1,600 years overall but 5,000 years for the 16189 'Saxon' sub-branch) in Forster et al. (2004). 7,000 years ($n = 58$, SD ±1,600) in Tambets et al. (2003), who note that the age of 7,000 years 'might be an overestimate: subtracting a putative sub-clade ['Saxon'] node at np 16,189, the coalescence age drops to about 5000 BP'.

27. 2% vs 4% (Forster et al. 2004).

28. In support of this hypothesis, the putative Anglo-Saxon homeland also possesses a female H gene type known as the 'Saxon' marker at rates of about 25%. 'Low-German-speaking ("Saxon") areas of the North German Plain harbour a "Saxon" mtDNA marker H/16189 at about 25 per cent (16/61, updated from Richards et al. 1995)' (Forster et al. 2004). This female Saxon marker is rare in England, which it would not have been had the 'Anglo-Saxon' invasion carried significant numbers of women.

29. Age of J1a in Europe: 5,000 years (SE ±3,000) according to Forster et al. (2004). Age of J1a in Northern Europe according to Tambets et al. (2003): 7,000 years (SE ±1,600). Age of J1a without an additional sub-clade 16189: 5,000 years (Tambets et al. 2003).

30. Forster gives several estimates for the age of J2 in Europe, using different approaches, but his favoured 7,000 years has an SE of ±2,000 (Forster et al. 2004).

31. J1b1-16192 is called 'Celtic' in Forster et al. (2004) on the basis of distribution, but this does not seem to be appropriate.

32. i.e. formerly Brythonic celtic-speaking.

33. Found at rates of around 3–6% in Scotland, Wales (5.4%) and Cornwall (3%), J1b1-16192 is notably absent from England and originally Goidelic-speaking areas such as Ireland. The

distribution of the immediate J1b1 ancestor (without the 16192 mutation) in the rest of Europe is diffuse, although it is rare in non-Scandinavian Germanic-speaking areas. Tantalizingly, there is no obvious Continental source for the specific British 16192 mutation on J1b1 *except* Norway, the only other part of the European mainland known to have the type (see Figure 5.3b in this book).

34. Forster et al. (2004).

35. Present study. I1a-5 is 5.2% of British I1a (12/233). Absolute rate in Western Isles 5.7% (5/88); age 5,700 years ($n = 9$, SD $\pm 2,010$). R1a1-2b: age in Norway 5,700 years (rooted on Ht. 87, $n = 37$, SD $\pm 2,160$), in British Isles 5,600 years (rooted on Ht. 87, $n = 38$, SD $\pm 2,650$).

36. Norwegian clusters founding in Scotland/Shetland/Orkney during the Neolithic are I1a-5 and R1a1-2. I1a-5: 5.7% in Western Isles, age in Britain 5,700 years ($n = 9$, SD $\pm 2,010$), age in Norway 8,800 years ($n = 18$, SD $\pm 3,760$). R1a1-2 (rooted on Ht. 87), age in Britain 5,590 years ($n = 38$, SD $\pm 2,650$), age in Norway 5,740 years ($n = 37$, SD $\pm 2,160$), frequency in Shetland 14.1%, Orkney 8.3%, Penrith 5.6%, Oban 4.8%, Stonehaven 4.5%.

37. Note Forster's 'Celtic' appellation (Forster et al. 2004).

38. Harding et al. (2000), Rees (2000).

39. Tambets et al. (2003).

40. Present study (mtDNA database), and Tambets et al. (2003).

41. This problem is acknowledged by Richards in an excellent review and update of the Richards et al. (2000) founder paper, 'The Neolithic invasion of Europe': 'Finally, the results are, at best, estimating the proportion of lineages in the present-day population that can be attributed to each founder event from the Near East (or to bottlenecks within Europe), rather than from the immediate source region' (Richards 2003, p. 152). On the expansion signalled initially by Cardial Ware, he notes that 'There are fewer Neolithic-derived lineages along the Mediterranean and the Atlantic west, at around 10%, again mainly from haplogroup J' (p. 153).

42. 'The Basque region, which was an outlier in the PC analyses of both mtDNA and classical markers, has the lowest Neolithic component, at around 7%. The Basque outlier status may therefore be partly the result of reduced Neolithic penetration, as well as considerable genetic drift due to isolation and small population size. They are little more of a Mesolithic relict than any other European population' (Richards 2003, p. 153).

43. Represented e.g. by mtDNA haplogroup H5a, which dates to 8,000 years ago ($n = 32$, SE $\pm 4,000$), see Pereira et al. (2005).

44. McEvoy et al. (2004).

45. Richards (2003), p. 153.

46. Percentage of Ivan in English: present study 19.5% (172/883). See also Rootsi et al. (2004), estimated on same dataset but including several larger islands around England: 18.4% (174/945). For the role of I in the post-glacial and Neolithic expansions from the Balkans, see also Passarino et al. (2002).

47. 8.3% (168/2,082), present study.

48. 'Gaelic Modal Haplotype' = Ht. 199 in present study. Gaelic-named male data from Hill et al. (2000).

49. The original publication on Rory's 'Gaelic connection' (Hill et al. 2000) suggested a Neolithic age for the whole cluster.

50. Rory re-expansions: Mesolithic: R1b-14b, distribution Wales and Western Scotland > Ireland, age: 8,440 years ($n = 40$, SD $+/4060$). Neolithic: R1b-14c, distribution Ireland > Scotland, age 6,541 years ($n = 53$, SD $\pm4,614$). Irish Neolithic: R1b-14a, mainly Ireland, age 5,450 years ($n = 83$, SD $\pm2,090$).

51. Age of R1b-11 to 13: 9,540 years ($n = 225$, SD $\pm4,740$); R1b-13, 7,790 years ($n = 79$, SD $\pm4,090$); R1b-11, 4,560 years ($n = 121$, SD $\pm3,370$); and R1b-12, 4,620 years ($n = 25$, SD $\pm4,140$). Rate of R1b11 plus R1b-12: 7.0% (146/2,082).

52. FMH is the modal haplotype (Ht. 151 in the present study and closely related to the AMH) of cluster R1b-8. The name is not intended to identify Frisia as the source, since it is also present in the Iberian refuge region.

53. One-step derivative: Ht. 150.

54. R1b-8 (Ht. 150–153 and 17; frequency as percentage of R1b in Durness 36.6% (15/41), as a percentage of total Durness 29.4% (15/51). R1b-8 founding age in British Isles: 4,674 years (rooted on Ht. 151, $n = 96$, SD $\pm4,674$). Total size of R1b-8 in UK: $n = 273$. R1b-8 age in Frisia (no uniquely derived lineages): 2,024 years ($n = 19$, SD $\pm1,510$).

55. R1b-7 (Ht. 145–8) (= small cluster). Founding age in UK 3,940 years ($n = 13$, SD $\pm3,119$).

56. Neolithic contributions by re-expansion of existing clusters in the British Isles: $n = 136$ (R1b-14) + 146 (R1b-11 and 12) + 109 (R1b-7 and 8) = 391; 391/2,082 = 18.8% of male lines (or 25.8% of R1b in British Isles).

57. Rate of group I in British Isles: 16.1% (336/2,082); age of haplogroup I in Europe: see Rootsi et al. (2004) and Semino et al. (2000).

58. Age of I in entire present dataset, rooted on Ht. 311: 54,100 years ($n = 627$, SD $\pm16,240$).

59. Present study (entire database): I1a age 14,940 years (rooted on Ht. 393, $n = 458$, SD $\pm6,428$); I1c, 20,830 years (rooted on Ht. 346, $n = 109$, SD $\pm4,550$); and I1b, 21,000 years (rooted on Ht. 299, $n = 41$, SD $\pm6,970$). See also Rootsi et al. (2004), table 3: times since sub-clade divergence: I1a, 15,900 ($\pm5,200$) years; I1c, 14,600 ($\pm3,800$) years. See also Semino et al. (2000), Eu7 and Eu8 inferred estimate c.22,000 years; and Inos in Oppenheimer (2003), pp. 146, 151 (inferred 33,000 years by analogy with age of mtDNA group HV).

60. One of the problems in working out the routes the Ivan subgroups took after the end of the Ice Age is in knowing where they had each already spread to *before* the LGM. While the main source of the Ivan group information (Semino et al. 2000, Rootsi et al. 2004) suggests that although Ivan may have already spread in a limited way with the Gravettian technology before the LGM, the main expansions of its subgroups occurred after the LGM. This is borne out in the divergence ages of each subgroup, which by their estimates are all post-glacial. Furthermore, if we look at which part of the Ivan geographical distribution holds all his ancestral as well as the filial diversity (i.e. including I* and I1a to c), we find that they are all in the same distribution as I1b*. In other words, south-east Europe, where Ivan

represents up to 45% of male lines, has every appearance of a homeland (see Figures 3.7, 4.11a and 4.11b in this book). I should make it clear that this Balkan homeland of all the main branches of Ivan is not what is proposed by Rootsi et al. (2004), but the movement of Ivan (and R1a1) into Germany and Norway during the Neolithic is supported to a certain extent by Passarino et al. (2002).

61. Present study. Age of I1b*: 23,000 years (rooted on Ht. 300, $n = 24$, SD ±7,870); I1b-2, 14,330 years (rooted on Ht. 289, $n = 17$, SD ±7,500). See also Rootsi et al. (2004), who give ages of, for I1b*, 10,700 (±4,800) years and, for I1b2, 9,300 (±7,600) years.

62. Another possibility is that the Balkan refuge extended a little farther to the east, around the north coast of the Black Sea, which during the LGM would have been a shrunken fresh-water lake.

63. Piggott (1965).

64. British ages of the three I1c clusters: I1c-1, 12,800 years ($n = 12$, SD ±7,400); I1c-2, 14,200 years ($n = 19$, SD ±3,880); and I1c-3, 11,600 years ($n = 25$, SD ±5,470). See also similar dates for I1c in Western Europe in Rootsi et al. (2004), table 3.

65. Ages: I1a-4, 7,700 years (rooted on Ht. 401, $n = 20$, SD ±4,440); Ib-2, 8,600 years ($n = 6$ unique descendants, SD ±4,270), dated in British expansion; total I1b-2 in British including types shared with Iberia = 14 in this study. Numbers used for founder calculations include only uniquely British derived haplotypes – i.e. less than total immigrants for each cluster (see Appendix C). For comparison, ages estimated by Rootsi et al. (2004) for I1a: time since sub-clade divergence (i.e. whole branch) 15,900 (±5,200) years, age of STR variation 8,800 (±3,200) years, time since population divergence 6,800 (±1,900) years.

66. For distribution and expansion ages for I1a, I1b* and I1b-2 in Europe, see Rootsi et al. (2004), figure 1 and table 3.

67. Gronenborn (2003).

68. Age of I1b*: see Rootsi et al. (2004), table 3. STR variation 7,600 (±2,700) years; time since population divergence 7,100 (±2,500) years. For the geographical distribution of I1b* and I1b-2, see figure 1 in Rootsi et al. (2004). From table 3 in Rootsi et al. (2004): I1b-2 age of STR variation, 8,000 (±4,000) years; time since population divergence, 7,900 (±3,600) years.

69. European rates of I1b-2: see Rootsi et al. (2004), table 1. Rates in British Isles: present study. Age of I1b-2 in British Isles in present study 8,600 years ($n = 6$, SD ±4,270).

70. Figure 1c in Rootsi et al. (2004).

71. Gronenborn (2003).

72. Percentages of I1a for Germany, the Netherlands, Switzerland and Normandy from Rootsi et al (2004), table 1; for Belgium inferred from Hg-2 in Rosser et al. (2000). In spite of the view of Rootsi et al. (2004) that 'the I1a data in Scandinavia are consistent with a post-LGM recolonization of north-western Europe from Franco-Cantabria', their own figures for the presence of I1a show low rates in southern France (5.3%) and Lyon/Poitiers (2%). Only Normandy, in the north, shows appreciable rates (11.9%).

73. I1a-2, 3 and 4: total 173/233 = 74% of British I1a. I1a-3: 30.4% of British I1a (71/233), British date 1,200 years (rooted on Ht. 398, $n = 68$, SD ±400). 'Bronze Age' of I1a-3 in southern Scandinavia 3,608 years (rooted on Ht. 398; $n = 164$, SD ±1,390).

74. I1a-3 is also found both farther north, in Norway and Sweden (I1a in southern Sweden 35.7% ($n = 168$), in Norway 38.9% ($n = 72$), Rootsi et al. (2004); in Oslo 33% (9/27), present study), and south across the Baltic in Denmark. It is less common, however, in northern Germany (I1a in Germany 25%, Rootsi et al. (2004); in northern Germany 7.5% (14/185), present study). Unlike all the other I1a branches, however, I1a-3 has a significantly lower and more focused representation in Britain than might be expected from its high frequency in Norway and the distribution of other I1a clusters there (I1a-3 rate in Norway: 11.6% (33/284), and 3.4% in the British Isles (71/2,082); odds ratio 0.29, i.e. I1a-3 is only about one-third of its expected frequency in the British Isles). This low ratio is consistent with the limited nature of the Viking invasions compared with the Neolithic Scandinavian entrants.

75. I1a-5 is 5.2% of British I1a (12/233). Absolute rate in Western Isles, 5.7% (5/88); age, 5,700 years ($n = 9$, SD ±2,010).

76. I1a-2, 27.0% of British I1a (63/233).

77. I1a-3 is 31.8% of British I1a (74/233); I1a-3 constitutes 9% of Danish I1a (9/100), 14% (12/85) of Schleswig-Holstein/northern Germany and 6.3% of Frisia (6/94) – all from present study.

78. I1a-4 rates: Oslo 9.1% (6/66), Bergen 7.8% (10/128), Denmark 4% (4/100), Frisia 4.3% (4/94); England overall rate 1.3% (28/2,082), Fakenham 7.5% (4/53); British age 7,700 years (rooted on Ht. 401, $n = 20$, SD ±4,440).

79. Nine unique derived British haplotypes; four others shared with the Continent.

80. I1a-7 rate among British I1a types 9.4% (22/233). While this is the British cluster nearest to Frisia in identity, it does not show evidence of a recent invasion: one sub-cluster, which shows a clear founding event in East Anglia, has a date suggestive of the early Bronze Age. Overall rate I1a-7 rate in British Isles 1% (22/2,082), age of I1a-7b cluster 3,940 years (rooted on Ht. 461, $n = 13$, SD ±2,420).

81. I1a-3 represents 30.4% of British I1a types (71/233) and 3.4% in Britain overall (71/2,082); frequency in Bergen 15.6%, in Oslo 13.6%, Denmark 12%, eastern England 8–11% (all figures present study). However, although it is the only dated late founder, I1a-3 is probably not the only I1a cluster to enter Britain from the Anglo-Saxon region during the Dark Ages, with I1a-2, I1a-4, I1a-6 and I1a-7 (in that order of importance) contributing overall a further 3–5% in eastern England.

82. For example, I1a forms a large part of haplogroup 2 in the Principal Components Analyses by Weale et al. (2002) and Hill et al. (2000).

83. E3b is most common in the southern Balkans and southern Italy (25%), and in southern Spain (10%). E3b is rare to absent in the Basque Country and the southern French Ice Age refuge regions, although it is common at 6–14% in other parts of southern Spain and the Spanish coast, both Atlantic and Mediterranean (Semino et al. 2004, and present study), whereas E3b is found in Galicia (11.3%) and Valencia (6.5%). Age of E3b in the British Isles: main cluster 4,500 years (rooted on Ht. 66, $n = 20$, SD ±2,480). Three smaller E3b clusters give dates of the same order – total British $n = 47$. Highest rate at Abergele 33.3% (6/18 – four haplotypes belonging to three clusters, i.e. *not* an extreme founder event); Southwell

5.7%. For European/African distribution and ages see Semino et al. (2004); see also Richards (2003).

84. However, J2 does seem to have moved partly through southern France to Britain rather than round it.

85. Overall J frequency in British Isles 2% (38/2,082), frequency in southern Britain 4–7%. Frequencies in Spain: Valencia 16.1%, Galicia 15.1%, Catalonia 3.7%. For European/African distribution and ages of J, see Semino et al. (2004); see also Richards (2003).

86. Age of E3b in the British Isles, main cluster: 4,500 years (rooted on Ht. 66, $n = 20$, SD ±2,480). Three smaller E3b clusters give dates of the same order – total British $n = 47$. Highest rate Abergele 33.3% (6/18 – 4 haplotypes belonging to 3 clusters i.e. *not* an extreme founder event); Southwell 5.7%. For European/African distribution and ages see Semino et al. (2004); see also Richards (2003).

87. Haak et al. (2005).

88. 'Neolithic skeletons from 16 sites of the LBK/AVK [Alföldi Vonaldiszes Kerámia] culture from Germany, Austria, and Hungary ... All human remains were dated to the LBK or AVK period (7,500 to 7,000 years ago) on the basis of associated cultural finds' (Haak et al. 2005). Identities of ancient maternal haplotypes in Haak's study: 'Eighteen of the sequences belonged to typical western Eurasian mtDNA branches; there were seven H or V sequences [4H, 2HV and 1V], five T sequences, four K sequences, one J sequence, and one U3 sequence.' It should be noted that, of these eighteen, eleven comprised 3T2, T3, T*, 4K, J* and U3, reflecting more of a Mesolithic complexion. For relative Mesolithic dating of T2 and K as European founders, see Richards et al. (2000), figure 1.

89. See e.g. figure 1 in Richards et al. (2000).

90. Haak et al. (2005). For Anatolia see Tambets et al. (2000).

91. Ricaut et al. (2004).

92. Herodotus, *Histories* 4.21

93. The film *King Arthur* (2004), directed by Antoine Fuqua.

94. Reviewed in Richards (2003).

95. 11/146 = 7.5% estimated on haplogroups I1a ($n = 5$), I*/I1* ($n = 3$), I1b/I1b* ($n = 1$), E3b ($n = 1$) and KxPN3 ($n = 1$).

96. Post-Mesolithic estimate in present study for Norfolk and Fens: 34.6% (46/133). Anglo-Saxon homeland and overall Scandinavian intrusive haplotypes in this region are estimated at 7.5% (10/133) and 13.5% (18/133), respectively. The Scandinavian types comprise a large proportion of Neolithic entrants.

97. McEvoy et al. (2004).

98. Richards et al. (2000).

99. British figures, present study, shown graphically in Figures 5.4a–c. Neolithic intrusion from north-west Europe estimated by summing the following gene clusters: I1a-2'1b, I1a-4, I1a-5, I1a-6b, I1a-7b, R1a1-1 and R1a1-2b. Neolithic intrusion from south-west Europe estimated by summing the following gene groups and gene clusters: E3, J, FxIJK, I* and I1b.

100. Present study.

101. See Chapter 6. See also e.g. Renfrew (1989), Diamond and Bellwood (2003).
102. Father clearly having a Gaelic name, see Hill et al. (2000) (and derived figures in the present study).
103. 'Surname subdivision reveals a cline in Irish samples, with exogenous samples clearly showing lower frequencies of *Hg1* [i.e. effectively of R1b] (English, 62.5%; Scottish, 52.9%; Norman/Norse, 83.0%) than Gaelic Irish samples (Leinster, 73.3%; Ulster, 81.1%; Munster, 94.6%), which almost reach fixation in the westernmost province (Connaught, 98.3%)' (Hill et al. 2000).
104. Hill et al. (2000).
105. i.e. those discussed so far in this book.
106. Richards et al. (2000), Pereira et al. (2005).
107. Forster et al. (2006).
108. Dyen et al. (1992).
109. Cunliffe (2004), p. 175ff.
110. Cunliffe (2004), p. 187.
111. Cunliffe (2004), p. 169.
112. Scarre (1995), p. 13.
113. Cunliffe (2004), p. 190.
114. Cunliffe (2004), pp. 187–9.
115. Cunliffe (2004), pp. 160–61.
116. See Chapter 6 and also Renfrew (1989).
117. Richards et al. (2000); for a review see Richards (2003).
118. A survey of European ponies and horses shows a geographically localized mitochondrial 'cluster C1, which is distinctive for northern European ponies' (Jansen et al. 2002), including Viking, Shetland and Highland ponies (see p. 239).
119. The confidence intervals for dates of this genetic expansion (see note 122 below) actually include the whole Neolithic period in Western Europe, so would not falsify a Kurgan migration either. European dates of I1a: table 3 in Rootsi et al. (2004); British dates: present study. See the earlier section 'Ian: the northern Neolithic line?' and Appendix C.
120. Tambets et al. (2004), table 3.
121. Rosser et al. (2000).
122. Present study, R1a1-2b. Age in Norway 5,700 years (rooted on Ht. 87, $n = 37$, SD $\pm 2{,}160$), in British Isles 5,600 years (rooted on Ht. 87, $n = 38$, SD $\pm 2{,}650$). This evidence suggests that a Neolithic Norwegian cluster expanded to Shetland, where a genetic founding event dates to the same period of the Mesolithic–Neolithic transition. For the Shetland Mesolithic–Neolithic transition see Melton and Nicholson (2004).
123. Gimbutas (1970).
124. Present study. R1a1-3 forms two founding clusters in Britain dating to the Late Neolithic/Bronze Age, 3a and 3c, rooted on haplotypes 95 and 93, respectively. Age of R1a1-3a in British Isles, 4,080 years (rooted on Ht. 95, $n = 22$, SD $\pm 2{,}260$); age of R1a1-3c in British Isles, 3,660 years (rooted on Ht. 93, $n = 7$, SD $\pm 3{,}660$). A third sub-cluster, R1a1-3b, is more consistent with a Dark Ages entry to the British Isles (see pp. 393–4): age in

Norway, 3,200 years (rooted on Ht. 96, $n = 20$, SD $\pm 1,920$); in British Isles 1,830 years (rooted on Ht. 96, $n = 21$, SD $\pm 1,831$).

125. Cunliffe (2004), pp. 255–60; and Scarre (1995), p. 117.

126. Cunliffe (2004), pp. 217–18.

127. Cunliffe (2004), p. 246.

128. Tom Higham, 'Radiocarbon dating', Wessex Archaelology, <http://www.wessexarch.co.uk/projects/amesbury/tests/radiocarbon.html>.

129. Carol Chenery, 'The Amesbury Archer: oxygen isotope analysis', Wessex Archaelology <http://www.wessexarch.co.uk/projects/amesbury/tests/oxygen_isotope.html>.

130. Cunliffe (2004), pp. 218–19.

131. Lewis (1996).

132. Caution should be exercised in the genetic interpretation, since the total sample size for Abergele is only $n = 18$, but the larger nearby Llangefni sample shares similar lines, and random chance does not explain their unique make-up, diversity and behaviour in Principal Components Analysis.

133. Clearly, the problems of dating and the difficulty in detecting repeated migrations from the same source would tend to blunt my assertion; see also the discussions in Richards (2003) and McEvoy et al. (2004).

134. Lineages derived from north-west Europe, putatively during the Bronze Age, contribute overall 3.1% (65/2,082) to modern British male lines (for details see note 142 below); and from Iberia: potentially E3b, J and I1b2 (but difficult to distinguish from Neolithic gene flow).

135. Cunliffe (2004), p. 287.

136. Cunliffe (2004), pp. 247–50, figure 6.18.

137. Cunliffe (2004), p. 290.

138. Cunliffe (2004), p. 281.

139. Cunliffe (2004), pp. 292–3.

140. Cunliffe (2004), pp. 291–2.

141. Clusters I1a-6a ($n = 21$), I1a-7a ($n = 13$), R1a1-3a ($n = 22$) and R1a1-3c ($n = 10$). Total $n = 66/2,082$ (3.2% of British gene pool). British age of I1a-6a cluster, 3,800 years (rooted on Ht. 424, $n = 10$, SD $\pm 2,870$); I1a-7a, 3,940 years (rooted on Ht. 461, $n = 13$, SD $\pm 2,420$); R1a1-3a, 4,080 years (rooted on Ht. 95, $n = 22$, SD $\pm 2,260$); and R1a1-3c, 3,660 years (rooted on Ht. 93, $n = 7$, SD $\pm 3,660$).

142. I1a-7a (East Anglia from northern Germany and Denmark): British age 3,940 years (rooted on Ht. 461, $n = 13$, SD $\pm 2,420$); I1a-6a (from Norway to north-eastern Britain, including the Orkneys and Shetland): British age 3,800 years (rooted on Ht. 424, $n = 10$, SD $\pm 2,870$).

143. From North Germany and Denmark to the east coast of England: R1a1-3a, age 4,080 years (rooted on Ht. 95, $n = 22$, SD $\pm 2,260$); R1a1-3c, 3,660 years (rooted on Ht. 93, $n = 7$, SD $\pm 3,660$; from Norway to the Scottish islands).

144. Concerning equine clusters C1 and E: 'A total of 17 of 19 documented horses with C1 are northern European ponies (Exmoor, Fjord, Icelandic, and Scottish Highland). Additionally, 14 of 27 undocumented horses [3] with C1 are ponies, including Connemara ponies … Furthermore, mtDNA cluster E ($n = 16$) consists entirely of Icelandic, Shetland, and Fjord

ponies' (Jansen et al. 2002). The distribution of C1 is restricted to the Balkans, the British Isles and Scandinavia, including Iceland.

145. Two alternative fossil calibrations gave genetic dates for C1 of 8,000 years and 2,000 years. 'The cluster is younger than perhaps 8,000 y … but definitely older than 1,500 y, because C1 was also found in two ancient Viking horses' (Jansen et al. 2002).

146. Raftery (1994).

147. Ó'Donnabháin (2000).

Chapter 6

1. For example, a single point mutation recently discovered in Europeans which is responsible for a large part of their skin bleaching (Lamason et al. 2005).

2. Adams and Otte (1999).

3. Martin Richards and his colleagues paid me a compliment by incorporating this perspective, with acknowledgement, in their milestone paper on European genetic Founder Analysis (Richards et al. 2000).

4. Diamond and Bellwood (2003), esp. p. 598. See also my critique (Oppenheimer 2004) and their response (Bellwood and Diamond 2005).

5. Gimbutas (1970), but see the discussions in Renfrew (1989), Oppenheimer (2004) and Richards (2003).

6. Renfrew (1989).

7. Cavalli-Sforza et al. (1994), pp. 292–3, figures 5.11.1 and 5.11.3. See also Cavalli-Sforza et al. (1993).

8. Renfrew (1989).

9. See e.g. Gamkrelidze and Ivanov (1995) and Rexova et al. (2003).

10. For example, resulting from different rates of language evolution, ascertainment bias and undetected lexical borrowing – see p. 257.

11. Dyen et al. (1992).

12. Gray and Atkinson (2003).

13. Atkinson et al. (2005).

14. Semino et al. (2000), Rootsi et al. (2004). Interestingly, the only European populations with I rates as low as the Greeks are in south-west Europe and on the Atlantic fringe, speaking insular-celtic languages.

15. As defined in Rootsi et al. (2004).

16. Cowgill (1970).

17. The potential fallacy in interpreting Cavalli-Sforza's First Principal Component (the south-east to north-west vector) as necessarily reflecting just a Near Eastern Neolithic expansion was pointed out not by a geneticist but by archaeologist Marek Zvelebil, who first suggested the analogy with a 'palimpsest'. The literature, both archaeological and genetic, is well reviewed by Richards (2003).

18. Adams and Otte (1999).

19. See the review article by Pringle (1998).

20. Balter (1998).
21. McMahon and McMahon (2003).
22. Forster and Toth (2003).
23. 10,100 (±1,900) years ago (Forster and Toth 2003).
24. McMahon and McMahon (2006).

Part 3

Introduction

1. Gildas, *De excidio Britanniae* 22–23.

Chapter 7

1. Strabo, *Geography* 4.5.2.
2. Tacitus, *Agricola* 11.
3. Harding et al. (2000), Rees (2000).
4. Julius Caesar, *Gallic Wars*, 5.12.
5. Julius Caesar, *Gallic Wars*, 5.14.
6. Sims-Williams (2006), esp. maps 11.1 and 11.2; Parsons and Sims-Williams (2000), pp. 169–78; Evans (1967); Rivet and Smith (1979).
7. Evans (1967), p. 16, quoting Caesar from *Gallic Wars*, 1.1–2.
8. Evans (1967), p. 16.
9. Julius Caesar, *Gallic Wars*, 1.1.
10. Julius Caesar, *Gallic Wars*, 2.3.
11. Julius Caesar, *Gallic Wars*, 2.4.
12. Julius Caesar, *Gallic Wars*, 2.4.
13. Treharne and Fullard (1976), p. 15.
14. Tacitus, *Germania* 28.
15. Julius Caesar, *Gallic Wars*, 2.29.
16. Sims-Williams (1998b), note 71 on p. 19.
17. Tacitus, *Germania* 28; see also Tacitus on Cimbri in *Germania* 37.
18. Evans (1967).
19. Sims-Williams (2002), p. 7.
20. Sims-Williams (1998), p. 19.
21. Kuhn (1962), pp. 105–8 and maps 9–16.
22. Kuhn (1962). As a cross-check there is a degree of congruity between Kuhn's micro-analysis of Belgic place-names and Patrick Sims-Williams' recent massive survey of Celtic place-names in Europe (Sims-Williams 2006). See Figure 2.1b in this book.
23. Kuhn (1962), map 13.
24. Jackson (1953), Evans (1967), Rivet and Smith (1979), Sims-Williams (2006).
25. Bragg (2003), p. 5.

26. Parsons and Sims-Williams (2000). While there is the advantage that numbers of different language derivations are compared using the same ancient geographer, unfortunately their ultimate aim of a standardized, quantitative, region-by-region comparison is not fulfilled quite yet.
27. Parsons (2000), p. 174.
28. Parsons (2000), p. 174.
29. Parsons (2000), p. 174.
30. Scheers (1972).
31. Cunliffe (1981b), Kent (1981); see also figures 39–44 in Cunliffe (1981a).
32. Sims-Williams (2002), pp. 14–15.
33. Sims-Williams (2002), pp.10–11.
34. Sims-Williams (2002), p. 14.
35. Celtic inscribed stones also continued to be set up in Brittany, Orkney and Ireland for that matter; see Figure 7.4 in this book.
36. Sims-Williams (2002), pp. 15–19.
37. For up-to-date reviews see Sims-Williams (2003, 2006).
38. CISP, <http://www.ucl.ac.uk/archaeology/cisp/database>.
39. Bragg (2003), p. 5.

Chapter 8

1. Tacitus, *Germania* 2. The linguist Peter Schrijver (1999) argues that the Ingvaeonic vowel changes might have resulted from contact with Celtic.
2. See e.g. Nielsen (1981). See also Kortlandt (1999), who suggests that the split between Old English and Old Frisian on the one hand and Saxon on the other effectively occurred before the Anglian invasion, indicating that a Saxon invasion preceded the Anglian one. This is an interesting view, since one might imagine that the time difference between the Saxon and Anglian invasions of England would have to have been sufficiently large to produce such an effect. A large gap is just what I argue later, though on other grounds.
3. Forster et al. (2006). For an alternative theory of pre-invasion splits and chronological separation of English dialects, see Kortlandt (1999). For *Heliand*, an early ninth-century (AD 825) epic Christian poem written in Old Low German, see Genzmer (1982).
4. Walter (1911). Such unique derived forms are known as lexical synapomorphies. See also Forster et al. (2006).
5. Dyen et al. (1992).
6. Gray and Atkinson (2003).
7. McMahon and McMahon (2003). The citation in the quote is to Embleton (1986), pp. 100–101. A Swadesh list is a list of basic vocabulary terms in two or more related languages used to identify the proportion of shared cognates.
8. Except possibly the Jutes, but this is not apparent in Old Kentish.
9. Genzmer (1982).
10. Gildas, *De excidio Britanniae* 23.3. See also Sims-Williams (1983), p. 22.

11. *OED*, 2nd edn, 1989. See also Forster et al. (2006).
12. Although in fact Middle Dutch had a cognate, *kiel*, meaning 'ship', which subsequently lost that meaning, taking on the common meaning of 'keel' instead.
13. Cronan (2004).
14. Forster et al. (2006).
15. Forster et al. (2006).
16. Forster et al. (2006).
17. *OED*, 2nd edn, 1989.
18. Adamnan, *Life of St Columba* 1.27, 2.33.

Chapter 9

1. See discussion in Pryor (2004), e.g. p. 131.
2. Bede, *Ecclesiastical History* 1.6.
3. Pryor (2004), pp. 135–43.
4. Pryor (2004), p. 135.
5. Pryor (2004), p. 139.
6. Pryor (2004), p.141.
7. Pryor (2004), p. 141.
8. Pryor (2004), p. 142.
9. Pryor (2004), p. 143.
10. Pryor (2004), p. 135.
11. It should be noted that this whole unresolved issue concerning the *Notitia dignitatum*, the 'forts' and the meaning of 'Saxon' in that context is not new, and discussion goes back to the 1930s. I use Pryor's book in the discussion because it is topical and many will have read it or seen the documentary.
12. See discussion/review in Rivet and Smith (1979), pp. 297–300.
13. Parsons (2000), p. 175.
14. See the discussion in Rivet and Ellis (1979), p. 281.
15. Bede, *Ecclesiastical History* 5.24.
16. *Epistola Cuthberti de obitu Bedae* in Opland (1980), pp. 140–41.
17. See the discussion in Sims-Williams (1983).
18. For example in *The Life of Gildas* by Caradoc of Llancarfan, c.1130–50.
19. Gildas, *De excidio Britanniae* 23.
20. Page (1999), p. 213.
21. Page (1999), pp. 16–17.
22. Page (1999), p. 23.
23. Bede, *Ecclesiastical History* 4.13.
24. Sims-Williams (1990), figure 2.
25. Page (1999), p. 18.
26. Page (1999), p. 19.
27. Page (1999), pp. 228–9.

28. To my reading, Procopius is ambiguous about who he means here – presumably the Franks on the Continent rather than the Frisians.
29. Procopius, *History of the Wars* 8.20.6–10.
30. Cruciform brooches are also absent from Lower Saxony: see e.g. Scarre (1995), p. 180, map 1.

Chapter 10

1. i.e. from the northern Germanic region, including Denmark and farther north.
2. e.g. *The Anglo-Saxon Chronicle*, which was originally compiled on the orders of King Ælfred the Great, c. AD 890, and continued into the Middle Ages. Part 1 (entry for the year 495) contains a list of the Saxon Kings of England, mostly lifted from Bede. Translation available at the Online Medieval and Classical Library: <http://omacl.org/Anglo/part1.html>.
3. There is also a genealogy mentioned in *Beowulf*. The date of *Beowulf* has yet to be determined.
4. *The Anglo-Saxon Chronicle*, entry for the year AD 449.
5. Heyerdahl and Lillieström (2001).
6. The issue of 'Old Saxony' during Roman times and its likely location at the base of the Cimbrian Peninsula and on the east bank of the Elbe has been addressed by several authors, who independently came to the same conclusion. There is textual evidence for this, both at the time of Augustus' death in AD 14 (Treharne and Fullard 1976, plates 10 and 15, map 2a), and in Ptolemy's *Geographia* (c. AD 150). See also Chadwick (1907), figure on p. 195; Forster et al. (2006), figure 11.2; and Forster (1995).
7. Chadwick (1907), p. 112; Treharne and Fullard (1976), Medieval section, p. 7.
8. Saxo Grammaticus, *Gesta Danorum*, late twelfth to early thirteenth century.
9. Saxo Grammaticus, Book 1.
10. Anderson (1736).
11. This formula (declaration) of renunciation is dated AD 743 (Vulpius 1826).
12. Wrenn and Bolton (1988), pp. 25–33.
13. Wrenn and Bolton (1988), p. 56.
14. Wrenn and Bolton (1988), p. 38.
15. The Winchester MS is also known as the *The Parker Chronicle* (Corpus Christi College, Cambridge, MS. 173).
16. Bede, *Ecclesiastical History* 2.15.
17. Clarke (1960), p. 138ff.
18. Newton (1992); see also Newton (1993).
19. Gildas, *De excidio Britanniae* 23.
20. Newton (2003).
21. Carver (1998), pp. 32, 36, 43, 47, 56, 128.
22. Carver (1998), p. 36.
23. Carver (1998), p. 38.
24. Hines (1992).
25. Myhre (1992).

26. Hills (1998), Weber (1998).
27. Dumville (1989).

Chapter 11

1. Traditional archaeological view of replacement: e.g. Myres (1969, 1986), Leeds (1912). Still in favour of significant demic impact: e.g. Härke (2002). Geneticists in favour of Anglo-Saxon 'replacement': e.g. Weale et al. (2002).
2. Higham (1992), Hamerow (1997); see also earlier sceptics: Arnold (1984) and Hodges (1989).
3. Crawford (1997).
4. See e.g. Härke (2002).
5. Hamerow (1997).
6. Various figures have been estimated, but no higher than 5%: e.g. 1% (Härke 2002) and 3–5% (Bragg 2003, p. 42).
7. Diamond (1997).
8. Pryor (2004), pp. 143–4.
9. Hamerow (1997), p. 37.
10. Hamerow (1997), p. 40.
11. Myres (1986), pp. 176–8.
12. Myres (1986), p. 14.
13. Cavalli-Sforza et al. (1994).
14. Watkin and Mourant (1952).
15. Watkin (1966).
16. Viereck (1998).
17. Kopeć (1970). Note that similar arguments have been put forward for Rhesus negative rates, which are high in the Basque Country. To keep things simple in this discussion of method, I do not refer to these.
18. Viereck (1998).
19. See e.g. Daniels (2002).
20. Weale et al. (2002).
21. For an accessible edition see Hinde (2004).
22. In more detail: 'Firstly, little genetic differentiation exists among the Central English towns … Secondly, in contrast to the Central English towns, the two North Welsh towns show highly significant differences, both from each other and from the five Central English towns … Thirdly, no significant differences in haplotype frequencies exist between Friesland and any of the Central English towns. Comparisons between Norway and the Central English towns, on the other hand, are all significant, apart from Bourne … Similar results were obtained using F_{ST} values based on haplogroup frequencies, but tests on F_{ST} values based on haplotype frequencies were not significant because of the large number of singletons at this level … Taken together, these results suggest considerable male-line commonality between Central England and Friesland' (Weale et al. 2002).

23. In spite of the claims by Weale et al. (2002) that they were using 150 highly differentiated gene types to make comparisons, the 'statistical significance' tests required to make the inferences of 'extremely high affinity' rested on calculations based on only seven haplogroups rather than many haplotypes.

24. Thomas et al. (2006).

25. Capelli et al. (2003).

26. Wilson et al. (2001).

27. The stated reason for amalgamating these samples was that there were 'no significant differences between them' (Capelli et al. 2003). However, the small number of samples (23) from the north-west Germany source (Lower Elbe river) may have been more pertinent. There were 100 samples from Denmark and 62 from Schleswig-Holstein.

28. To improve on the haplogroup number and resolution, Capelli et al. (2003) added several more UEP (unique event polymorphisms – see below) markers and also created three STR-defined clusters of one-step relatives around the three of the more common haplotypes: the Atlantic Modal Haplotype and two others (Ht. 157, 393 and 96 in this study). In this context, haplogroup is defined by unambiguous markers in the Y chromosome, referred to by Capelli et al. as UEPs and more generally in the literature as 'bi-allelic' (i.e. dichotomous) markers: see Underhill et al. (2000) and Semino et al. (2000).

    The three extra clusters are defined by STR (single tandem repeat polymorphisms), which are less stable and not 'unambiguous'. They correspond, respectively, to the clusters R1b-10, I1a-4 and R1a1-3, within haplogroups R1b, I and R1a1, which I have used in this book. The difference is that while Capelli et al. have chosen only three clusters, thus adding to their 11 UEP-defined haplogroups to make 14 groups, I have systematically broken up all the large haplogroups into clusters in this way (see Appendix C), thus adding 30 clusters to 11 haplogroups and making 41 groups.

29. Plot based on Principal Components Analysis using their 14 groups and displayed their results in the form of a two-dimensional genetic distance map using first two Principal Components (Capelli et al. 2003).

30. They also note that 'All continental populations, however, show significant differences from the indigenous group ($p < 0.01$)' (Capelli et al. 2003).

31. In a separate Principal Components plot using simulated populations based on various theoretical admixtures of Norwegian, NGD and 'indigenous British gene groups', Capelli et al. (2003) show a very similar outcome which illustrates where each level of admixture would place the simulated populations. By analogy, this essentially places the majority of the sampled British market towns on the 'indigenous' side of the plot (i.e. less than 50% intrusion).

32. Capelli et al. (2003) do address this point: 'With regard to source populations, we note that Weale et al. ... recently used Friesland as an Anglo-Saxon representative source population and suggested a substantial replacement of pre-Anglo-Saxon paternal lineages in central England. We therefore compared Frisians to our North German/Danish sample and found that the two sets are not significantly different from each other ($p = 0.3$, data not shown). When included in the PC analysis, the Frisians were more "Continental" than any of the British samples, although they were somewhat closer to the British ones than the North German/Denmark sample.'

I repeated the analysis and was able to confirm that the Frisians were indeed intermediate and closer to the British than to the NGD sample. However, the lack of significant difference between Frisians and the NGD sample in haplogroup frequency is to be expected, given the overall genetic similarity and shallow regional geographical gradients between *all* the north-west European populations at this level of haplogroup resolution. For 'shallow geographic gradients' at the haplotype level, see figure 4a–c in Roewer et al. (2005), who look at STR variation geographically, but do not use it to resolve the relevant haplogroups.

33. Rosser et al. (2000). While using slightly different UEP markers and 12 haplogroups, and defining slightly fewer haplogroups than in the Weale or Capelli studies, all the important divisions are present in the Rosser database and additionally they can be closely matched from the Weale and Capelli English databases (Hg1 = R1b and PxR1a1; Hg2 = I and FxJK; Hg3 = R1a1; Hgs 4,8,12 = E3; Hg9 = J; Hg16 = N3; Hg26 = KxPN3); in addition they include five British locations. See also Roewer et al. (2005); they include only two British sample points, have a different selection of UEPs and STRs and avoid using the phylogenetic information, but their analysis still, like Rosser's, shows three poles of variation (represented as three 'dimensions').

34. Amos et al. (2006) state that: 'Two previous studies have used large datasets to explore the male genetic ancestry of Britain [Weale et al. (2002), Capelli et al. (2003)] … Both studies reveal interesting patterns, but both are limited by a combination of using a small number of extrinsic continental outgroups (three), uncertainties about the timing of events and the perennial question of who is mixing with whom. To examine the effectiveness of the mixing profile approach to elucidate complicated population histories, we therefore reanalysed the larger dataset of Capelli et al., augmenting it with further data from other continental populations.'

35. Amos et al. (2006).

36. i.e with a sufficient number of new, mutationally novel genetic haplotypes related to an identifiable founding line.

37. The numerous (278) singleton haplotypes were all assigned to their statistically/phylogeographically most likely source. Haplotypes that were unique to the British Isles were assigned as such, and in nearly every case could be assigned to uniquely British clusters over 3,000 years old. The identification of the latter separate class is the main reason why the overall degree of intrusion from Continental sources is greater than that for the haplotype matches. I tabulated 490 haplotypes belonging to 40 clusters from 11 haplogroups. There were 278 singleton haplotypes, leaving 212 types with potential for matching. Statistical testing has not been performed in this analysis, since error calculation would be difficult.

38. Present study: overall 29.6% (319/1,077); lowest is Faversham, Kent with 14.5% (8/55); highest is Fakenham, Norfolk with 41.5% (22/53). Admixture proportion estimate for England from Capelli et al. (2003) 'over 40%', from North Germany/Denmark 37.5%, from Norway 24.3%. These figures are not mutually exclusive, and the overall intrusion from north-west Europe would not be as much as their sum, but certainly more than 40%.

39. 'Anglo-Saxon intrusions' to British Isles: England (including Cornwall) 5.5% (59/1,077), British Isles overall 3.8% (79/2,082), Isle of Man 4.5% (2/44), Scottish Isles 2.9% (10/344), Scotland 1.7% (3/178), Wales 1.5% (3/201), Ireland 0.8% (2/238).

40. Härke (2002), p.150.

41. Procopius, *History of the Wars* 8.20.9–10.

42. J1a (J-16231); see Forster et al. (2004).

43. To dissociate Dark Ages Continental 'Anglo-Saxon' women from English women, Forster et al. (2004) note that 'the Low-German-speaking ("Saxon") areas of the North German Plain harbour a "Saxon" mtDNA marker H/16189 at about 25 per cent ... which is rare in England where there is a frequency of only 3.5 per cent ... This low proportion indicates a contribution of zero to maximally 25 per cent of north German women to the native population of England.'

Chapter 12

1. Cunliffe (2004), p. 489.

2. Cunliffe (2004), pp. 488–93.

3. i.e. as opposed to Neolithic contacts, which can be inferred from the genetic record discussed in this book.

4. Wrenn and Bolton (1988), pp. 41–5.

5. The three historic clusters are I1a-3, British age 1,200 years (rooted on Ht. 398, $n = 68$, SD ±400); R1b-8a, age 2,137 years (rooted on Ht. 152, $n = 12$, SD ±1,510); and R1a1-3b, British age 1,830 years (rooted on Ht. 96, $n = 21$, SD ±1,830). Overall rate in British Isles 4.9% (101/2,082).

6. Danish matches 4.4% (91/2,082), Anglo-Saxon matches 3.8% (79/2,082).

7. Fakenham has the highest percentage intrusion (20%) of I1a-3 of all British geographical samples.

8. I1a-3 British age 1,200 years (rooted on Ht. 398, $n = 68$, SD ±400). Distribution: particularly in Norfolk, with six individuals of Ht. 398 in Fakenham.

9. The root type of I1a-3, haplotype 398, is identical to the common haplotype 2.49 in the dataset of Wilson et al. (2001).

10. Neither of the two founding clusters of the historical period from Scandinavia (total $n = 93/2,082$, 4.5%) is particularly associated with Norfolk except for I1a-3 root haplotype 398, which has six representatives in Fakenham and one in Sheringham (R1a1-3b has one representative in Sheringham as well). I1a-3 otherwise tends to follow the full Viking dispersal, while R1a1-3b is distributed more to the north of Britain and its islands and specifically with Norwegian Viking dispersals.

11. 18.9% (10/53). For haplotype matching, Ht. 398 is assigned to Denmark rather than Norway on the basis of overall frequency; this, however, splits cluster I1a-3.

12. (1) R1a1-3c (total in cluster $n = 10$), age in Britain 3,660 years (rooted on Ht. 93, $n = 7$, SD ±3,660); (2) R1a1-3a, 4,080 years (rooted on Ht. 95, $n = 22$, SD ±2,260); (3) R1a1-1 (total $n = 10$), undated and derives from northern Gemany/Denmark and Frisia; (4) R1a1-2 (total in cluster $n = 52$), 5,600 years (rooted on Ht. 87, $n = 38$, SD ±2,650); (5) R1a1-2a, age in Britain 7,700 years (rooted on Ht. 86, $n = 10$, SD ±5,440), derives from Norway. (6) R1a1-2b, age in Norway 5,700 years (rooted on Ht. 87, $n = 38$, SD ±2,160).

'Two Bronze Age clusters', respectively (1) and (2) above: R1a1-3c centres on the Isle of Man; R1a1-3a centres on Shetland, Orkney and the Western Isles.

13. R1a1-3b (total $n = 22$), age in British Isles 1,830 years (rooted on Ht. 96, $n = 21$, SD ±1,831). Source cluster in Norway: age 3,200 years (rooted on Ht. 96, $n = 20$, SD ±1,920).

14. 17.5% (22/126).

15. R1b-8a, age 2,137 years (rooted on Ht. 152, $n = 12$, SD ±1,510).

16. R1b haplogroup in Shetland 63.7% (86/135); 'indigenous/British' Shetland haplotypes 71.8% (97/135). Other studies have arrived at similar figures: for instance, Gooadcre et al. (2005) found a patrilineal ancestry admixture in Shetland of 55.5% Irish/British. They also found an equally balanced male–female Scandinavian admixture in Shetland and Orkney, unlike the Western Isles and Iceland, in each of which the male component was twice as large. They suggest that this implies that Orkney and Shetland may have been settled from Scandinavia by families (see p. 398). See also table 4 in Helgason et al. (2001), who estimate a 64.5% 'Gaelic' mtDNA component in Orkney and a 62.5% 'Gaelic' component in Iceland.

17. Admixture estimates from Goodacre et al. (2005), table 3. See also Helgason et al. (2001), table 4, where the Norse : Gaelic ancestral mtDNA ratio in Iceland is given as 3 : 5. For earlier estimates of male/female, British/Scandinavian components in Iceland, see Helgason et al. (2000a,b).

18. R1b in Iceland: 37.0% (67/181).

19. R1b in Trondheim: 28.9% (26/90); I1a rates are 24.4% (22/90) in Trondheim and 29.3% (53/181) in Iceland. More specifically, I1a-3 constitutes 76.9% (40/52) of I1a in Iceland and only 18.2% (4/22) of I1a in Trondheim. The highest equivalent figure elsewhere in Scandinavia is in Bergen, farther south: 52.6% (20/38). These figures suggest both a greater contribution from southern Scandinavia and a strong I1a-3 founder event in Iceland. Orkney and Shetland show no such I1a-3 founding event; rather they have a diverse mix of I1a and R1a1 haplotypes (in addition to indigenous R1b) belonging to lineages which arrived earlier in Britain. The same applies to a lesser extent to the Western Isles, where Viking founder events (I1a-3 and R1a1-3b) account collectively for 45.4% (10/22) of I1a and R1a1 types, but there are also roughly equal representatives of nine Neolithic Scandinavian founders, in particular I1a-5, which has its main British presence there (present study). For genetic drift in Iceland, see also Helgason et al. (2003).

20. Danish haplotype matches in Shetland and Orkney: 3% (8/256). For a graphic display of the relative importance of mtDNA from northern Britain and southern Norway in the colonization of Iceland, see Forster et al. (2004), figure 8.4.

21. 13.3% (18/135).

22. Goodacre et al. (2005), table 3.

23. Cunliffe (2004), p. 502; but see Smith (2001), Crawford (1981/2) and Wainwright (1962).

24. *Orkneyinga Saga* (12th century), translation in Cunliffe (2004), p. 493.

25. Present study.

26. Goodacre et al. (2005). The change in sex ratio in the Western Isles may, of course, partly reflect distance from Scandinavia. The Western Isles have roughly equal Neolithic and Viking male intrusions from Scandinavia – see note 16 in this chapter.

27. e.g. Smith (2001).
28. But see also Forsyth (1995).
29. Norwegian haplotype intrusion: to Shetland 20% (27/135), Orkney 17% (21/121), Western Isles 20% (18/88), Isle of Man 16% (10/62), Channel Islands 6.2% (8/128).
30. Norwegian haplotype intrusion based on exact matches: Oban 12% (5/42), Durness 13.7% (7/51). For the western British locations (excluding the Channel Islands) I have combined southern and northern Scandinavian haplotype matches as 'Norwegian' since, owing to the distribution of I1a-3, these would otherwise appear as 'Danish', which is historically less likely. Results are based on the present study and are generally lower than estimated by Goodacre et al. (2005), which is to be expected, given the use of exact matching.
31. 6.2% (129/2,082) (ignoring the Danish-to-Norwegian adjustment made in the previous note).
32. Härke (2002) has suggested 2–4%, invasion based on a population denominator of 1–2 million. The latter figure is rather larger than estimated elsewhere for the late Anglo-Saxon period, so 2–4% could be nearer 4–6%, although archaeological estimates during this period are really guestimates. An alternative estimate for the ninth-century Anglo-Saxon population is 750,000–850,000: see Coleman and Salt (1992).
33. Two estimates: 1.75 and 1.1–2.6 million: see Coleman and Salt (1992), table 1.1 and pp. 2–8.
34. Härke (2002).
35. Bragg (2003), p. 42.
36. Golding (2001), pp. 61–2.
37. Hinde (2004).
38. There are other sources of demographic measurement and estimation from this period.
39. Bragg (2003).
40. Douglas (1942).
41. Cunliffe (2004), p. 498.

Epilogue

1. Much of the best detail of colonization of the British Isles comes from my re-analysis of the Y-chromosome evidence; so most of the figures summarized here relate to that evidence. But published work on mitochondrial DNA inherited from our mothers generally supports that picture.
2. British Isles 72.6% (1,511/2,082), Irish 87.9% (233/265), Welsh 81.1% (180/222), Cornwall 78.8% (41/52), Scotland and islands 70.1% (366/522), England 67.7% (691/1,021).
3. Pryor (2004), pp. 143–4.
4. Forster et al. (2006).
5. National Statistics, 2001 Census, <http://www.statistics.gov.uk/census2001/profiles/commentaries/ethnicity.asp>.
6. Gildas, *De excidio Brittaniae* 3.27.

Appendices

1. If one takes the HVS 2 segment of the control region normally studied – see methods in Forster et al. (1996).

2. See methods in Forster et al. (1996) and Saillard et al. (2000).

3. See e.g. Forster et al. (2000), Zhivotovsky (2001) and Zhivotovsky et al. (2004).

4. See methods and calibration in Forster et al. (2000).

5. For the consensus nomenclature, see The Y Chromosome Consortium (2002).

6. See figure 3.3 in Oppenheimer (2003).

7. Additional dates from Richards et al. (2000), Pereira et al. (2005) and Achilli et al. (2004, 2005).

8. Oppenheimer (2003).

9. Gamble et al. (2004), Torroni et al. (2001).

10. Pereira et al. (2005), Achilli et al. (2004).

11. Capelli et al. (2003).

12. A couple of the other datasets lack the M170 UEP, which allows unambiguous differentiation of some Ivan types (IxI1b2 in Figure A3) from rare relatives. This potentially affects haplogroup assignment in sixteen samples from 3/32 of the UK sites that I use (in Weale et al. 2002), the Irish Gaelic sample (Hill et al. 2000) and Galicia and Valencia in Spain (Brion et al. 2003, 2004). In practice, the correct Ivan gene group assignment can be inferred using the STR gene type data (see below) by referring to the detailed analysis of UEPs and STRs carried out by Siiri Rootsi and colleagues (Rootsi et al. 2004). The probability of mis-assignment or ambiguity is very low in any case. This method of inference was also adequate to assign gene sub-groups within the IxI1b2 group (namely I*, I1*, I1b*, I1c and I1a). The Gaelic sample also lacks the UEP '12f2' identifying group J in the same part of the tree; but again assignment could be made using STR types, and group J is in any case absent from both the other Irish samples.

13. Ivan (I), in Weale's, Brion's and Hill's additional data, is inferred from STR type as described in the previous note. Severe drift of Ruisko in Iberian refuge is inferred from the overall Late Glacial age of the Capelli haplogroup RxR1a1 both in Iberia and Western Europe.

14. See e.g. Hurles et al. (2002), Capelli et al. (2001).

15. To do this, rank priority for microsatellites was given in the order used elsewhere (e.g. Wilson et al. 2001): DYS388, DYS393, DYS392, DYS19, DYS390, DYS391.

16. Capelli et al. (2003), Wilson et al. (2001).

17. Network 4.200, available from Fluxus Technology Ltd at [http://www.fluxus-engineering.com/sharenet.htm].

18. Forster et al. (2000).

19. Methods used for Founder Analysis and identifying and dating founder clusters are described in Richards et al. (2000).

20. The software used was Surfer 8, available from Golden Software Inc. [http://www.goldensoftware.com/products/surfer/surfer.shtml].

BIBLIOGRAPHY

Achilli, A., Rengo, C., Magri, C., Battaglia, V.A. et al. (2004), 'The molecular dissection of mtDNA haplogroup H confirms that the Franco-Cantabrian glacial refuge was a major source for the European gene pool', *American Journal of Human Genetics*, 75: 910–18.

Achilli, A., Rengo, C., Battaglia, V., Pala, M. and Olivieri, A. et al. (2005), 'Saami and Berbers – An unexpected mitochondrial DNA link', *American Journal of Human Genetics*, 76: 883–6.

Adams, J.M. and Otte, M. (1999), 'Did Indo-European languages spread before farming?', *Current Anthropology*, 40: 73–7.

Ammerman, A.J. and Cavalli-Sforza, L.L. (1984), *The Neolithic Transition and the Genetics of Populations in Europe* (Princeton University Press).

Amos, W., Jow, H. and Burroughs, N.J. (2006), 'Uncovering the male history of Britain', in Shuichi Matsumura, Peter Forster and Colin Renfrew (eds), *Simulations, Genetics and Human Prehistory – A Focus on Islands*, McDonald Institute Monograph Series (Cambridge: McDonald Institute for Archaeological Research) (in press).

Anderson, James (1736), *Royal Genealogies. The genealogical tales of emperors, kings and princes from Adam to these times*.

Anderson, J.M. (1988), *Ancient Languages of the Hispanic Peninsula* (Lanham, MD: University Press of America).

Anglo-Saxon Chronicle [Savage, Anne (ed.) (1982), *The Anglo-Saxon Chronicles* (London: Heinemann)].

Arnold, C.J. (1984), *From Roman Britain to Saxon England* (London: Croom Helm).

Atkinson, Q., Nicholls, G., Welch, D. and Gray, R. (2005), 'From words to dates: Water into wine, mathemagic or phylogenetic inference?', *Transactions of the Philological Society*, 103(2): 193–219.

Bailey, R.C., Head, G., Jenike, M., Owen, B. et al. (1989), 'Hunting and gathering in tropical rain forest: is it possible?', *American Anthropologist*, 91: 59–82.

Balter, M. (1998), 'Why settle down? The mystery of communities', *Science*, 282: 1442–4.

Bandelt, H.J., Forster, P. and Röhl, A. (1999), 'Median-joining networks for inferring intraspecific phylogenies', *Molecular Biology and Evolution*, 16: 37–48.

Barton, N. and Roberts, A. (2004), 'The Mesolithic period in England: Current perspectives and new research', in Alan Saville (ed.), *Mesolithic Scotland and its Neighbours: The Early Holocene Prehistory of Scotland, Its British and Irish Context, and Some Northern European Perspectives* (Edinburgh: Society of Antiquaries of Scotland), pp. 339–58.

Barton, R.N.E. (1999), 'Colonisation and resettlement of Europe in the Late Glacial: A view from the western periphery', *Folia Quaternaria*, 70: 71–86.

Barton, R.N.E., Jacobi, R.M., Stapert, D. and Street, M.J. (2003), 'The Late-Glacial reoccupation of the British Isles and the Creswellian', *Journal of Quaternary Science*, 18: 631–43.

Bede, *Ecclesiastical History* [Jane, L.C. (trans.) (1910) *The Ecclesiastical History of the English Nation*, with an introduction by Vida D. Scudder (London: J.M. Dent; New York: E.P. Dutton)].

Bellwood, P. and Diamond, J. (2005), 'On explicit "replacement" models in Island Southeast Asia: A reply to Stephen Oppenheimer', *World Archaeology*, 37: 503–6.

Bosch, E., Calafell. F., Comas, D., Oefner, P.J. et al. (2001), 'High-resolution analysis of human Y-chromosome variation shows a sharp discontinuity and limited gene flow between Northwestern Africa and the Iberian Peninsula', *American Journal of Human Genetics*, 68: 1019–29.

Bragg, Melvyn (2003), *The Adventure of English: The Biography of a Language* (London: Sceptre).

Brion, M., Salas, A., Gonzalez-Neira, A., Lareu, M.V. and Carracedo, A. (2003a), 'Insights over the Iberian population origin through the construction of highly informative Y-chromosome haplotypes using biallelic markers, STRs and the MSY1 minisatellite', *American Journal of Physical Anthropology*, 122: 147–61.

Brion, M., Quintans, B., Zarrabeitia, M. and Gonzalez-Neira, A. (2004), 'Microgeographical differentiation in Northern Iberia revealed by Y-chromosomal DNA analysis', *Gene*, 329: 17–25.

Buchanan, George (1582), *Rerum Scoticarum historia* (Edinburgh: Alexander Arbuthnet).

Campbell, Ewan (2001), 'Were the Scots Irish?', *Antiquity*, 75: 285–92.

Capelli, C., Wilson, J.F., Richards, M., Gratrix, F. et al. (2001), 'A predominantly indigenous paternal heritage for the Austronesian-speaking peoples of Insular Southeast Asia and Oceania', *American Journal of Human Genetics*, 68: 432–43.

Capelli, C., Redhead, N., Abernethy, J.K., Gratrix, F. et al. (2003), 'A Y chromosome census of the British Isles', *Current Biology*, 13: 979–84.

Carver, Martin (1998), *Sutton Hoo: Burial Ground of Kings?* (London: British Museum Press).

Cavalli-Sforza, L.L., Menozzi, P. and Piazza, A. (1993), 'Demic expansions and human evolution', *Science*, 259: 639–46.

Cavalli-Sforza, L. Luca, Menozzi, Paolo and Piazza, Alberto (1996), *The History and Geography of Human Genes* (Princeton University Press).

Chadwick, H.M. (1907), *The Origin of the British Nation* (Cambridge University Press).

Clarke, R.R. (1960), *East Anglia*, Ancient People and Places (London: Thames & Hudson).

Cohen, Mark Nathan (1989), *Health and the Rise of Civilization* (New Haven: Yale University Press).

Cohen, Mark Nathan and Armelagos, George J. (eds) (1984), *Paleopathology at the Origins of Agriculture* (New York: Academic Press).

Coleman, David and Salt, John (1992) *The British Population: Patterns, Trends and Processes* (Oxford University Press).

Collis, John (2003), *The Celts: Origins, Myths, Inventions* (Stroud: Tempus).

Cowgill, Warren (1970), 'Italic and Celtic superlatives and the dialects of Indo-European', in G. Cardona et al. (eds), *Indo-European and Indo-Europeans*, Papers presented at the Third Indo-European Conference (Philadelphia: University of Pennsylvania Press), pp. 113–53.

Crawford, I.A. (1981/2), 'War or peace – Viking colonisation in the Northern and Western Isles of Scotland reviewed', *Medieval Scandinavia*, Supplements (Proceedings of the 8th International Viking Congress), pp. 259–69.

Crawford, Sally (1997), 'Britons, Anglo-Saxons and the Germanic burial ritual', in John Chapman and Helena Hamerow (eds), *Migrations and Invasions in Archaeological Explanation* (Oxford: Archaeopress), pp. 45–72.

Cronan, D. (2004), 'Poetic words, conservatism and the dating of Old English poetry', *Anglo-Saxon England*, 33(Dec.): 23–50.

Cunliffe, Barry (ed.) (1981a), *Coinage and Society in Britain and Gaul: Some Current Problems*, British Research Report No. 38 (London: Council for British Archaeology).

Cunliffe, Barry (1981b), 'Money and society in pre-Roman Britain' in Cunliffe (1981a), pp. 29–39.

Cunliffe, Barry W. (1988), *Greeks, Romans and Barbarians: Spheres of Interaction* (London: Batsford).

Cunliffe, Barry (1997), *The Ancient Celts* (Oxford University Press).

Cunliffe, Barry (2002), *The Extraordinary Voyage of Pytheas the Greek: The Man Who Discovered Britain* (London: Penguin).

Cunliffe, Barry (2003), *The Celts: A Very Short Introduction* (Oxford University Press).

Cunliffe, Barry (2004) *Facing the Ocean: The Atlantic and Its Peoples* (Oxford University Press).

Daniels, G. (2002), *Human Blood Groups* (Oxford: Blackwell).

Diamond, J. (1997), *Guns, Germs and Steel: The Fates of Human Societies* (London: Jonathan Cape).

Diamond, J. and Bellwood, P. (2003), 'Farmers and their languages: the first expansions', *Science*, 300: 597–603.

Domesday Book [Hinde, T. (ed.) (2004), *The Domesday Book: England's Heritage, Then and Now* (London: Salamander)].

Douglas, D.C. (1942), 'Rollo of Normandy', *English Historical Review*, 57: 414–36.

Dumville, D. (1989), 'The Tribal Hidage: An introduction to its texts and their history', in Steven Bassett (ed.), *The Origins of Anglo-Saxon Kingdoms: Studies in the Early History of Britain* (Leicester University Press), pp. 225–30.

Dyen, Isidore, Kruskal, Joseph B. and Black, Paul (1992), 'An Indoeuropean classification: A lexicostatistical experiment', *Transactions of the American Philosophical Society*, 82: Part 5.

Embleton, Sheila (1986), *Statistics in Historical Linguistics* (Bochum: Brockmeyer).

Evans, David Ellis (1967), *Gaulish Personal Names* (Oxford University Press).

Forster, P. (1995), 'Einwanderungsgeschichte Norddeutschlands' (Immigration history of northern Germany), *North-Western European Language Evolution*, Suppl. vol. 12: 141–63. (English translation available at <http://mcdonald. com.ac.uk/genetics/forster1995english.pdf>.)

Forster, P., Harding, R., Torroni, A. and Bandelt, H.-J. (1996), 'Origin and evolution of Native American mtDNA variation: A reappraisal', *American Journal of Human Genetics*, 59: 935–45.

Forster, P., Röhl, A., Lünnemann, P., Brinkmann, C. et al. (2000), 'A short tandem repeat–based phylogeny for the human Y chromosome', *American Journal of Human Genetics*, 67: 182–96.

Forster, Peter and Toth, Alfred (2003), 'Toward a phylogenetic chronology of ancient Gaulish, Celtic, and Indo-European', *Proceedings of the National Academy of Sciences of the USA*, 100: 9079–84.

Forster, P., Romano, V., Calì, F., Röhl, A. and Hurles, M. (2004), 'MtDNA markers for Celtic and Germanic language areas in the British Isles', in Martin Jones (ed.), *Traces of Ancestry: Studies in Honour of Colin Renfrew*, McDonald Institute Monograph Series (Cambridge: McDonald Institute for Archaeological Research), pp. 99–111.

Forster, P., Polzin, T. and Röhl, A. (2006), 'Evolution of English basic vocabulary within the network of Germanic languages', in Peter Forster and Colin Renfrew (eds), *Phylogenetic Methods and the Prehistory of Languages*, McDonald Institute Monograph Series (Cambridge: McDonald Institute for Archaeological Research), pp. 131–7.

Forsyth K. (1995) 'The ogham-inscribed spindle-whorl from Buckquoy: Evidence for the Irish language in pre-Viking Orkney?', *Proceedings of the Society of Antiquaries of Scotland*, 125: 677–96.

Gamble, C., Davies, W., Pettitt, P. and Richards, M. (2004), 'Climate change and evolving human diversity in Europe during the last glacial', *Philosophical Transactions of the Royal Society of London B*, 359: 243–54.

Gamkrelidze, T. V. and Ivanov, V. V. (1995), *Indo-European and the Indo-Europeans: A Reconstruction and Historical Analysis of a Proto-Language and Proto-Culture*, Trends in Linguistics Vol. 80 (Berlin: Mouton de Gruyter).

Genzmer, F. (1982), *Heliand und die Bruchstücke der Genesis* (Stuttgart: Reclam).

Gildas, *De excidio Britanniae* [Williams, Hugh (trans.) (1899) *Two Lives of Gildas by a Monk of Ruys and Caradoc of Llancarfan* (Cymmrodorion Record Series; facsimilie reprint by Llanerch Publishers, Felinfach, 1990)].

Gimbutas, M. (1970), 'Proto-Indo-European culture: the Kurgan culture during the fifth, fourth and third millennia B.C.', in G. Cardona et al. (eds), *Indo-European and Indo-Europeans* (Philadelphia: University of Pennsylvania Press), pp. 155–95.

Golding, Brian (2001), *Conquest and Colonisation: The Normans in Britain, 1066–1100*, rev. edn (Basingstoke: Palgrave).

Goodacre, S., Helgason, A., Nicholson, J. and Southam, L. (2005), 'Genetic evidence for a family-based Scandinavian settlement of Shetland and Orkney during the Viking periods', *Heredity*, 95(2): 129–35.

Gray, Russell and Atkinson, Quentin (2003), 'Language-tree divergence times support the Anatolian theory of Indo-European origin', *Nature*, 426: 435–9.

Gronenborn, D. (2003), 'Migration, acculturation and culture change in western temperate Eurasia, 6500–5000 cal BC', *Documenta Praehistorica*, XXX: 79–91.

Gusma, L., Sánchez-Diz, P., Alves, C., Beleza, S. et al. (2003), 'Grouping of Y-STR haplotypes discloses European geographic clines', *Forensic Science International*, 134: 172–9.

Haak, W., Forster, P., Bramanti, B., Matsumura, S. et al. (2005), 'Ancient DNA from the first European farmers in 7500-year-old Neolithic sites', *Science*, 310: 1016–18.

Hachmann, R., Kossack, G. and Kuhn, H. (1962), *Völker zwischen Germanen und Kelten: Schriftquellen, Bodenfunde und Namengut zur Geschichte des nördlichen Westdeutschlands um Christi Geburt* (Neumunster: Karl-Wachholtz-Verlag).

Haetta, O.M. (1996), *The Sami: An Indigenous People of the Arctic* (Karasjok: Davvi Girji).

Hamerow, Helena (1997), 'Migration theory and the Anglo-Saxon "identity crisis"', in John Chapman and Helena Hamerow (eds), *Migrations and Invasions in Archaeological Explanation* (Oxford: Archaeopress), pp. 33–44.

Harding, R.M., Healy, E., Ray, A.J., Ellis, N.S. et al. (2000), 'Evidence for variable selective pressures at the human pigmentation locus, MC1R', *American Journal of Human Genetics*, 66: 1351–61.

Härke, H. (2002), 'Kings and warriors: Population and landscape from post-Roman to Norman Britain', in Paul Slack and Ryk Ward (eds), *The Peopling of Britain: The Shaping of a Human Landscape. The Linacre Lectures* (Oxford University Press), pp. 145–75.

Harrison, R. (1980), *The Beaker Folk* (London: Thames & Hudson).

Helgason, A., Sigurðadóttir, A., Gulcher, J., Ward, R. and Stefánsson, K. (2000a), 'MtDNA and the origin of the Icelanders: Deciphering signals of recent population history', *American Journal of Human Genetics*, 66: 999–1016.

Helgason, A., Sigurðadóttir, A., Nicholson, J., Sykes, B. et al. (2000b), 'Estimating Scandinavian and Gaelic ancestry in the male settlers of Iceland', *American Journal of Human Genetics*, 67: 697–717.

Helgason, A., Hickey, E., Goodacre, S. and Bosnes, V. (2001), 'mtDNA and the islands of the North Atlantic: Estimating the proportions of Norse and Gaelic ancestry', *American Journal of Human Genetics*, 68: 723–37.

Helgason, A., Nicholson, G., Stefánsson, K. and Donnelly, P. (2003), 'A reassessment of genetic diversity in Icelanders: Strong evidence from multiple loci for relative homogeneity caused by genetic drift', *Annals of Human Genetics*, 67: 281–97.

Herodotus, *Histories* [Macaulay, G.C. (trans.) (2001) *An Account of Egypt* (440–430 BC), Project Gutenberg, etext 2131].

Heyerdahl, Thor and Lillieström, Per (2001), *Jakten på Odin: På sporet av vår fortid* ('The Search for Odin') (Oslo: J.M. Stenersens).

Higham, N. (1992), *Rome, Britain, and the Anglo-Saxons* (London: Seaby).

Hill, E.W., Jobling, M.A. and Bradley, D.G. (2000). 'Y-chromosome variation and Irish origins', *Nature*, 404: 351–2.

Hills, Catherine (1998), 'Did the people of Spong Hill come from Schleswig-Holstein?' *Sonderdruck aus Studien zur Sachsenforschung*, 11: 145–54.

Hinde (2004) *see under* Domesday Book.

Hines, John (1992), 'The Scandinavian character of Anglian England: An update', in Martin Carver (ed.), *The Age of Sutton Hoo: The Seventh Century in North-western Europe* (Woodbridge, Suffolk: Boydell Press), chapter 22.

Hladik, C.M., Hladik, A., Linares, O.F., Pagezy, H. et al. (eds) (1993), *Tropical Forests, People and Food. Biocultural Interactions and Applications to Development* (Paris: UNESCO Publications).

Hodges, Richard (1989), *The Anglo-Saxon Achievement* (London: Duckworth).

Hoppa, R.D. and FitzGerald, C.M. (eds) (1999), *Human Growth in the Past* (Cambridge University Press).

Hurles, M.E., Nicholson, J., Bosch, E., Renfrew, C. et al. (2002), 'Y chromosomal evidence for the origins of Oceanic-speaking peoples', *Genetics*, 160: 289–303.

Jackson, Kenneth (1953), *Language and History in Early Britain* (Edinburgh University Press).

Jackson, K.H. (1955), 'The Pictish language', in F.T. Wainwright (ed.), *The Problem of the Picts* (Edinburgh: Nelson), pp. 129–66.

James, Simon (1999), *The Atlantic Celts: Ancient People or Modern Invention?* (London: British Museum Press).

Jansen, T., Forster, P., Levine, M.A., Oelke, H. et al. (2002), 'Mitochondrial DNA and the origins of the domestic horse', *Proceedings of the National Academy of Science of the USA*, 99: 10905–10.

Julius Caesar, *The Gallic Wars* [McDevitte, W.A. and Bohn, W.S. (trans.) (1869) *De Bello Gallico* (New York: Harper & Brothers)].

Kent, J.P.C. (1981), 'The origins of coinage in Britain', in Cunliffe (1981a), pp. 40–42.

Kopeć, Ada C. (1970), *The Distribution of the Blood Groups in the United Kingdom* (London: Oxford University Press).

Kortlandt, F. (1999), 'The origin of the Old English dialects revisited', *Amsterdamer Beiträge zur älteren Germanistik*, 51, 45–51.

Kozlowski, J. and Bandi, H.G. (1984), 'The paleohistory of circumpolar arctic colonization', *Arctic*, 37: 359–72.

Kuhn, H. (1962), 'Das Zeugnis der Namen', in Hachmann et al. (1962), pp. 105–35.

Lamason, R.L., Mohideen, M.-A.P., Mest, J.R., Wong, A.C. et al. (2005), 'SLC24A5 affects pigmentation in zebrafish and man', *Science*, 310: 1782–6.

Lambert, P.-Y. (1994) *La Langue Gauloise* (Paris: Editions Errance).

Leeds, E.T. (1912), 'The distribution of the Anglo-Saxon saucer brooch in relation to the Battle of Bedford AD 571, *Archaeologia* 2ND SER, 13: 159–202.

Lejeune, M. (1988), *Recueil des inscriptions Gauloises*, Supplément à Gallia 45, Vol. 2 (Paris: CNRS Editions).

Lewis, C. Andrew (1996), 'Prehistoric mining at the Great Orme: Criteria for the identification of early mining', M.Phil thesis, University of Wales, Bangor Agricultural and Forest Sciences. (available on-line at <http://www.greatorme.freeserve.co.uk/MPhil.htm>).

Lhuyd, Edward (1707), *Archaeologia Britannica: Giving some account Additional to what has been hitherto publish'd, of the Languages, Histories, and Customs of the Original Inhabitants of Great Britain* (Oxford).

Macalister, R.A. Stewart (ed. & trans.) (1938–56), *Lebor Gabála Érenn* (1st extant recension *The Book of Leinster*, c.1150), Vols 34, 35, 39, 41, 44 (London: Irish Texts Society).

McCone, K. (1996), *Towards a Relative Chronology of Ancient and Medieval Celtic Sound Change* (Maynooth: Department of Old and Middle Irish, St. Patrick's College).

McEvoy, B., Richards, M., Forster, P. and Bradley, D.G. (2004), 'The *Longue Durée* of genetic ancestry: Multiple genetic marker systems and Celtic origins on the Atlantic facade of Europe', *American Journal of Human Genetics*, 75: 693–702.

McMahon, April and McMahon, Robert (2003), 'Finding families: Quantitative methods in language classification', *Transactions of the Philological Society*, 101: 7–55.

McMahon, A. and McMahon, R. (2006), 'Why linguists don't do dates: Evidence from Indo-European and Australian languages', in Peter Forster and Colin Renfrew (eds), *Phylogenetic Methods and the Prehistory of Languages*, McDonald Institute Monograph Series (Cambridge: McDonald Institute for Archaeological Research), pp. 153–60.

Marichal, R. (1988), *Les Graffites de la Graufesenque*, Supplément à Gallia 47 (Paris: CNRS Editions).

Melton, N.D. and Nicholson, R.A. (2004), 'The Mesolithic in the Northern Isles: The preliminary evaluation of an oyster midden at West Voe, Sumburgh, Shetland, U.K.', *Antiquity* 78(299), <http://antiquity.ac.uk/ProjGall/nicholson>.

Myhre, Bjørn (1992), 'The royal cemetery at Borre, Vestfold: A Norwegian centre in a European periphery', in Martin Carver (ed.), *The Age of Sutton Hoo: The Seventh Century in North-western Europe* (Woodbridge, Suffolk: Boydell Press), pp. 301–13.

Myres, J.N.L. (1969), *Anglo-Saxon Pottery and the Settlement of England* (Oxford University Press).

Myres, J. N. L. (1986), *The English Settlements*, The Oxford History of England Vol. 1b (Oxford: Clarendon Press).

Newton, Sam (1992), 'Beowulf and the East Anglian royal pedigree', in Martin Carver (ed.), *The Age of Sutton Hoo: The Seventh Century in North-Western Europe* (Woodbridge, Suffolk: Boydell Press), pp. 65–74.

Newton, Sam (1993), *The Origins of Beowulf and the Pre-Viking Kingdom of East Anglia* (Woodbridge, Suffolk: Boydell & Brewer).

Newton, Sam (2003), *The Reckoning of King Rædwald:The Story of the King Linked to the Sutton Hoo Ship-Burial* (Woodbridge, Suffolk: Boydell & Brewer).

Nielsen, H.F. (1981), 'Old Frisian and Old English dialects', *Us Wurk* 30(2) 49–66.

Nygaard, S. (1989), 'The stone age of Northern Scandinavia: A review', *Journal of World Prehistory*, 3: 71–116.

Ó'Donnabháin, B. (2000), 'An appalling vista? The Celts and the archaeology of later prehistoric Ireland', in Angela Desmond et al. (eds), *New Agendas in Irish Prehistory* (Bray, Ireland: Wordwell), pp. 189–96.

O'Donovan, John (ed. & trans.) (1848–51), *Annals of the Four Masters, edited from MSS in the Library of the Royal Irish Academy and of Trinity College Dublin with a translation and copious notes*, 7 vols (Dublin); available on-line as *Annala Rioghachta Eireann: Annals of the Kingdom of Ireland by the Four Masters, from the earliest period to the year 1616* at <http://www.ucc.ie/celt/online/T100005A>.

Olofsson, Anders (2003), 'Pioneer settlement in the Mesolithic of northern Sweden', D.Phil dissertation, Umeå University; pdf free on-line at <www.diva-portal.org/diva/getDocument?urn_nbn_se_umu_diva-108–1__fulltext.pdf>.

Opland, J. (1980), *Anglo-Saxon Oral Poetry: A Study of the Traditions* (New Haven and London: Yale University Press).

Oppenheimer, Stephen (1998), *Eden in the East* (London: Weidenfeld & Nicolson).

Oppenheimer, Stephen (2003), *Out of Eden: The Peopling of the World* (London: Constable).

Oppenheimer, Stephen (2004), 'The "express train from Taiwan to Polynesia": On the congruence of proxy lines of evidence', *World Archaeology*, 36: 591–600.

O'Rahilly, T.F. (1946), *Early Irish History and Mythology* (Dublin Institute for Advanced Studies).

Page, R.I. (1999), *An Introduction to English Runes*, 2nd edn (Woodbridge, Suffolk: Boydell Press).

Parsons, David (2000), 'Classifying Ptolemy's English place-names', in Parsons and Sims-Williams (2000), pp. 169–78.

Parsons, David N. and Sims-Williams, Patrick (eds) (2000), *Ptolemy: Towards a Linguistic Atlas of the Earliest Celtic Place-names of Europe* (Aberystwyth: CMCS Publications); also available on CD.

Passarino, G., Cavalleri, G.L., Lin, A.A., Cavalli-Sforza, L.L. et al. (2002), 'Different genetic components in the Norwegian population revealed by the analysis of mtDNA and Y chromosome polymorphisms', *European Journal of Human Genetics*, 10: 521–9.

Pereira, L., Richards, M., Goios, A., Alonso, A. et al. (2005), 'High-resolution mtDNA evidence for the late-glacial resettlement of Europe from an Iberian refugium', *Genome Research*, 15: 19–24.

Pezron, Paul-Yves (1703), *L'Antiquité de la nation et de la langue des Celtes, autrement appelés Gaulois*.

Pezron, Paul-Yves (1706), *The Antiquities of Nations, More particularly of the Celtae or Gauls, Taken to be Originally the same People as our Ancient Britains* (London).

Piggott, S. (1965) *Ancient Europe from the Beginnings of Agriculture to Classical Antiquity* (Edinburgh University Press).

Pitkänen, K (1994), 'Suomen väestön historialliset kehityslinjat', in S. Koskinen et al. (eds), *Suomen Väestö* (Hämeenlinna: Gaudeamus), pp. 19–63.

Pringle, H. (1998), 'Neolithic agriculture: The slow birth of agriculture', *Science*, 282: 1446–50.

Pryor, Francis (2004), *Britain AD: A Quest for Arthur, England and the Anglo-Saxons* (London: HarperCollins).

Raftery, B. (1994), *Pagan Celtic Ireland: The Enigma of the Irish Iron Age* (London: Thames & Hudson).

Rankin, David (1996), *Celts and the Classical World* (London: Routledge; 1st edn 1987, London, Areopagitica Press).

Rees, J.L. (2000), 'The melanocortin 1 receptor (MC1R): More than just red hair', *Pigment Cell Research*, 13: 135–40.

Renfrew, Colin (1989), *Archaeology and Language: The Puzzle of Indo-European Origins* (London, Penguin; 1st edn 1987, London, Jonathan Cape).

Rexova, K., Frynta, D. and Zrzavy, J. (2003), 'Cladistic analysis of languages: Indo-European classification based on lexicostatistical data', *Cladistics*, 19: 120–27.

Ricaut, F.-X., Keyser-Tracqui, C., Bourgeois, J., Crubezy, E. and Ludes, B. (2004), 'Genetic analysis of a Scytho-Siberian skeleton and its implications for ancient Central Asian migrations', *Human Biology*, 76: 109–25.

Richards, M. (2003), 'The Neolithic invasion of Europe', *Annual Review of Anthropology*, 32: 135–62.

Richards, M.B., Forster, P., Tetzner, S., Hedges, R. and Sykes, B.C. (1995), 'Mitochondrial DNA and the Frisians', *North-Western European Language Evolution*, Suppl. vol. 12: 141–63.

Richards, M., Côrte-Real, H., Forster, P., Macaulay, V. et al. (1996), 'Paleolithic and neolithic lineages in the European mitochondrial gene pool', *American Journal of Human Genetics*, 59: 185–203.

Richards, M., Macaulay, V., Hickey, E., Vega, E. et al. (2000), 'Tracing European founder lineages in the Near Eastern mitochondrial gene pool', *American Journal of Human Genetics*, 67: 1251–76.

Richards, M., Macaulay, V., Torroni, A. and Bandelt, H.-J. (2002), 'In search of geographical patterns in European mitochondrial DNA', *American Journal of Human Genetics*, 71: 1168–74.

Ripoll, S., Muñoz, F., Pettitt, P. and Bahn, P. (2004), 'New discoveries of cave art in Church Hole, Creswell Crags, England', *INORA – International Newsletter on Rock Art*, 40: 1–6.

Rivet, Albert and Smith, Colin (1979), *The Place-names of Roman Britain* (London: Batsford).

Roewer, L., Croucher, P.J.P., Willuweit, S., Lu, T.T. et al. (2005), 'Signature of recent historical events in the European Y-chromosomal STR haplotype distribution', *Human Genetics*, 116: 279–91.

Rootsi, S. (2004), 'Human Y-chromosomal variation in European populations', PhD dissertation, Tartu University, pp. 30, 46.

Rootsi, S., Magri, C., Kivisild, T, Benuzzi, G. et al. (2004), 'Phylogeography of Y-chromosome haplogroup I reveals distinct domains of prehistoric gene flow in Europe', *American Journal of Human Genetics*, 75: 128–37.

Rosser, Z.H., Zerjal. T., Hurles, M.E., Adojaan, M. et al. (2000), 'Y-chromo-somal diversity in Europe is clinal and influenced primarily by geography, rather than language', *American Journal of Human Genetics*, 67: 1526–43.

Saillard, J., Forster, P., Lynnerup, N., Bandelt, H.-J. and Nørby, S. (2000), 'mtDNA variation among Greenland Eskimos: The edge of the Beringian expansion', *American Journal of Human Genetics*, 67: 718–26.

Sather, C. (1995), 'Sea nomads and rainforest hunter-gatherers: Foraging adaptations in the Indo-Malaysian archipelago', in P. Bellwood (ed.), *The Austronesians* (Canberra: Australian National University), pp. 229–68.

Saxo Grammaticus, *Gesta Danorum* [Elton, Oliver (trans.) (1905), *The Nine Books of the Danish History of Saxo Grammaticus* (New York: Norroena Society). See also Davidson, Hilda Ellis (2002) *History of the Danes: Books I–IX*, trans. Peter Fisher (Rochester, NY: Boydell & Brewer)].

Scarre, Chris (ed.) (1995), *Past Worlds: The Times Atlas of Archaeology* (London: Times Books).

Scheers, S. (1972), 'Coinage and currency of the Belgic tribes during the Gallic war', *British Numismatic Journal*, 41: 1–6.

Schirrmeister, L., Siegert, C., Kuznetsova, T., Kuzmina, S. et al. (2002), 'Paleoenvironmental and paleoclimatic records from permafrost deposits in the Arctic region of Northern Siberia', *Quaternary International*, 89: 97–118.

Schmidt, K.H. (1988). 'On the reconstruction of Proto-Celtic', in G. W. MacLennan (ed.), *Proceedings of the First North American Congress of Celtic Studies*, Ottawa, pp. 231–48.

Schrijver, P. (1999) 'The Celtic contribution to the development of the North Sea Germanic vowel system, with special reference to Coastal Dutch', *North-Western European Language Evolution* 35: 3–47.

Scozzari, R., Cruciani, F., Pangrazio, A., Santolamazza, P. et al. (2001), 'Human Y-chromosome variation in the western Mediterranean area: Implications for the peopling of the region', *Human Immunology*, 62: 871–84.

Semino, O., Passarino, G., Oefner, P.J., Lin, A.A. et al. (2000), 'The genetic legacy of Paleolithic *Homo sapiens sapiens* in extant Europeans: A Y chromosome perspective', *Science*, 290: 1155–9.

Semino, O., Magri, C., Benuzzi, G., Lin, A.A. et al. (2004), 'Origin, diffusion, and differentiation of Y-chromosome haplogroups E and J: Inferences on the Neolithization of Europe and later migratory events in the Mediterranean area', *American Journal of Human Genetics*, 74: 1023–34.

Siddall, M., Rohling, E.J., Almogi-Labin, A., Hemleben, Ch. et al. (2003), 'Sea-level fluctuations during the last glacial cycle', *Nature*, 423: 853–8.

Sims-Williams, P. (1983), 'Gildas and the Anglo-Saxons', *Cambrian Medieval Celtic Studies*, 6: 1–23.

Sims-Williams, P. (1998a), 'Genetics, linguistics, and prehistory: Thinking big and thinking straight', *Antiquity*, 72: 505–27.

Sims-Williams, P. (1998b), 'Celtomania and Celtoscepticism', *Cambrian Medieval Celtic Studies*, 36: 1–35.

Sims-Williams, P. (1990), *Religion and Literature in Western England, 600–800* (Cambridge University Press).

Sims-Williams, P. (2002), 'The five languages of Wales in the pre-Norman inscriptions', *Cambrian Medieval Celtic Studies*, 44: 1–36.

Sims-Williams, Patrick (2003), *The Celtic Inscriptions of Britain: Phonology and Chronology, c.400–1200* (Oxford: Blackwell).

Sims-Williams, P. (2006), *Ancient Celtic Place-Names in Europe and Asia Minor*, Publications of the Philological Society, Vol. 39 (Oxford: Blackwell).

Smith, Brian (2001), 'The Picts and the martyrs: Or did the Vikings kill the native population of Orkney and Shetland?', *Northern Studies*, 36: 7–32.

Strabo, *Geography* [Jones, H.L. (ed. and trans.) (1917–32), *The Geography of Strabo*. Vols 1–8, containing Books 1–17, Loeb Classical Library (London: Heinemann). See also <http://penelope.uchicago.edu/Thayer/E/Roman/Texts/Strabo/home.html>].

Stuart, A.J. and Lister A.M. (2001), 'The Late Quaternary extinction of woolly mammoth (*Mammuthus primigenius*), straight-tusked elephant (*Palaeoloxodon antiquus*) and other megafauna in Europe', in G. Cavaretta et al. (eds), *The World of Elephants, Proceedings of the 1st International Congress* (Rome: Consiglio Nazionale delle Ricerche), pp. 571–6.

Sumkin, V. J. (1990), 'On the ethnogenesis of the Saami: An archaeological view', *Acta Borealia*, 7: 3–20.

Tacitus, *Agricola* [Church, Alfred John, and Brodribb, William Jackson (eds) *(1942) The Life of Cnæus Julius Agricola* (New York: Random House)].

Tambets, K., Kivisild, T., Metspalu, E., Parik, J. et al. (2000), 'The topology of the maternal lineages of the Anatolian and Trans-Caucasus populations and the peopling of the Europe: Some preliminary considerations', in C. Renfrew and K. Boyle (eds), *Archaeogenetics: DNA and Population Prehistory of Europe*, McDonald Institute Monograph Series (Cambridge: McDonald Institute for Archaeological Research), pp. 219–35.

Tambets, K., Tolk, H.-V., Kivisild, T., Metspalu, E. et al. (2003). 'Complex signals for population expansions in Europe and beyond' in P. Bellwood and C. Renfrew (eds), *Examining the Farming/Language Dispersal Hypothesis*, McDonald Institute Monograph Series (Cambridge: McDonald Institute for Archaeological Research), pp. 449–57.

Tambets, K., Rootsi, S., Kivisild, T., Help, H. et al. (2004), 'The Western and Eastern roots of the Saami: The story of genetic "outliers" told by mitochondrial DNA and Y chromosomes', *American Journal of Human Genetics*, 74: 661–82.

Thomas, M., Stumpf, M.P.H. and Härke, H. (2006), 'Evidence for an apartheid social structure and differential reproductive success in early Anglo-Saxon England', in Shuichi Matsumura, Peter Forster and Colin Renfrew (eds), *Simulations, Genetics and Human Prehistory – A Focus on Islands*, McDonald Institute Monograph Series (Cambridge: McDonald Institute for Archaeological Research) (in press).

Thomas, M.G., Bradman, N. and Flinn, H.M. (1999), 'High throughput analysis of 10 microsatellite and 11 diallelic polymorphisms on the human Y-chromosome', *Human Genetics*, 105: 577–81.

Torroni, A., Bandelt, H.-J., D'Urbano, L., Lahermo, P. et al. (1998), 'mtDNA analysis reveals a major late Paleolithic population expansion from south-western to north-eastern Europe', *American Journal of Human Genetics*, 62: 1137–52.

Torroni, A., Bandelt, H.-J., Macaulay, V., Richards, M. et al. (2001), 'A signal, from human mtDNA, of post-glacial recolonization in Europe', *American Journal of Human Genetics*, 69: 844–52.

Treharne, R.F. and Fullard, H. (eds) (1976), *Muir's Historical Atlas: Ancient Medieval & Modern* (London, George Philip & Son).

Ulijaszek, S.J. (2000), 'Interpreting patterns of growth and development among past populations', *Perspectives in Human Biology*, 5: 1–11.

Underhill, P.A., Shen, P., Lin, A.A., Jin, L. et al. (2000), 'Y chromosome sequence variation and the history of human populations', *Nature Genetics*, 26: 358–61.

Underhill, P.A., Passarino, G., Lin, A.A., Shen, P. et al. (2001), 'The phylogeography of Y chromosome binary haplotypes and the origins of modern human populations', *Annals of Human Genetics*, 65: 43–62.

Untermann, J. (1961), *Sprachräume und Sprachbewegungen in vorrömischen Hispanien* (Wiesbaden: Harrassowitz).

Untermann, J. (1997), *Monumenta Linguarum Hispanicarum*, Vol. IV: *Die tartessischen, keltiberischen und lusitanischen Inschriften* (Wiesbaden: Reichelt).

Viereck, W. (1998), 'Geolinguistics and haematology: The case of Britain', *Links & Letters* (Departament de Filologia Anglesa i de Germanística, Universitat Autònoma de Barcelona), 5: 167–79.

Villar, F. (1997), 'The Celtiberian language', *Zeitschrift für Celtische Philologie*, 49/50: 898–941.

Villar, F. and Pedrero, R. (2001), 'La nueva inscripción lusitana: Arroyo de la Luz III', in F. Villar and M.P. Fernández Alvarez (eds), *Religión, lengua y cultura prerromanas de Hispania* (Universidad de Salamanca), pp. 663–98.

Vulpius, Christian August (1826), *Handwörterbuch der Mythologie der deutschen, verwandten, benachbarten und nordischen Völker* (Leipzig).

Wainwright F.T. (1962), 'Picts and Scots' and 'The Scandinavian settlement' in F.T. Wainwright (ed.), *The Northern Isles* (Edinburgh: Nelson), pp. 116, 162.

Walter, G. (1911), 'Der Wortschatz des Altfriesischen', *Münchener Beiträge zur romanischen und englischen Philologie*, 53.

Watkin, I.M. (1966), 'An anthropological study of eastern Shropshire and south-western Cheshire: ABO blood-groups', *Man* NS 1: 375–85.

Watkin, I.M. and Mourant, A.E. (1952), 'Blood groups, anthropology and language in Wales and the western countries', *Heredity*, 6: 13–36.

Weale, M.E., Weiss, D.A., Jager, R.F., Bradman, N. and Thomas, M. (2002), 'Y chromosome evidence for Anglo-Saxon mass migration', *Molecular Biology and Evolution*, 19: 1008–21.

Weber, Martin (1998), 'Das Gräberfeld von Issendorf, Niedersachsen: Ausgangangspunkt für Wanderungen nach Brittanien', *Sonderdruck aus Studien zur Sachsenforschung*, 11: 199–212.

Wells, R.S., Yuldasheva, N., Ruzibakiev, R., Underhill, P.A. et al. (2001), 'The Eurasian heartland: A continental perspective on Y-chromosome diversity', *Proceedings of the National Academy of Sciences of the USA*, 98: 10244–9.

Wilson, J.F., Weiss, D.A., Richards, M., Thomas, M.G. et al. (2001), 'Genetic evidence for different male and female roles during cultural transitions in the British Isles', *Proceedings of the National Academy of Sciences of the USA*, 98: 5078–83.

Wrenn, C.L. and Bolton, W.F. (eds) (1988), *Beowulf* (University of Exeter Press).

The Y Chromosome Consortium (2002), 'A nomenclature system for the tree of human Y-chromosomal binary haplogroups', *Genome Research*, 12: 339–48.

Yeats, W.B. (1902), *The Celtic Twilight* (London: A.H. Bullen).

Zerjal, T., Dashnyam, B., Pandya, A., Kayser, M. et al. (1997), 'Genetic relationships of Asians and Northern Europeans, revealed by Y-chromosomal DNA analysis'. *American Journal of Human Genetics*, 60: 1174–83.

Zhivotovsky, L.A. (2001), 'Estimating divergence time with use of microsatellite genetic distances: Impacts of population growth and gene flow', *Molecular Biology and Evolution*, 18: 700–709.

Zhivotovsky, L.A., Underhill, P.A., Cinnioğlu, C., Kayser, M. et al. (2004), 'On the effective mutation rate at Y-chromosome STRs with application to human population divergence time', *American Journal of Human Genetics*, 74: 50–61.

Zohary, Daniel and Hopf, Maria (2000), *Domestication of Plants in the Old World: The Origin and Spread of Cultivated Plants in West Asia, Europe, and the Nile Valley*, 2nd edn (New York: Oxford University Press).

Illustrations

Note Unless otherwise stated, all ages / dates derived from radiocarbon dating in captions to figures and plates and in the main text are given as 'years ago', meaning that they are already corrected or calibrated in primary cited literature sources. There are self-identified exceptions, e.g. Figure 6.1a, where raw radiocarbon dates were used in the source reference.

INDEX

Secondary colonization of the British Isles

The last 7,500 years – farmers, Anglo-Saxons and Vikings

Neolithic

Viking

Neolithic

Neolithic

Viking

Viking

Neolithic

Neolithic

Viking

Viking

Neolithic

Viking

Anglo-Saxon

Neolithic

Neolithic

Neolithic

Neolithic

Mediterranean Sea